Electrical Modeling and Design for 3D System Integration

IEEE Press
445 Hoes Lane
Piscataway, NJ 08854

Electrical Modeling and Design for 3D System Integration

3D Integrated Circuits and Packaging, Signal Integrity, Power Integrity and EMC

Er-Ping Li, BSc, MSc, PhD, IEEE Fellow
Zhejiang University
Hangzhou, China

IEEE PRESS

A John Wiley & Sons, Inc., Publication

Published by John Wiley & Sons, Inc., Hoboken, New Jersey.
Published simultaneously in Canada.

For general information on our other products and services or for technical support, please contact our Customer Care Department within the United States at (800) 762-2974, outside the United States at (317) 572-3993 or fax (317) 572-4002.

Wiley also publishes its books in a variety of electronic formats. Some content that appears in print may not be available in electronic formats. For more information about Wiley products, visit our web site at www.wiley.com.

Library of Congress Cataloging-in-Publication Data:

Li, Er-Ping.
   Electrical modeling and design for 3D system integration : 3D integrated circuits and packaging, signal integrity, power integrity and EMC / Er-Ping Li.
        p. cm.
   ISBN 978-0-470-62346-6 (hardback)
1. Three-dimensional integrated circuits.   I. Title.
   TK7874.893.L53 2011
   621.3815–dc23

                                                                    2011028946

Printed in the United States of America.

10 9 8 7 6 5 4 3 2 1

Contents

Foreword

Today, the modeling of electrical interconnects and packages is very important from both a practical and a theoretical point of view. High performance and high speed especially require a great deal of skill. An ever-increasing number of practical designs fall into this class.

The fact that we now have powerful design tools increases our ability to solve a larger number of real-world problems for many different issues. This greatly helps solve most of the important problems for a large class of geometries. However, the ever-increasing performance of the technology requires a continuous evolution of the skills in modeling techniques. A key performance issue is the reduction in effort and computing time for very large problems. Clearly, better design tools and techniques lead to better designs. Over the years, we also could observe that the opposite is true, namely that the more challenging problems lead to improved tools as well as better technical solutions. A consequence of this process is the continuous bootstrapping of the tools and techniques as well as the designers' skills.

This book represents an educational tool for modelers as well as for tool designers. It offers an unusual combination of the latest techniques for the electromagnetic (EM) modeling of packages and signal interconnections, including the challenging via problems. In fact, it is much more detailed than some of the introductory texts which are available today on the subject. It considers all aspects such as the analysis methods for the construction of macromodels which are stable, causal, and passive. Such models are widely in use today, and the passivity issue impacts the accuracy in both the frequency and time domains, while instability is unacceptable in the time domain. Also, key aspects of the modeling are the noise interactions between the multitude of wires and signal planes which are present in a typical design. All these aspects are considered in detail from an electromagnetic point of view, and sophisticated solution techniques are given. It is evident from this book that addressing the modern 3D packaging technology is an integral part of what makes the book relevant.

   We are fortunate to find in this book the contributions of an author who is both experienced and knowledgeable in this field. Dr. Er-Ping Li is an internationally well-known contributor to the field of electromagnetic solutions in the area of interest. He has been a Principal Scientist and Director of Electronics and Photonics at A*STAR (Agency for Science, Technology and Research) Institute of High Performance Computing in Singapore. From 2010 he holds an appointment as Chair Professor in Zhejiang University, China. He is a Fellow of the IEEE and a Fellow of the Electromagnetic Academy, USA. He received numerous international awards and honors in recognition of his professional work.

ALBERT E. RUEHLI, *PhD, Life Fellow of IEEE*
Emeritus, IBM T. J. Watson Research Center, Yorktown, NY, USA
Adjunct Professor, EMC Lab,
Missouri University of Science and Technology, Rolla, MO, USA

Preface

The requirements of higher bandwidth and lower power consumption of electronic systems render the integration of circuits and electronic packages more and more complex. In particular, the introduction of three-dimensional (3D) structures based on through-silicon via (TSV) technology provides a potential solution to reduce the size and to increase the performance of these systems. As a consequence, the electromagnetic compatibility (EMC) between circuits, signal integrity (SI), and power integrity (PI) in electronic integration are of vital importance. For this reason, the electronic circuits and packaging systems must be designed by taking into account the trade-offs between cost and performance. This requires ever more accurate modeling techniques and powerful simulation tools to achieve these goals. Incredible progress in electromagnetic field modeling has been achieved in the world. My research group has invested considerable efforts to develop novel simulation techniques over the last decade. Nevertheless, the present modeling techniques may be still far from perfect; for example, the modeling of multiphysics relevant to 3D integration is still far behind the requirements of the available technology.

This book presents the material that results from many years of our collective research work in the fields of modeling and simulation of SI, PI, and EMC in electronic package integration and multilayered printed circuit boards. It represents the state-of-the-art in electronic package integration and printed circuit board simulation and modeling technologies. I hope this book can serve as a good basis for further progress in this field in both academic research and industrial applications. The book consists of six chapters: Chapter 1 is written by Er-Ping Li, Chapter 2 by Enxiao Liu and Er-Ping Li, Chapter 3 by Zaw-Zaw Oo and Er-Ping Li, Chapter 4 by Xingchang Wei and Er-Ping Li, Chapter 5 by Yaojiang Zhang, and Chapter 6 by En-Xiao Liu.

Chapter 1 provides a review of progress in modeling and simulation of SI, PI, and EMC scenarios; Chapter 2 focuses on the macromodeling technique used in the electrical and electromagnetic modeling and simulation of complex interconnects in 3D integrated systems; Chapter 3 presents the semianalytical scattering matrix method (SMM) based on the N-body scattering theory for modeling of 3D electronic package and multilayered printed circuit boards with multiple vias. In Chapter 4, 2D and 3D integral equation methods are employed for the analysis of power distribution networks in 3D package integration. Chapter 5 describes the physics-based algorithm for extracting the equivalent circuit of a complex power distribution network in 3D integrated systems and printed circuit boards; Chapter 6 presents an equivalent-circuit model of through-silicon vias (TSV) and addresses the metal-oxide-semiconductor (MOS) capacitance effects of TSVs.

I gratefully acknowledge the technical reviewers of this book, Dr. Albert Ruehli, Emeritus of the IBM Watson Research Center, Yorktown, New York, USA; Prof. Wolfgang Hoefer, A*STAR, Singapore, and Prof. Zhongxiang Shen, Nanyang Technological University, Singapore, who donated their time and effort to review the manuscript. Also acknowledged are the contributors of the book, Dr. Xingchang Wei, Dr. Enxiao Liu, Dr. Zaw Zaw OO, and Dr. Yaojiang Zhang, who did the really hard work. I also wish to express my gratitude to Mary Hatcher at Wiley/IEEE Press for her great help in keeping us on schedule. Finally, I am grateful to my wife and the contributors' wives, for without their continuing support and understanding, this book would have never been published.

I hope that this book will serve as a valuable reference for engineers, researchers, and postgraduate students in electrical modeling and design of electronic packaging, 3D electronic integration, integrated circuits, and printed circuit boards. Even though much work has been accomplished in this field, I anticipate that many more exciting challenges will arise in this area, particularly in 3D integrated circuits and systems.

ER-PING LI
West Lake, Hangzhou, China

CHAPTER 1

Introduction

1.1 INTRODUCTION OF ELECTRONIC PACKAGE INTEGRATION

The rapid growth and convergence of digital computers and wireless communication have been driving semiconductor technology to continue its evolution following Moore's law in today's nanometer regime. Future electronic systems require higher bandwidth with lower power consumption to handle the massive amount of data, especially for large memory systems, high-definition displays, and high-performance microprocessors. Electronic packaging is one of the key technologies to realize a wider bus architecture with high bandwidth operating at higher frequencies. Various packages have been developed toward a higher density structure. In particular, a three-dimensional (3D) integration based on through-silicon via (TSV) [1] arrays technology provides a potential solution to reduce the size and to increase the performance of the systems. Furthermore, nano-interconnects to replace the Cu-based interconnects provides a promising solution for long-term application.

There is a great challenge for further increasing of the signal speed in electronic systems due to the serious electromagnetic compatibility (EMC) problem. Figure 1.1 plots the technology trends versus actuals and survey, and Figure 1.2 shows the trends of microprocessors predicted by the International Technology Roadmap for Semiconductors (ITRS) [2, 3]. From these figures one can see that

Electrical Modeling and Design for 3D System Integration: 3D Integrated Circuits and Packaging, Signal Integrity, Power Integrity and EMC, First Edition. Er-Ping Li.
© 2012 Institute of Electrical and Electronics Engineers. Published 2012 by John Wiley & Sons, Inc.

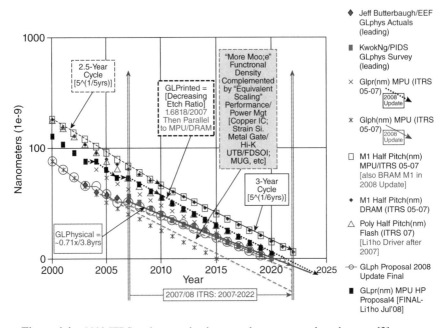

Figure 1.1   2008 ITRS update—technology trends versus actuals and survey [2].

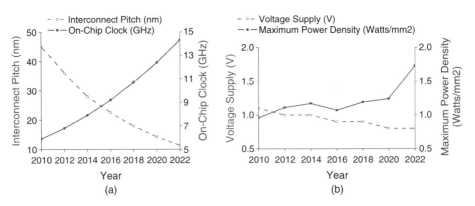

Figure 1.2   The trends of microprocessor predicted by the International Technology Roadmap for Semiconductors (ITRS).

- Interconnect pitch will continue to decrease to 11.3 nanometer, while the on-chip clock frequency will be increased to 14.3 GHz by 2022. Due to the reduction of the feature size and pitch, more and more circuits are integrated into one electronic package, such as the system in package (SIP) and the 3D integration. This results in a complex and high-density environment inside the electronic systems. At the same time, with the ever-increasing clock frequency (also its high-frequency harmonics), the physical size of the small electronic package becomes electrically large, and so the electromagnetic wave propagation inside such a small structure must be considered.

- Until 2011, the voltage supply of the microprocessor is continually reduced with an increased power density. The electromagnetic noise will be pronounced due to the increased power density, which then makes the decreased voltage supply unstable. To design a high-speed and stable electronic system, we need better understand the electromagnetic interactions and the EMC issues inside the electronic package.

The EMC researches related to the high-speed circuit systems have a long history, which can be classified into different levels according to the size of the interested objects, which includes the system level, printed circuit board (PCB) level, electronic package level, and component level. The increasing clock frequency makes the size of tiny structures on the chip be comparable with the wavelength of interest. The fluctuation of electromagnetic wave cannot be ignored any more. Therefore, we must accurately model the electromagnetic wave behavior for all scales of the high-speed circuit systems. In the near future, the nanoscale integrated circuits (ICs) will be characterized by using the *electric and magnetic fields* instead of the conventional *voltage and current*. *EM in micro-E* is becoming a hot topic in both academic community and industrial applications.

The EMC analysis for high-speed electronics includes lots of issues, such as the ground bounce, cross talk, conducted emission, radiated emission, conducted immunity, and radiated immunity. The interaction between on-board capacitance and on-chip capacitance causes an antiresonance which induces a peak in the total power distribution network (PDN) impedance as shown in Figure 1.3. Figure 1.4 shows a typical multilayered advanced electronic package which consists of two

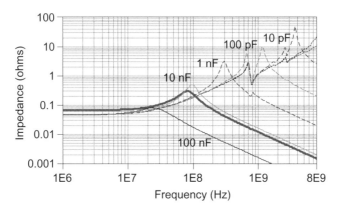

Figure 1.3 Example of antiresonances in total PDN impedances for various on-chip capacitance values [3]. (See color insert.)

Figure 1.4 A schematic diagram of a multilayered electronic package [37].

main electrically functional systems: the PDN and the signal distribution network (SDN). The passive structures are composed of three main categories: (1) traces or transmission lines, typically microstrip lines or striplines, (2) vias used as vertical interconnections, and (3) conductor plates serving as power or ground planes. Because of the complexity of an advanced package, it is difficult to model the entire SDN or PDN simultaneously. Yet, we need to consider the impact of the PDN on the SDN in order to characterize the SDN more accurately. Many researchers have proposed various approaches to study the electrical properties of the above passive structures [4–44].

A typical EMC problem residing in this PDN of the electronic package is illustrated in Figure 1.5. In Figure 1.5, the power and ground planes are used to supply DC power for the circuits integrated in the electronic package. The signal traces are often laid out in different

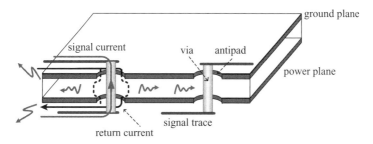

Figure 1.5   Noise coupling inside and emission from the power-ground planes.

layers of power-ground planes. Their return currents flow on the power-ground planes just below them. When the traces pass through different layers, their return currents also exchange from one plane to another plane, as shown in Figure 1.5. Accordingly, a vertical displacement current is induced between different planes for the continuity of the return currents. This displacement current will excite electromagnetic field noise, which then propagates inside the power-ground planes and couples to other signal traces passing through the same layer. At the same time, this noise also leaks to the surrounding area of the electronic package through the periphery and gaps of the power-ground planes. These interferences will be further amplified if the noise's spectrum covers any inherent resonant frequency of the cavity-like power-ground planes.

To achieve first-pass design success, we must employ an advanced modeling and simulation technique to analyze the electrical performance of the 3D electronic packages, PCB, and chips at the system level. However, both industry and academia communities face the great challenges in developing the electrical design and simulation tools due to the multiscale nature of the problem, the strong local and global electromagnetic coupling, and the complexity of 3D integration systems. ITRS has summarized the state of the art of current semiconductor industry development, where the major challenges for simulation and modeling are listed as [2] mixed-signal co-design and simulation environment, rapid turnaround modeling and simulation, electrical (power disturbs, electromagnetic interference (EMI), signal and power integrity associated with higher frequency/current and lower voltage switching), system-level co-design, electronic design automation (EDA) for "native" area array to meet the roadmap projections, and models for reliability prediction. Therefore, advanced modeling techniques, which stand up to the challenges imposed by the complexity of nanoscale silicon chips and their interconnections including 3D ICs, 3D packaging, and PCB [45–47], are in great demand.

1.2 REVIEW OF MODELING TECHNOLOGIES

Modeling of transmission lines has a long history and is well documented in many textbooks [4]. So in the following, we will mainly review the modeling of vias and power-ground planes for electronic packaging and PCBs. Such modeling methods can be roughly classified into three categories: (1) lumped circuit approaches, (2) full-wave approaches, and (3) hybrid circuit coupled full-wave approaches.

For its simplification and ease of understanding, at the beginning of the research, lumped circuit approaches have been used for the electrical modeling of electronic packages. Such examples are shown in Figures 1.6 and 1.7. Empirical and analytical formulae for via capacitance and inductance can be easily found in many handbooks. Quasi-static numerical methods have also been introduced to calculate the

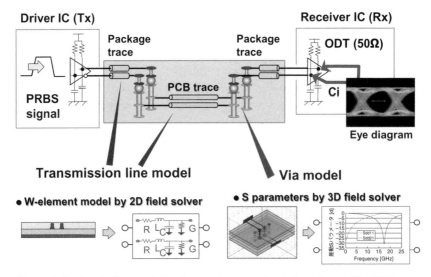

Figure 1.6 A typical transmission line model on a printed circuit board [3]. PRBS: pseudo-random binary sequence. ODT: on die termination.

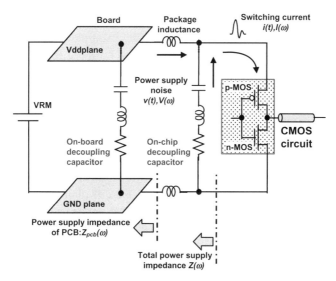

Figure 1.7   Power distribution network [3].

lumped circuit values in T or PI types of via models [5, 6]. These methods allow rapid computation, but often suffer from accuracy problems. The authors in Reference 7 proposed a model of a magnetic-frill array and utilized microwave network theory for analysis of vias in multilayered packages. But it is a single via model which is difficult to be generalized to multiple vias. The equivalent circuits of much complex via array can be extracted by using de-embedding method [8]. Distributed circuit approaches have also been widely used for package modeling, such as the partial element equivalent circuit (PEEC) method [9, 10].

Second, full-wave methods both in the time and frequency domains have been employed to study the packaging problems. The commonly used full-wave commercial simulators include Ansoft HFSS and CST Microwave Studio, which are based on finite element method (FEM) [11, 12] and finite integral technique, respectively. Other full-wave algorithm includes the finite-difference time-domain (FDTD) method [13] and the transmission line matrix method (TLM) [14]. Recently, the integral equation-based full-wave method begins to attract more attention and had been employed. The advantage of the integral equation method is that it can use the suitable Green's function to present the effect of the complex environment, so that the unknowns are only placed on discontinuities inside the PDN. This can give an efficient simulation. According to the different Green's functions used, the integral equation methods can be classified into (a) two-dimensional (2D) integral equations, including 2D mode method [15] and image method for rectangular power and ground planes, and 2D transverse magnetic (TM) integral equation for arbitrarily shaped power and ground planes [16–18]; (b) 3D integral equations, including 3D cavity mode method for rectangular power and ground planes [19] and parallel plate mode method for arbitrarily shaped power and ground planes [20, 21]. For most real applications the parallel plates have regular shapes, such as rectangles, circles, or triangles, a closed form of the Green's functions can be formulated which results in an impedance formula in terms of the summation of infinite number of resonant modes [22–24]. This 2D integral equation method is sometimes called the cavity resonator method. Segmentation techniques may be applied to extend the cavity resonator method for parallel plates with irregular shapes.

These full-wave methods are versatile and able to solve a wide range of problems, but at the expense of large memory usage and long CPU time, especially for those 3D full-wave methods. Although the overall size of the electronic packages is small enough to apply these full-wave methods, the high aspect ratio of the power and ground planes and the tiny structures, such as the signal traces and narrow slots, result in a huge number of meshing. This makes these full-wave methods very expensive in terms of computing time and memory requirement.

Third, to avoid the computational cost of these full-wave methods and the geometrical limitations of the analytical methods, a more efficient approach is to combine both methods together, so that we can benefit from both analytic and numerical techniques. The coupled circuit-field approaches are also widely used to model the electronic packages in order to leverage the advantages of both circuit and full-wave approaches. An important approach under this category is rooted in the theory of modal decomposition and the salient features of electronic packages. The transmission lines and power-ground planes in an electronic package convey different modes, that is, transmission line modes and parallel plate modes. Modal decomposition can be used to decouple these two modes, which are then solved independently. These two modes are finally recombined to reflect the original problem. The coupling between the transmission line mode and the parallel-plate mode often occurs due to the vias. The current flowing in the via excites the parallel-plate mode field, while the transmission line experiences the loading effect of the power-ground plane in the presence of the via.

The complete modal decomposition and recombination approach has been demonstrated by several researchers. Current or voltage controlled sources are used to link these two modes. A general modal recombination approach was presented in Reference 26 for coupled striplines and nonideal power-ground planes, while the parallel-plate mode associated with the power-ground planes has been studied by many researchers. 2D full-wave methods have been extensively employed in the literature to model the power-ground planes. The 2D integral equation method is also called the contour integral method and has been used in Reference 16 to study general parallel-plate structures with arbitrary shapes. Another 2D approach, called the 2D FDTD, has also been used to model parallel plates [27]. Discretization of the metal plates by the finite-difference method was interpreted as a 2D distributed LC circuit, and a rigorous derivation is given in Reference [27]. The 2D distributed $RLCG$ (resistance, inductance, capacitance, and conductance) circuit network, which is widely used in the literature to represent the power and ground planes, can be considered as an extension of the LC network derived from the finite-difference method. Instead of using Simulation Program with Integrated Circuit Emphasis (SPICE)-like solvers to simulate the large equivalent circuit network of power-ground planes, the latency insertion method is proposed in Reference 28 to perform fast transient simulation of large RLC networks. Moreover, a transmission matrix method reported in Reference 29 divides the 2D distributed $RLCG$ circuit network into many interdependent blocks, and each block is formulated as a transmission ($ABCD$) matrix. Cascading those transmission matrices produces a fast way to obtain the desired impedance of the power-ground plane. A multilayered finite-difference method (MFDM) was recently proposed in Reference 26. The 2D finite element method (2D FEM) is also used to simulate power-ground planes [30] and had been integrated into the commercial software Ansoft SIWave. In addition, the radial transmission line theory has been applied to derive an admittance matrix to account for the effect of the parallel plates [31]. However, image theory [32] is needed to model the reflection from the edges of finite-sized substrates. Image theory is elegant for modeling the boundary with a regular shape but is cumbersome for modeling arbitrary shapes of the edges of PCBs or packages. In the model decomposition and recombination approach, a single via can be represented by a PI type of equivalent circuit. The capacitance and inductance in the PI circuit are usually computed by analytical formulae or quasi-static solvers. Recently, an elegant analytical formula was derived for the via barrel-plate capacitance [6].

1.3 ORGANIZATION OF THE BOOK

This book is organized in six chapters. Chapter 1 provides an overview of the state-of-art of electrical modeling and simulation techniques for electrical packaging systems. Chapter 2 focuses on the macromodeling technique widely used in the electrical and electromagnetic modeling and simulation of complex interconnects in 3D integrated systems. Macromodels are generated by employing the vector fitting (VF) method to perform rational-function approximation of scattering or admittance network parameters of high-speed complex interconnects and passive circuits. Subsequently, the macromodel can be synthesized as an equivalent circuit, which is compatible with the SPICE circuit simulator and can be combined with other external linear or nonlinear circuits to perform signal and power integrity analysis or other electrical performance analysis of electronic systems. The stability, causality, and passivity assessment and enforcement of the macromodel are also discussed in detail. Finally, numerical examples of macromodeling are presented and discussed.

In Chapter 3, the semianalytical scattering matrix method (SMM) based on the N-body scattering theory is presented for modeling of 3D electronic package and multilayered PCBs with multiple vias. Using the modal expansion of fields in a parallel-plate waveguide, the formula derivation of the SMM is presented in detail. In the conventional SMM, the power-ground planes are assumed to be infinitely large so they cannot capture the resonant behavior of the real-world packages. In particular, the SMM method has been extended to solve the finite domain of power-ground planes in coupling with a novel boundary modeling method proposed by the author's group. This method has demonstrated its unique features which is capable to efficiently handle the complex real-world 3D package integration and PCB structures.

In Chapter 4, 2D and 3D integral equation methods are employed for the analysis of PDN in 3D package integration. The 2D integral equation method provides a comprehensive way for one to quickly extract the equivalent circuits of the PDN, and then substitute them into a SPICE-like simulator to perform the signal and power integrity analysis. The 3D integral equation method provides a more accurate solution for both the emission and susceptibility issues of the PDN. Both of the 2D and 3D integral equation methods are optimized by making a full use of the structural features of the PDN.

Chapter 5 is based on the physical-based algorithm to extract the equivalent circuit of the complex PDN in 3D integrated systems and PCBs. An intrinsic via circuit model is first derived through rigorous electromagnetic analysis for an irregular plate pair with multiple vias in a PCB. The derivation of the intrinsic via circuit model naturally leads to a new impedance definition of plate pair or power-bus, which is expressed in terms of cylindrical waves. The new plate pair impedance has clear physical meaning and makes possible signal/power integrity co-simulations. Numerical and measurement examples have indicated that while the new impedance gives almost the same results to the conventional one in a plate pair with few vias, it can correctly predict the resonant frequency shift in the case of a plate pair with a large amount of vias.

Chapter 6 presents a compact wideband equivalent-circuit model for electrical modeling of TSVs and addresses the metal-oxide-semiconductor (MOS) capacitance effects of TSVs.

REFERENCES

[1] P. GARROU, C. BOWER, and P. RAMM, *Handbook of 3D Integration*, Wiley-VCH Verlag GmbH&Co., Weinheim, 2008.
[2] International Technology Roadmap for Semiconductors (ITRS), http://www.itrs.net/.
[3] E.-P. LI, X. C. WEI, A. C. CANGELLARIS, E. X. LIU, Y. J. ZHANG, M. D'AMORE, J. KIM, and T. SUDO, Progress review of electromagnetic compatibility analysis technologies for packages, PCB and novel interconnects, *IEEE Trans. Electromagn. Compat.*, vol. 52, no. 2, pp. 248–265, 2010.
[4] C. PAUL, *Analysis of Multiconductor Transmission Lines*, 2nd. ed., Wiley, Hoboken, NJ, 2007.
[5] P. A. KOK and D. D. ZUTTER, Prediction of the excess capacitance of a via-hole through a multilayered board including the effect of connecting microstrips or

striplines, *IEEE Trans. Microw. Theory Tech.*, vol. 42, no. 12, pp. 2270–2276, 1994.

[6] Q. GU, Y. E. YANG, and M. A. TASSOUDJI, Modeling and analysis of vias in multilayered integrated circuits, *IEEE Trans. Microw. Theory Tech.*, vol. 41, no. 2, pp. 206–214, 1993.

[7] Y. J. ZHANG, J. FAN, G. SELLI, M. COCCHINI, and D. P. FRANCESCO, Analytical evaluation of via-plate capacitance for multilayer packages or PCBs, *IEEE Trans. Microw. Theory Tech.*, vol. 56, no. 9, pp. 2118–2128, 2008.

[8] X. C. WEI and E. P. LI, Integral-equation equivalent-circuit method for modeling of noise coupling in multilayered power distribution networks, *IEEE Trans. Microw. Theory Tech.*, vol. 58, no. 3, pp. 559–565, 2010.

[9] A. E. RUEHLI, Equivalent circuit models for three-dimensional multiconductor systems, *IEEE Trans. Microw. Theory Tech.*, vol. 22, pp. 216–221, 1974.

[10] A. E. RUEHLI, G. ANTONINI, J. ESCH, J. EKMAN, A. MAYO, and A. ORLANDI, Non-orthogonal PEEC formulation for time and frequency domain EM and circuit modeling, *IEEE Trans. Electromagn. Compat.*, vol. 45, no. 2, pp. 167–176, 2003.

[11] J. G. YOOK, N. I. DIB, and L. P. B. RATEHI, Characterization of high frequency interconnects using finite difference time domain and finite element methods, *IEEE Trans. Microw. Theory Tech.*, vol. 42, no. 9, pp. 1727–1736, 1994.

[12] J. M. JIN, *The Finite Element Method in Electromagnetics*, John Wiley & Sons, New York, 2002.

[13] S. MAEDA, T. KASHIWA, and I. FUKAI, Full wave analysis of propagation characteristics of a through hole using the finite-difference time-domain method, *IEEE Trans. Microw. Theory Tech.*, vol. 39, no. 12, pp. 2154–2159, 1991.

[14] P. B. JOHNS, A symmetrical condensed node for the TLM method, *IEEE Trans. Microw. Theory Tech.*, vol. 35, no. 4, pp. 370–377, 1987.

[15] T. OKOSHI, *Planar Circuits for Microwaves and Lightwave*, Springer-Verlag, Munich, Germany, 1984.

[16] X. C. WEI, E. P. LI, E. X. LIU, and X. CUI, Efficient modeling of re-routed return currents in multilayered power-ground planes by using integral equation, *IEEE Trans. Electromagn. Compat.*, vol. 50, no. 3, pp. 740–743, 2008.

[17] X. C. WEI, E. P. LI, E. X. LIU, and R. VAHLDIECK, Efficient simulation of power distribution network by using integral equation and modal decoupling technology, *IEEE Trans. Microw. Theory Tech.*, vol. 56, no. 10, pp. 2277–2285, 2008.

[18] M. STUMPF and M. LEONE, Efficient 2-D integral equation approach for the analysis of power bus structures with arbitrary shape, *IEEE Trans. Electromagn. Compat.*, vol. 51, no. 1, pp. 38–45, 2009.

[19] X. C. WEI, E. P. LI, E. X. LIU, E. K. CHUA, Z. Z. OO, and R. VAHLDIECK, Emission and susceptibility modeling of finite-size power-ground planes using a hybrid integral equation method, *IEEE Trans. Adv. Packag.*, vol. 31, no. 3, pp. 536–543, 2008.

[20] X. C. WEI, G. P. ZOU, E. P. LI, and X. CUI, Extraction of equivalent network of arbitrarily shaped power-ground planes with narrow slots using a novel integral equation method, *IEEE Trans. Microw. Theory Tech.*, vol. 58, no. 11, pp. 2850–2855, 2010.

[21] M. R. ABDUL-GAFFOOR, H. K. SMITH, A. A. KISHK, and A. W. A. G. A. W. GLISSON, Simple and efficient full-wave modeling of electromagnetic coupling in realistic RF multilayer PCB layouts, *IEEE Trans. Microw. Theory Tech.*, vol. 50, no. 6, pp. 1445–1457, 2002.

[22] C. WANG, J. MAO, G. SELLI, S. LUAN, L. ZHANG, J. FAN, D. J. POMMERENKE, R. E. DUBROFF, and J. L. DREWNIAK, An efficient approach for power delivery network design with closed-form expressions for parasitic interconnect inductances, *IEEE Trans. Adv. Packag.*, vol. 29, no. 2, pp. 320–334, 2006.

[23] J. KIM, Y. JEONG, J. KIM, J. LEE, C. RYU, J. SHIM, M. SHIN, and J. KIM, Modeling and measurement of interlevel electromagnetic coupling and fringing effect in a hierarchical power distribution network using segmentation method with resonant cavity method, *IEEE Trans. Adv. Packag.*, vol. 31, no. 3, pp. 544–557, 2008.

[24] R. L. CHEN, J. CHEN, T. H. HUBING, and W. M. SHI, Analytical model for the rectangular power-ground structure including radiation loss, *IEEE Trans. Electromagn. Compat.*, vol. 47, pp. 10–16, 2005.

[25] A. E. RUEHLI and A. C. CANGELLARIS, Progress in the methodologies for the electrical modeling of interconnects and electronic packages, *Proc. IEEE*, vol. 89, pp. 740–771, 2001.

[26] A. E. ENGIN, W. JOHN, G. SOMMER, W. MATHIS, and H. REICHL, Modeling of striplines between a power and a ground plane, *IEEE Trans. Adv. Packag*, vol. 29, no. 3, pp. 415–426, 2006.

[27] W. K. GWAREK, Analysis of an arbitrarily shaped planar circuit: A time-domain approach, *IEEE Trans. Microw. Theory Tech.*, vol. 33, no. 10, pp. 1067–1072, 1985.

[28] J. E. SCHUTT-AINE, Latency insertion method (LIM) for the fast transient simulation of large networks, *IEEE Trans. Circuits Syst. I*, vol. 48, no. 1, pp. 81–89, 2001.

[29] J. H. KIM and M. SWAMINATHAN, Modeling of irregular shaped power distribution planes using transmission matrix method, *IEEE Trans. Adv. Packag.*, vol. 24, no. 3, pp. 334–346, 2001.

[30] J. E. BRACKEN, S. POLSTYANKO, I. BARDI, A. MATHIS, and Z. J. CENDES, Analysis of system-level electromagnetic interference from electronic packages and boards, in *Proc. 14th Elect. Performance Electron. Packag. Conf.*, 2005, pp. 183–186.

[31] R. ABHARI, G. V. ELEFTHERIADES, and E. van DEVENTER-PERKINS, Physics-based CAD models for the analysis of vias in parallel-plate environments, *IEEE Trans. Microw. Theory Tech.*, vol. 49, no. 10, pp. 1697–1707, 2001.

[32] R. ITO, R. W. JACKSON, and T. HONGSMATIP, Modeling of interconnections and isolation within a multilayered ball grid array package, *IEEE Trans. Microw. Theory Tech.*, vol. 47, no. 9, pp. 1819–1825, 1999.

[33] L. TSANG, H. CHEN, C. C. HUANG, and V. JANDHYALA, Modeling of multiple scattering among vias in planar waveguides using Foldy-Lax equations, *Microw. Opt. Technol. Lett.*, vol. 31, pp. 201–208, 2001.

[34] L. TSANG and D. MILLER, Coupling of vias in electronic packaging and printed circuit board structures with finite ground plane, *IEEE Trans. Adv. Packag.*, vol. 26, pp. 375–384, 2003.

[35] C. C. HUANG, K. L. LAI, L. TSANG, X. X. GU, and C. J. ONG, Transmission and scattering on interconnects with via structures, *Microw. Opt. Technol. Lett.*, vol. 46, pp. 446–452, 2005.

[36] C. J. ONG, D. MILLER, L. TSANG, B. WU, and C. C. HUANG, Application of the Foldy-Lax multiple scattering method to the analysis of vias in ball grid arrays and interior layers of printed circuit boards, *Microw. Opt. Technol. Lett.*, vol. 2007, pp. 225–231, 2007.

[37] Z. Z. OO, E. X. LIU, E. P. LI, X. C. WEI, Y. ZHANG, M. TAN, L. W. LI, and R. VAHLDIECK, A semi-analytical approach for system-level electrical modeling of electronic packages with large number of vias, *IEEE Trans. Adv. Packag.*, vol. 31, no. 2, pp. 267–274, 2008.

[38] E. X. LIU, E. P. LI, Z. Z. OO, X. WEI, Y. ZHANG, and R. VAHLDIECK, Novel methods for modeling of multiple vias in multilayered parallel-plate structures, *IEEE Trans. Microw. Theory Tech.*, vol. 57, no. 7, pp. 1724–1733, 2009.

[39] Z. Z. OO, E. P. LI, X. C. WEI, E. X. LIU, Y. J. ZHANG, and L. W. LI, Hybridization of the scattering matrix method and modal decomposition for analysis of signal traces in a power distribution network, *IEEE Trans. Electromagn. Compat.*, vol. 51, no. 3, pp. 784–791, 2009.

[40] C. SCHUSTER, Y. KWARK, G. SELLI, and P. MUTHANA, Developing a "physical" model for vias, in *Proc. IEC DesignCon Conf.*, Santa Clara, CA, February 6–9, 2006, pp. 1–24.

[41] G. SELLI, C. SCHUSTER, Y. H. KWARK, M. B. RITTER, and J. L. DREWNIAK, Developing a physical via model for vias—Part II: Coupled and ground return vias, in *Proc. IEC DesignCon Conf.*, Santa Clara, CA, January 29–February 1, 2007, pp. 1–22.

[42] G.-T. LEI, R. W. TECHENTIN, P. R. HAYES, D. J. SCHWAB, and B. K. GILBERT, Wave model solution to the ground/power plane noise problem, *IEEE Trans. Instrum. Meas.*, vol. 44, no. 2, pp. 300–303, 1995.

[43] Y. J. ZHANG and J. FAN, An intrinsic via circuit model for multiple vias in an irregular plate pair through rigorous electromagnetic analysis, *IEEE Trans. Microw. Theory Tech.*, vol. 58, no. 8, pp. 2251–2265, 2010.

[44] Y. J. ZHANG, Z. Z. OO, X. C. WEI, E. X. LIU, E. P. LI, and J. FAN, Systematic microwave network analysis for multilayer printed circuit boards with vias and decoupling capacitors, *IEEE Trans. Electromagn. Compat.*, vol. 52, no. 2, pp. 401–409, 2010.

[45] Y. XIE, J. CONG, and S. SAPATNEKAR, *Three-Dimensional Integrated Circuit Design*, Springer, New York, 2010.

[46] Y. DENG and W. P. MALY, *3-Dimensional VLSI: A 2.5 Dimensional Integrated Scheme*, Springer, New York, 2010.

[47] S. H. HALL and H. L. HECK, *Advanced Signal Integrity for High-Speed Digital Design*, John Wiley & Sons, New Jersey, 2009, pp. 274–279.

[48] S. MCMORROW and C. HEARD, The impact of PCB laminate weave on the electrical performance of differential signaling at multi-gigabit data rates, *DesignCon*, 2005.

[49] T. BANDYOPADHYAY, R. CHATTERJEE, D. CHUNG, M. SWAMINATHAN, and R. TUMMALA, Electrical modeling of annular and co-axial TSVs considering MOS

capacitance effects, in *IEEE 18th Conference on Electrical Performance of Electronic Packaging and Systems*, Portland, OR, October 2009, pp. 117–120.

[50] J. Kim, E. Song, J. Cho, J. S. Pak, H. Lee, K. Park, and J. Kim, Through silicon via equalizer, in *IEEE 18th Conference on Electrical Performance of Electronic Packaging and Systems*, Portland, OR, October 2009, pp. 13–16.

[51] R. Schmitt, X. Huang, L. Yang, and C. Yuan, System level power integrity analysis and correlation for multi-gigabit designs, DesignCon, 5-WA2, 2004.

[52] N. Hirano, M. Miura, Y. Hiruta, and T. Sudo, Characterization and reduction of simultaneous switching noise for multilayer package, in *Proceedings of 44th ECTC*, 1994, pp. 949–956.

[53] T. Sudo, Characterization of simultaneous switching noise and electromagnetic radiation associated with chip and package properties, *IEICE Trans. Electron.*, vol. J89-C, no. 7, pp. 429–439, 2006.

[54] M. Swaminathan, J. Kim, I. Novak, and J. Libous, Power distribution networks for system-on-package: Status and challenges, *IEEE Trans. Adv. Packag.*, vol. 27, pp. 286–300, 2004.

[55] J. Qin, O. M. Ramahi, and V. Granatstein, Novel planer electromagnetic bandgap structures for wideband noise suppression and EMI reduction in high speed circuits, *IEEE Trans. Electromagn. Compat.*, vol. 49, no. 3, pp. 661–669, 2007.

[56] T.-K. Wang, T.-W. Han, and T.-L. Wu, A novel power/ground layer using artificial substrate EBG for simultaneously switching noise suppression, *IEEE Trans. Microw. Theory Tech.*, vol. 56, no. 5, pp. 1164–1171, 2008.

[57] M. Swaminathan and A. Ege Engin, *Power Integrity Modeling and Design for Semiconductors and Systems*, Prentice Hall, Englewood Cliffs, NJ, 2007, pp. 415–445.

CHAPTER 2

Macromodeling of Complex Interconnects in 3D Integration

2.1 INTRODUCTION

Complex passive interconnects widely used in printed circuit boards (PCBs), electronic packages, and other electronic systems, are best characterized in the frequency domain. The frequency-dependent characteristics of interconnects require descriptions by either scattering (S) parameters, admittance (Y) parameters, or impedance (Z) parameters. Compared to the admittance parameters, the scattering parameters are well defined to represent networks consisting of high-speed interconnects because they are bounded quantities. Thus, the scattering parameters are primarily used in this chapter. Although the dispersive nature of interconnects is amicable to a frequency-domain representation, nonlinear circuits coexisting with the interconnect structures in electronic systems require time-domain description. System-level simulation for signal and power integrity analysis and electrical performance verification must address the problem of mixed time-frequency domain simulation required by an electronic system with both nonlinear circuits and linear passive interconnects.

Electrical Modeling and Design for 3D System Integration: 3D Integrated Circuits and Packaging, Signal Integrity, Power Integrity and EMC, First Edition. Er-Ping Li.
© 2012 Institute of Electrical and Electronics Engineers. Published 2012 by John Wiley & Sons, Inc.

Many approaches have been proposed in the literatures to address the mixed frequency-time domain problem. A straightforward approach to solve this problem was to employ the inverse fast Fourier transform (IFFT) and convolution method [1]. However, this approach suffers from excessive computational cost in the convolution process. Another approach to solve this mixed domain problem was the complex frequency hopping (CFH) method by moment matching [2, 3]. The difficulty faced by this approach is that for every moment, a corresponding derivative of each parameter must be computed using numerical integration across the entire time domain. This process has to be repeated on multiple frequency expansion points, which can be cumbersome for high-order approximation, or a network with a large number of ports.

An efficient macromodeling approach based on sampled frequency-domain data has been discussed in References 4–10. The macromodeling approach uses direct rational function approximation instead of moment matching to tackle the mixed domain problem. The macromodel obtained by rational function approximation can be used in conjunction with recursive convolution [11] to simulate interconnects together with nonlinear devices efficiently. Alternatively, the resultant macromodel can be converted to equivalent lumped circuits, which can be incorporated into the industrial standard SPICE (Simulation Program with Integrated Circuit Emphasis) circuit simulator to perform electrical analysis [12].

Many researchers have applied different methods to performing the rational function approximation. A section-by-section approximation approach was proposed in Reference 4, which partitioned the frequency band of the data into small sections to avoid numerically ill-conditioning problems. The drawback of this approach is that the final model has an artificially large number of poles accumulated from the approximation of each section. The matrix equations in Reference 6 introduced unnecessary ill conditioning to the approximation by using ω^2 terms in the polynomials at the numerator and denominator. An improved approach was proposed in Reference 5 to compute pole-zero pairs recursively. But this method is only valid for real poles, which restricts its application to resistance-inductance (RL) and resistance-capacitance (RC) circuits.

A robust method for rational function approximation is the vector fitting (VF) method developed by Gustavsen and Semlyen in Reference

13. The VF method has advantages over other fitting methodologies [14]. Most conventional fitting methods rely on nonlinear optimization algorithms that are complex and may converge to a local minimum. Conversely, the VF method relies on the solution of two linear least-squares problems and thus obtains the optimal solution rather directly. At the same time, the VF method does not suffer much from the numerical stability problem, even when the bandwidth of interest is wide. Furthermore, one single run of the VF method can achieve the rational function approximation of all the elements in a transfer function matrix with a common set of poles. Therefore, the VF method has been widely adopted to generate stable macromodels.

This chapter focuses on the macromodeling techniques which are widely used in the electrical and electromagnetic modeling and simulation of complex interconnects. Macromodels are generated by employing the VF method to perform rational-function approximation of scattering or admittance network parameters of high-speed complex interconnects and passive circuits. Subsequently, the macromodel can be synthesized as an equivalent circuit, which is compatible with the SPICE circuit simulator and can be combined with other external linear or nonlinear circuits to perform signal integrity and power integrity analysis or other electrical performance analysis of electronic systems. The stability, causality, and passivity assessment and enforcement of the macromodel are also discussed in detail. Finally, numerical examples of macromodeling are presented and discussed.

2.1.1 Scope of Macromodeling

This chapter focuses on the modeling of linear time invariant (LTI) systems of high-speed passive interconnects, packages, and components used in digital, mixed signal systems, and radio frequency (RF) systems as well. A system is said to be linear if its response to a linear combination of two or more inputs is a superposition of the response due to each individual input. And the time invariant property of a system means that the output of the system, except for a time difference, is identical regardless of when the input is applied. Under such a context, macromodeling is defined as a process where the system identification approach is employed to generate a black-box continuous transfer-function model, which is often expressed in the form of partial fractions, from discrete frequency- or time-domain sampled data obtained through measurements or numerical simulation. Note that a review can be found in Reference 15 on the recent progress of electrical and electromagnetic modeling of interconnects, electronic packages, and PCBs). Thereafter, a set of first-order differential equations in the state-space form is derived through macromodel synthesis. The macromodeling technique differs from the model-order-reduction (MOR) method [12] which deals with known transfer functions. However, macromodeling may be followed by a MOR if the macromodel has a large number of poles. This chapter will limit the discussion to only linear macromodeling.

2.1.2 Macromodeling in the Picture of Electrical Modeling of Interconnects

The significance of the macromodeling technique is revealed in Figure 2.1, which shows a big picture of electrical modeling of interconnects, packages, and PCBs for signal, power integrity, and electromagnetic compatibility (SI/PI/EMC). The macromodeling technique bridges the frequency domain and the time domain, and thus resolves the mixed time-frequency simulation issue for system-level modeling of electronic systems.

2.2 NETWORK PARAMETERS: IMPEDANCE, ADMITTANCE, AND SCATTERING MATRICES

Network parameters describe the behavior of an electronic system observed externally. The commonly used network parameters in the electrical and electronic engineering include Z, Y, S, transfer (T), and chain/transmission (ABCD) parameters. This section will summarize the definitions and key properties of the impedance, admittance, and scattering network parameters.

The network studied in this chapter is assumed to be linear, passive, and reciprocal. A passive network contains no source of energy. A reciprocal network is one in which the power loss or transmission between any two ports is independent of the direction of signal or wave propagation. A network is reciprocal if it is passive and contains no

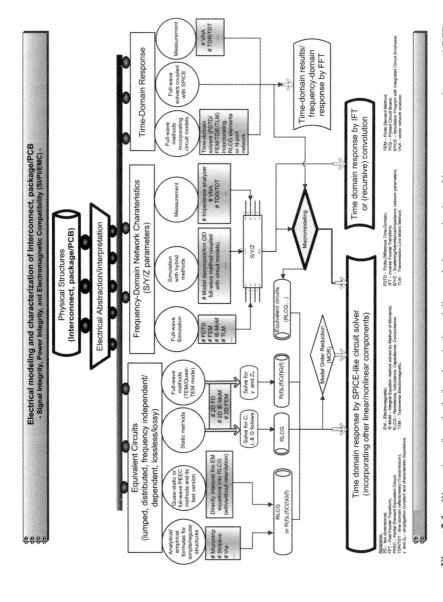

Figure 2.1 Illustration of methods for electrical modeling and characterization of interconnects, packages and PCBs, for signal, power integrity and electromagnetic compatibility (SI/PI/EMC). The macromodeling technique resolves the mixed time-frequency simulation issue.

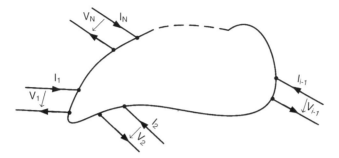

Figure 2.2 Schematic of an N-port network with voltages and currents shown at the ports.

active devices, and no nonreciprocal media such as ferrites, and plasmas. A network is symmetrical if its input is exchangeable with its output. Symmetrical networks are not necessarily physically symmetrical. A lossless network does not consist of energy-dissipative elements like resistors.

2.2.1 Impedance Matrix

For a network with N ports (see Fig. 2.2), an impedance matrix can be used to find the voltages at each port, if the currents at each port are known:

$$
\begin{bmatrix} V_1 \\ V_2 \\ \vdots \\ V_N \end{bmatrix} = \begin{bmatrix} Z_{11} & Z_{12} & \cdots & Z_{1N} \\ Z_{21} & Z_{22} & \cdots & Z_{2N} \\ \vdots & \vdots & \ddots & \vdots \\ Z_{N1} & Z_{N2} & \cdots & Z_{NN} \end{bmatrix} \begin{bmatrix} I_1 \\ I_2 \\ \vdots \\ I_N \end{bmatrix},
\tag{2.1}
$$

which is written in the following compact form,

$$
\{V\} = [Z]\{I\},
\tag{2.2}
$$

where V_i and I_i are port voltages and currents, respectively. Z_{ij} are the impedance parameters:

$$
Z_{ii} = \left. \frac{V_i}{I_i} \right|_{I_k = 0 (k \neq i)},
\tag{2.3}
$$

for self impedance parameters, and

$$Z_{ij} = \frac{V_i}{I_j}\bigg|_{I_k=0(k\neq j)}, \qquad (2.4)$$

for mutual/transfer impedance parameters. Note that open-circuit conditions are used to derive the impedance parameters.

The impedance parameters have the following properties:

$$
\begin{aligned}
Z_{ij} &= Z_{ji} \ (i \neq j), && \text{for a reciprocal network} \\
Z_{ii} &= Z_{jj}, && \text{for a symmetrical network} \quad (2.5) \\
Z_{ij} &= \pm j\psi_{ij} \ (\text{purely imaginary}), && \text{for a lossless network.}
\end{aligned}
$$

2.2.2 Admittance Matrix

An admittance matrix for an N-port network can be used to find the currents at each port, if the voltages at each port are known:

$$
\begin{bmatrix} I_1 \\ I_2 \\ \vdots \\ I_N \end{bmatrix} =
\begin{bmatrix} Y_{11} & Y_{12} & \cdots & Y_{1N} \\ Y_{21} & Y_{22} & \cdots & Y_{2N} \\ \vdots & \vdots & \ddots & \vdots \\ Y_{N1} & Y_{N2} & \cdots & Y_{NN} \end{bmatrix}
\begin{bmatrix} V_1 \\ V_2 \\ \vdots \\ V_N \end{bmatrix}. \qquad (2.6)
$$

It is written in the following compact form,

$$\{I\} = [Y]\{V\}, \qquad (2.7)$$

where Y_{ij} are the admittance parameters:

$$Y_{ii} = \frac{I_i}{V_i}\bigg|_{V_k=0(k\neq i)}, \qquad (2.8)$$

for self-admittance parameters, and

$$Y_{ij} = \frac{I_i}{V_j}\bigg|_{V_k=0(k\neq j)}, \qquad (2.9)$$

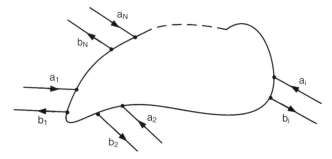

Figure 2.3   Schematic of an *N*-port network with incident and reflected waves shown at the ports.

for mutual/transfer admittance parameters. Note that short-circuit conditions are used to derive admittance parameters.

The admittance parameters have the following properties:

$$Y_{ij} = Y_{ji} \ (i \neq j), \qquad\qquad \text{for a reciprocal network}$$
$$Y_{ii} = Y_{jj}, \qquad\qquad \text{for a symmetrical network} \qquad (2.10)$$
$$Y_{ij} = \pm j\varphi_{ij} \ (\text{purely imaginary}), \quad \text{for a lossless network.}$$

2.2.3 Scattering Matrix

A scattering matrix for an *N*-port network (see Fig. 2.3) determines the relation between the incident and reflected waves at the ports:

$$\begin{bmatrix} b_1 \\ b_2 \\ \vdots \\ b_N \end{bmatrix} = \begin{bmatrix} S_{11} & S_{12} & \cdots & S_{1N} \\ S_{21} & S_{22} & \cdots & S_{2N} \\ \vdots & \vdots & \ddots & \vdots \\ S_{N1} & S_{N2} & \cdots & S_{NN} \end{bmatrix} \begin{bmatrix} a_1 \\ a_2 \\ \vdots \\ a_N \end{bmatrix}. \qquad (2.11)$$

Its compact form is

$$\{b\} = [S]\{a\}, \qquad\qquad (2.12)$$

where S_{ij} are the scattering parameters

$$S_{ii} = \frac{b_i}{a_i}\bigg|_{a_k=0(k\neq i)}, \tag{2.13}$$

for reflected scattering parameters, and

$$S_{ij} = \frac{b_i}{a_j}\bigg|_{a_k=0(k\neq j)}, \tag{2.14}$$

for transmitted scattering parameters. Note that the matched-load conditions are used to derive scattering parameters.

The scattering parameters have the following properties: (1) for a reciprocal network, the scattering matrix is symmetrical, i.e., $S_{ij} = S_{ji}(i \neq j)$; (2) for a lossless network, the scattering matrix is unitary:

$$[S]^T[S]^* = [U]$$

$$\sum_{k=1}^{N} S_{ki}S_{kj}^* = \delta_{ij} \quad (i, j = 1, \cdots, N), \tag{2.15}$$

where $[U] = diag\{1 \quad \cdots \quad 1\}$.

2.2.4 Conversion between Z, Y, and S Matrices

This section presents the formulae used for conversions between the Z, Y, and S matrices of a general N-port network.

The Z and Y matrices are converted to each other by

$$[Z] = [Y]^{-1}, \quad [Y] = [Z]^{-1}. \tag{2.16}$$

If different normalizing impedances are used at different ports, the following general formulae are used to convert from S to Z or Y matrices:

$$[Z] = \left[\sqrt{Z_0}\right]([U]+[S])([U]-[S])^{-1}\left[\sqrt{Z_0}\right], \tag{2.17}$$

$$[Y] = \left[\sqrt{Z_0}\right]^{-1}([U]-[S])([U]+[S])^{-1}\left[\sqrt{Z_0}\right]^{-1}, \tag{2.18}$$

where $[Z_0] = diag\{z_{01}, \cdots, z_{0N}\}$, and $\left[\sqrt{Z_0}\right] = diag\left\{\sqrt{z_{01}}, \cdots, \sqrt{z_{0N}}\right\}$. z_{0i} are the normalizing impedances at each port. Vice versa, the formulae for conversion from Z or Y to S matrices are given by

$$[S] = \left[\sqrt{Z_0}\right]^{-1}([Z]-[Z_0])([Z]+[Z_0])^{-1}\left[\sqrt{Z_0}\right], \qquad (2.19)$$

$$[S] = \left[\sqrt{Z_0}\right]([Y_0]-[Y])([Y_0]+[Y])^{-1}\left[\sqrt{Z_0}\right]^{-1}. \qquad (2.20)$$

2.3 RATIONAL FUNCTION APPROXIMATION WITH PARTIAL FRACTIONS

2.3.1 Introduction

Previous discussion shows that the interconnect network can be characterized by the S or Y parameters at M discrete frequency points ω_i over the frequency range of interest. An example for an N-port network is given by

$$H(\omega_i) = \begin{bmatrix} H_{11}(\omega_i) & H_{12}(\omega_i) & \cdots & H_{1N}(\omega_i) \\ H_{21}(\omega_i) & H_{22}(\omega_i) & \cdots & H_{2N}(\omega_i) \\ \vdots & \vdots & \ddots & \vdots \\ H_{N1}(\omega_i) & H_{N2}(\omega_i) & \cdots & H_{NN}(\omega_i) \end{bmatrix}, \quad i = 1, \cdots, M \quad (2.21)$$

where $H(\omega_i)$ can be discrete scattering parameters $S(\omega_i)$ or admittance parameters $Y(\omega_i)$.

In order to facilitate the time-domain analysis of an electronic system with interconnects, the frequency-dependent data in Equation (2.21) can be approximated by rational functions to obtain a macro-model of the interconnect network. The idea of rational-function approximation is to fit the frequency response of a network by the ratio of two polynomials with real coefficients in the Laplace domain,

$$H_{ij}(s) = \frac{g_0 + g_1 s + g_2 s^2 + \cdots + g_n s^n}{1 + d_1 s + d_2 s^2 + \cdots + d_m s^m} = \frac{\sum_{i=0}^{n} g_i s^i}{1 + \sum_{i=1}^{m} d_i s^i}, \qquad (2.22)$$

where g_i denotes real coefficients for the numerator polynomial of degree n, and d_i represents real coefficients for the denominator polynomial of degree m. d_0 is normalized to one. $H_{ij}(s)$ can be $Y_{ij}(s)$ or $S_{ij}(s)$.

Asymptotically, Equation (2.22) can be written as $H_{ij}(s) \approx \left(\dfrac{g_n}{d_m}\right) s^{n-m}$. Therefore, the following expressions hold

$$H_{ij}(\infty) = \begin{cases} 0. & \text{if } n < m, \\ \dfrac{g_n}{d_m}, & \text{if } n = m, \\ \pm\infty, & \text{if } n > m. \end{cases} \qquad (2.23)$$

If $n \le m$, then the rational function in Equation (2.22) is proper; otherwise, it is improper. The case of $n > m$ usually does not occur in the circuit analysis [16].

A common way of performing rational function approximation is to multiply both sides of Equation (2.22) by its denominator. For M discrete frequency data, the resultant linear equations regarding the unknowns g_i and d_i are written as follows,

$$\begin{bmatrix} 1 & j\omega_1 & (j\omega_1)^2 & \cdots & (j\omega_1)^n & -j\omega_1 H(\omega_1) & (-j\omega_1)^2 H(\omega_1) \\ \vdots & \vdots & \vdots & \vdots & \vdots & \vdots & \vdots \\ 1 & j\omega_M & (j\omega_M)^2 & \cdots & (j\omega_M)^n & -j\omega_M H(\omega_M) & (-j\omega_M)^2 H(\omega_M) \end{bmatrix}$$

$$\begin{matrix} & & \\ \cdots & (-j\omega_0)^m H(\omega_0) \\ \cdots & (-j\omega_1)^m H(\omega_1) \\ \vdots & \vdots \\ \cdots & (-j\omega_M)^m H(\omega_M) \end{matrix} \begin{bmatrix} g_0 \\ g_1 \\ \vdots \\ g_n \\ d_1 \\ d_2 \\ \vdots \\ d_m \end{bmatrix} = \begin{bmatrix} H(\omega_1) \\ \vdots \\ H(\omega_M) \end{bmatrix}, \qquad (2.24)$$

where the subscripts ij for $H(s)$ are dropped for brevity.

Note that when a higher degree of the polynomial is needed for rational function approximation over a wide frequency range, Equation (2.24) may suffer from the numerical stability problem, which is attributed to the large discrepancy among the entries of the matrix in the left-hand side of Equation (2.24). The numerical stability problem can be overcome by the robust VF method, which will be discussed in detail later.

2.3.2 Iterative Weighted Linear Least-Squares Estimator

Many approaches have been proposed for the model generation and system identification especially in the field of control and systems [17]. The most widely used techniques are based on rational approximation with least-square fitting [18].

Rational transfer function in Equation (2.22) can be written in a general form as

$$H(s) \approx \frac{N(s)}{D(s)} = \frac{\sum_{n=1}^{Q} r_n \phi_n(s)}{\tilde{r}_0 + \sum_{n=1}^{Q} \tilde{r}_n \phi_n(s)}, \tag{2.25}$$

where r_n and $\tilde{r}_n$ are real coefficients and Q is the order of the macro-model. $\phi_n(s)$ denote the basis functions. The VF method uses rational functions as basis functions, which will be shown later.

An optimal model of a system can be obtained by minimizing the following nonlinear cost (objective) function:

$$\min_{(\tilde{r}, r)} \sum_{m=1}^{M} \left| \frac{N(s_m)}{D(s_m)} - H(s_m) \right|^2 = \min_{(\tilde{r}, r)} \sum_{m=1}^{M} \left| \frac{N(s_m) - D(s_m) H(s_m)}{D(s_m)} \right|^2. \tag{2.26}$$

A nonlinear optimization technique can be used to find the minimum norm solutions, but with high computation cost. Nonlinear optimization can be avoided by minimizing the following Levy's cost function:

$$\min_{(\tilde{r}, r)} \sum_{m=1}^{M} \left| N(s_m) - D(s_m) H(s_m) \right|^2. \tag{2.27}$$

The above Levy's cost function is obtained by multiplying Equation (2.26) with the denominator $D(s)$. However, this linearization has two major drawbacks: the overemphasizing of high frequency errors in Equation (2.27), and a large dynamic range of the terms in the normal equations. The former drawback may result in poor low-frequency fits, and the latter one may cause ill-condition problems, when systems with a large dynamic frequency range [17] are identified.

But the Sanathanan–Koerner iteration [19] can be used to overcome the problem of lacking of sensitivity to the low-frequency error in Equation (2.27). The τth step in the iteration minimizes the cost function in the following form:

$$\min_{(\tilde{n},n)} \sum_{m=1}^{M} \left| \frac{N^{\tau}(s_m)}{N^{\tau-1}(s_m)} - \frac{D^{\tau}(s_m)}{D^{\tau-1}(s_m)} H(s_m) \right|^2. \tag{2.28}$$

Most of the modified Sanathanan–Koerner methods fall into the framework of the following iterative weighted linear least squares [17]:

$$\min_{(\tilde{n},n)} \sum_{m=1}^{M} \frac{w_e^2(s_m)}{\left| D^{\tau-1}(s_m) \right|^2} \left| N^{\tau}(s_m) - D^{\tau}(s_m) H(s_m) \right|^2, \tag{2.29}$$

where $w_e(s_m)$ is an external weighting function. The weighted iterative cost function can be minimized by solving the following linear system of equations for all complex frequencies s:

$$\frac{w_e(s)}{D^{\tau-1}(s)} N^{\tau}(s) - \frac{w_e(s)}{D^{\tau-1}(s)} D^{\tau}(s) H(s) = 0. \tag{2.30}$$

The VF method, proposed by Gustavsen and Semlyen in Reference 13, resembles a modified Sanathanan–Koerner method [20]. It can be considered as a reformulation of Equation (2.30), which obviates solving the coefficients of the numerator and denominator polynomials in Equation (2.25), and thus directly produces partial fractions of a transfer function.

$$w_e(s) \underbrace{\frac{N^{\tau}(s)}{D^{\tau}(s)} \frac{D^{\tau}(s)}{D^{\tau-1}(s)}}_{(\lambda H)^{\tau}(s)} - w_e(s) \underbrace{\frac{D^{\tau}(s)}{D^{\tau-1}(s)}}_{(\lambda)^{\tau}(s)} H(s) = 0$$

$$\Updownarrow \tag{2.31}$$

$$w_e(s) \left(\hat{c} + \sum_{k=1}^{Q} \frac{\hat{r}_k}{s - \tilde{p}_k} \right) - w_e(s) \left(1 + \sum_{k=1}^{Q} \frac{\tilde{r}_k}{s - \tilde{p}_k} \right) H(s) = 0,$$

where $\lambda(s)$ is the scaling factor which is instrumental to the VF method. More details of the method are presented in the following sections.

2.4 VECTOR FITTING (VF) METHOD

With the assumption of no multiplicity of the poles, Equation (2.22) or Equation (2.25) is usually expanded into a pole-residue form as follows:

$$H(s) = c + \sum_{k=1}^{Q} \frac{r_k}{s - p_k}, \qquad (2.32)$$

where c, a real number, is a direct coupling constant. r_k and p_k represent the residues and poles of the transfer function, respectively, which can either be real or complex conjugate pairs. Note that the subscripts ij for $H(s)$ in Equation (2.22) are dropped for brevity, which should cause no confusion. Such a pole-residue form of the transfer function will facilitate the subsequent macromodel synthesis and time-domain simulation, because it has a direct interpretation in the time domain through inverse Laplace transform:

$$h(t) = c\delta(t) + \sum_{k=1}^{Q} r_k \cdot e^{p_k \cdot t}, \qquad (2.33)$$

where $\delta(t)$ is the Dirac unit impulse function, and t denotes time. Equation (2.32) is a nonlinear problem since the unknowns p_k are located in the denominator. However, the VF method solves it as a linear problem (cf. Eq. 2.31) in two steps: the first step is to compute the poles by scaling and iterative procedures, and the second step the residues.

2.4.1 Two Steps in Vector Fitting Method

2.4.1.1 Pole Identification

Instead of computing the poles p_k in Equation (2.32) directly, the first step of the VF method computes them via a scaling process, and converts the problem of calculating the poles into a problem of computing zeros.

First, a set of initial poles $\tilde{p}_k$ is chosen as an initial guess of the actual poles in Equation (2.32). And an unknown scaling function $\lambda(s)$ is expanded using the initial poles:

$$\lambda(s) = 1 + \sum_{k=1}^{Q} \frac{\tilde{r}_k}{s - \tilde{p}_k}, \qquad (2.34)$$

where $\tilde{r}_k$ are the residues of the scaling function corresponding to the poles, and the direct coupling term c is normalized to one.

Then, the original function $H(s)$ in Equation (2.32) is multiplied by the scaling function $\lambda(s)$ defined in Equation (2.34). The product of the two functions is denoted as $\theta(s)$, which is also approximated by the same set of initial poles $\tilde{p}_k$:

$$\theta(s) = \lambda(s)H(s) = \hat{c} + \sum_{k=1}^{Q} \frac{\hat{r}_k}{s - \tilde{p}_k}. \tag{2.35}$$

Substituting $\lambda(s)$ in Equation (2.35) with Equation (2.34) gives

$$\left(1 + \sum_{k=1}^{Q} \frac{\tilde{r}_k}{s - \tilde{p}_k}\right) H(s) = \hat{c} + \sum_{k=1}^{Q} \frac{\hat{r}_k}{s - \tilde{p}_k}. \tag{2.36}$$

Because the starting poles $\tilde{p}_k$ at both sides of Equation (2.36) are known, Equation (2.36) is now linear regarding the unknowns $\tilde{r}_k$, $\hat{r}_k$, and $\hat{c}$, and can be easily rearranged as

$$\left(\hat{c} + \sum_{k=1}^{Q} \frac{\hat{r}_k}{s - \tilde{p}_k}\right) - \left(\sum_{k=1}^{Q} \frac{\tilde{r}_k}{s - \tilde{p}_k}\right) H(s) = H(s). \tag{2.37}$$

For a given frequency s_l, Equation (2.37) becomes

$$A_l X_l = H(s_l), \tag{2.38}$$

where

$$A_l = \left\{1 \quad (s_l - \tilde{p}_1)^{-1} \quad \cdots \quad (s_l - \tilde{p}_Q)^{-1} \quad \frac{-H(s_l)}{(s_l - \tilde{p}_1)} \quad \cdots \quad \frac{-H(s_l)}{(s_l - \tilde{p}_Q)}\right\}, \tag{2.39}$$

$$X_l = \{\hat{c} \quad \hat{r}_1 \quad \hat{r}_2 \quad \cdots \quad \hat{r}_Q \quad \tilde{r}_1 \quad \tilde{r}_2 \quad \cdots \quad \tilde{r}_Q\}^T. \tag{2.40}$$

Because the coefficients of the numerator and denominator in Equation (2.22) are real, any complex poles and residues will present in conjugate pairs. A modification to Equation (2.39) is made in Reference 13 in order to preserve the conjugate property of the complex residues. Assuming that the kth and $(k+1)$th terms in the partial fraction expansion in Equation (2.37) contain complex conjugate pole and residue pairs, that is,

$$\tilde{p}_{k+1} = \tilde{p}_k^* = \text{Re}(\tilde{p}_k) - j\,\text{Im}(\tilde{p}_k), \quad \tilde{r}_{k+1} = \hat{r}_k^* = \text{Re}(\hat{r}_k) - j\,\text{Im}(\hat{r}_k). \quad (2.41)$$

The real and imaginary parts, $\text{Re}(r_k)$ and $\text{Im}(r_k)$ are used as the new unknowns in Equation (2.40) instead of using the original complex variables $\hat{r}_k$ and $\hat{r}_{k+1}$. This can be done by simply rearranging the two corresponding elements in the row vector A_l into the following form:

$$A_{l,k} = \frac{1}{s - \tilde{p}_k} + \frac{1}{s - \tilde{p}_k^*}, \quad A_{l,k+1} = \frac{j}{s - \tilde{p}_k} + \frac{j}{s - \tilde{p}_k^*}. \quad (2.42)$$

For all the M sampled frequencies, Equation (2.38) forms an over-determined linear matrix equation if $M > 2Q + 1$,

$$\begin{bmatrix} A_1 \\ A_2 \\ \vdots \\ A_l \\ \vdots \\ A_{M-1} \\ A_M \end{bmatrix}_{M \times (2Q+1)} \begin{bmatrix} \hat{c} \\ \hat{r}_1 \\ \vdots \\ \hat{r}_Q \\ \tilde{r}_1 \\ \vdots \\ \tilde{r}_Q \end{bmatrix}_{(2Q+1) \times 1} = \begin{bmatrix} H(s_1) \\ H(s_2) \\ \vdots \\ H(s_l) \\ \vdots \\ H(s_{M-1}) \\ H(s_M) \end{bmatrix}_{M \times 1}, \quad (2.43)$$

or

$$AX = B. \quad (2.44)$$

The problem of finding a vector X which minimize L_2 norm of the residue, that is, $\|B - AX\|_2$, is called a least-squares problem. Because only positive frequencies are used in the fitting process, the entries in A and B of Equation (2.44) is again formulated by using real quantities to retain the conjugate property of the solutions,

$$\begin{bmatrix} \text{Re}(A) \\ \text{Im}(A) \end{bmatrix} \{X\} = \begin{bmatrix} \text{Re}(B) \\ \text{Im}(B) \end{bmatrix}. \quad (2.45)$$

The least-squares solution X with the smallest norm $||X||$ is unique and is given by Reference 21:

$$A^T A X = A^T B, \tag{2.46}$$

or equivalently

$$X = \left(A^T A\right)^{-1} A^T B = A^\dagger B \tag{2.47}$$

where the superscript T denotes the transpose operation. And the Moore–Penrose pseudoinverse $A^\dagger$, which can be regarded as a generalization of matrix inversion to nonsquare matrices, may be computed from the singular value decomposition (SVD) [21, 22]. The SVD decomposes the matrix $A_{M \times (2Q+1)}$ of rank ρ into the product of two orthogonal matrices $U_{M \times M}$, $V_{(2Q+1) \times (2Q+1)}$, and a diagonal matrix $\overline{\Sigma}_{\rho \times \rho} = diag\left(\sigma_1, \sigma_2, \cdots \sigma_\rho\right)$,

$$A = U \begin{pmatrix} \overline{\Sigma} & 0 \\ 0 & 0 \end{pmatrix}_{M \times (2Q+1)} V^T. \tag{2.48}$$

The σ_i are called the nonzero singular values of A, and are sorted in decreasing order, that is, $\sigma_1 \geq \sigma_2 \geq \cdots \geq \sigma_\rho > 0$. Note that if σ_i is negative, one can make it positive by simply multiplying either the ith column of U or the ith column of V by -1. The factorization in Equation (2.48) is called the SVD of A, and the columns in U and V are called left-hand and right-hand singular vectors, respectively.

The pseudoinverse $A^\dagger$ can now be written as

$$\begin{aligned}
A^\dagger &= \left(A^T A\right)^{-1} A^T \\
&= \left[V \begin{pmatrix} \overline{\Sigma} & 0 \\ 0 & 0 \end{pmatrix} U^T U \begin{pmatrix} \overline{\Sigma} & 0 \\ 0 & 0 \end{pmatrix}^{-1} V \begin{pmatrix} \overline{\Sigma} & 0 \\ 0 & 0 \end{pmatrix} U^T \right] \\
&= V \begin{pmatrix} \overline{\Sigma} & 0 \\ 0 & 0 \end{pmatrix} U^T,
\end{aligned} \tag{2.49}$$

where the orthogonal property of matrices U and V is used in the derivation of Equation (2.49), and $\overline{\Sigma}^{-1} = diag\left(1/\sigma_1, 1/\sigma_2, \cdots 1/\sigma_\rho\right)$.

After Equation (2.45) is solved, one can proceed to compute the poles of the original problem. Notice that $\lambda(s)$ and $\theta(s)$ can be expanded into the following pole-zero form,

$$\lambda(s) = \frac{\prod_{k=1}^{Q}(s - \tilde{z})}{\prod_{k=1}^{Q}(s - \tilde{p}_k)}, \tag{2.50}$$

and

$$\theta(s) = \kappa \frac{\prod_{k=1}^{Q}(s - \hat{z}_k)}{\prod_{k=1}^{Q}(s - \tilde{p}_k)}, \tag{2.51}$$

where $\tilde{z}_i$ and $\hat{z}_i$ are the zeroes of $\lambda(s)$ and $\theta(s)$, respectively, and κ is a real constant.

Therefore, one can get

$$H(s) = \frac{\lambda(s)H(s)}{\lambda(s)} = \frac{\theta(s)}{\lambda(s)} = \kappa \frac{\prod_{k=1}^{Q}(s - \hat{z}_k)}{\prod_{k=1}^{Q}(s - \tilde{z}_k)}. \tag{2.52}$$

Equation (2.52) reveals that the poles of the original transfer function $H(s)$ are equal to the zeros of $\lambda(s)$, due to the same set of initial poles used for both expansions.

Following the solution of the linear Equation (2.45), the zeros of $\lambda(s)$ can be calculated as eigenvalues of the following matrix [13]:

$$\Upsilon = G - \Lambda R^T. \tag{2.53}$$

In the case of only real poles involved, G is a diagonal matrix containing the staring poles $\tilde{p}_k$, Λ is a column vector of ones, and R^T is a row vector comprising the residues of $\lambda(s)$. They are given as

$$G = \begin{bmatrix} \tilde{p}_1 & 0 & \cdots & 0 \\ 0 & \tilde{p}_2 & \vdots & 0 \\ \vdots & \vdots & \ddots & \vdots \\ 0 & 0 & \cdots & \tilde{p}_Q \end{bmatrix}, \tag{2.54}$$

$$\Lambda = [1 \quad 1 \quad \cdots \quad 1]_{1 \times Q}^T, \tag{2.55}$$

$$R^T = [\tilde{r}_1 \quad \tilde{r}_2 \quad \cdots \quad \tilde{r}_Q]. \tag{2.56}$$

Therefore, in the case of real poles, Equation (2.53) can be written explicitly as

$$
\Upsilon = \begin{bmatrix} \tilde{p}_1 & 0 & \cdots & 0 \\ 0 & \tilde{p}_2 & \vdots & 0 \\ \vdots & \vdots & \ddots & \vdots \\ 0 & 0 & \cdots & \tilde{p}_Q \end{bmatrix} - \begin{bmatrix} \tilde{r}_1 & \tilde{r}_2 & \cdots & \tilde{r}_Q \\ \tilde{r}_1 & \tilde{r}_2 & \cdots & \vdots \\ \vdots & \vdots & \cdots & \vdots \\ \tilde{r}_1 & \tilde{r}_2 & \cdots & \tilde{r}_Q \end{bmatrix}. \tag{2.57}
$$

It is obvious that the product ΛR^T is a Q by Q matrix, but its rank has only one. Equation (2.53) or Equation (2.57) can be considered as a special case of a generalized companion matrix used for root finding of a polynomial with an eigenvalue method [23].

If a complex pair of poles, for example, $\tilde{p}_k$ and $\tilde{p}_{k+1}$ with $\tilde{p}_{k+1} = \tilde{p}_k^*$, is present, the matrix block in Equation (2.53) corresponding to the conjugate complex poles is modified through a similarity transformation similar to Equation (2.80). Finally, the submatrices corresponding to the complex-pole pair is given by

$$
G_{2\times2}' = \begin{bmatrix} \mathrm{Re}(\tilde{p}_k) & \mathrm{Im}(\tilde{p}_k) \\ -\mathrm{Im}(\tilde{p}_k) & \mathrm{Re}(\tilde{p}_k) \end{bmatrix}, \quad \Lambda_{2\times2}' = \begin{bmatrix} 2 & 0 \end{bmatrix}^T,
$$
$$
\text{and} \quad R_{2\times1}'^T = \begin{bmatrix} \mathrm{Re}(\tilde{r}_k) & \mathrm{Im}(\tilde{r}_k) \end{bmatrix}. \tag{2.58}
$$

Again, the modification keeps Υ in Equation (2.53) as a real matrix even in the presence of complex poles, which ensure that its complex eigenvalues will be computed as conjugate pairs. Clearly, if both real poles and complex conjugate poles are present, the matrix G in Equation (2.53) is a tridiagonal matrix. The eigenvalues of Υ in Equation (2.53) can be computed by a QR-decomposition algorithm [22]. These eigenvalues are the new poles to replace the previously estimated poles in Equation (2.37). This is also called pole relocation. The iterative process proceeds until convergence is achieved, which implies that the scaling function $\lambda(s)$ in Equation (2.36) approaches unity at all frequencies.

2.4.1.2 Residue Identification

Substituting the new poles obtained in the first step into Equation (2.32) and writing it at the sampled frequencies, one can obtain an overdetermined linear problem similar to Equation (2.44), but with unknowns c

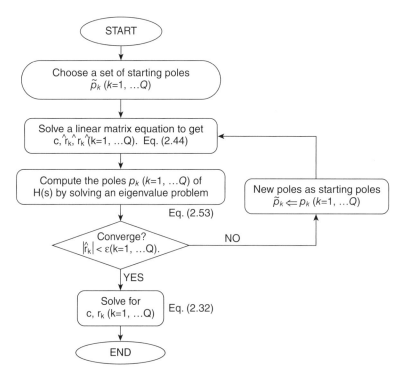

Figure 2.4   Procedures for rational function approximation via the vector fitting method.

and r_k. Solving the linear least-squares problem can produce the new residues corresponding to the new poles.

Steps 1 and 2 may have to be repeated several times with the new poles obtained in step 1 as initial poles until the problem converges. It can be observed from Equations (2.34) and (2.35) that if the convergence is achieved, that is, the actual poles and residues of $H(s)$ are obtained, $\theta(s)$ degenerates into unity, or equivalently all $\tilde{r}_k$'s become zero. Usually less than five iterations are needed for the applications in Reference 13. The procedures for the VF method are summarized in Figure 2.4.

2.4.2 Fitting Vectors with Common Poles

To this end, the VF algorithm is applied to a scalar function. As suggested by its name, the VF method is also applicable to a vector. A common set of poles is produced for all the elements in the vector. When it is applied to fitting the transfer matrix of a multiport network, the property of all the elements in the vector sharing a common set of poles will in general reduce the order of the model for the multiport network, and subsequently facilitate the model synthesis process.

It is straightforward for the preceding process of the VF method to be generalized from fitting a scalar function to a vector. Consider a vector with N elements

$$\mathbf{H} = \begin{bmatrix} H_1(s) \\ H_2(s) \\ \vdots \\ H_N(s) \end{bmatrix}. \tag{2.59}$$

Since all the elements have the same set of poles, the initial poles and the scaling function remain the same as those in Equation (2.34) for the scalar case. However, Equation (2.37) is modified accordingly as

$$\begin{bmatrix} \hat{c}^1 + \sum_{k=1}^{Q} \dfrac{\hat{r}_k^1}{s - \tilde{p}_k} \\ \hat{c}^2 + \sum_{k=1}^{Q} \dfrac{\hat{r}_k^2}{s - \tilde{p}_k} \\ \vdots \\ \hat{c}^N + \sum_{k=1}^{Q} \dfrac{\hat{r}_k^N}{s - \tilde{p}_k} \end{bmatrix} \begin{bmatrix} \left(\sum_{k=1}^{Q} \dfrac{\hat{r}_k}{s - \tilde{p}_k}\right) H_1(s) \\ \left(\sum_{k=1}^{Q} \dfrac{\hat{r}_k}{s - \tilde{p}_k}\right) H_2(s) \\ \vdots \\ \left(\sum_{k=1}^{Q} \dfrac{\hat{r}_k}{s - \tilde{p}_k}\right) H_N(s) \end{bmatrix} = \begin{bmatrix} H_1(s) \\ H_2(s) \\ \vdots \\ H_N(s) \end{bmatrix}. \tag{2.60}$$

The linear matrix equation can be written as

$$AX = B, \tag{2.61}$$

where

$$A = \begin{bmatrix} A_a & 0 & & 0 & A_b^1 \\ 0 & A_a & 0 & \vdots & A_b^2 \\ \vdots & 0 & \ddots & 0 & \vdots \\ 0 & \cdots & 0 & A_a & A_b^N \end{bmatrix}, \tag{2.62}$$

$$X = \begin{bmatrix} \hat{c}_1 & \hat{r}_1^1 & \cdots & \hat{r}_Q^1 & \cdots & \hat{c}^M & \hat{r}_1^M & \cdots & \hat{r}_Q^M & \tilde{c}^M & \tilde{r}_1^M & \cdots & \tilde{r}_Q^M \end{bmatrix}^T,$$
(2.63)

$$B = \begin{bmatrix} H_1(s) & H_2(s) & \cdots & H_N(s) \end{bmatrix}.$$
(2.64)

The submatrices A_a and A_b^i in Equation (2.62) are given by

$$A_a = \left\{ \begin{matrix} 1 & (s_1 - \tilde{p}_1)^{-1} & \cdots & (s_1 - \tilde{p}_Q)^{-1} \\ \vdots & \vdots & \vdots & \vdots \\ 1 & (s_M - \tilde{p}_1)^{-1} & \cdots & (s_M - \tilde{p}_Q)^{-1} \end{matrix} \right\},$$
(2.65)

$$A_b^i = \left\{ \begin{matrix} \dfrac{-H_i(s_1)}{(s_1 - \tilde{p}_1)} & \cdots & \dfrac{-H_i(s_1)}{(s_1 - \tilde{p}_Q)} \\ \vdots & & \vdots \\ \dfrac{-H_i(s_M)}{(s_M - \tilde{p}_1)} & & \dfrac{-H_i(s_M)}{(s_M - \tilde{p}_Q)} \end{matrix} \right\}, \quad i = 1, 2, \cdots, N.$$
(2.66)

Solving the linear system of equations in Equation (2.61) via Equation (2.47), one can find the residues $\tilde{r}_k$ of the scaling function $\lambda(s)$. Then the new poles can be computed as the eigenvalues of Υ in Equation (2.53). After several iterations, the accurate poles for the vector in Equation (2.59) can be obtained. Therefore, the corresponding residues for each element in the vector can be computed independently by solving linear least-squares problems with the poles as known quantities. The following section will discuss some practical problems about the implementation of the VF algorithm.

2.4.3 Selection of Initial Poles

Initial poles affect the convergence speed of the VF method [13]. The VF method, in general, needs iterations to relocate the initial poles into their final positions in the Laplace domain. The farther the initial poles are from their final positions, the more iterations need to be performed. Therefore, selection of appropriate initial poles $\tilde{p}_k$ in the first step of the VF method is important for a successful and fast rational function approximation.

If a transfer function to be approximated is a smooth one over its frequency range of interest, a set of real poles, which is linearly or logarithmically spaced as a function of frequencies, can be applied as initial poles. In contrast, for transfer functions with many resonant peaks, the initial poles should be chosen as complex conjugate pairs. Furthermore, the imaginary parts of these conjugate pairs shall be linearly distributed over the frequency range of interest and at least one hundred times larger than the real parts, that is,

$$\left| \mathrm{Im}\left\{ \tilde{p}_k \right\} \right| \geq 100 \cdot \left| \mathrm{Re}\left\{ \tilde{p}_k \right\} \right|. \tag{2.67}$$

Choosing complex pole pairs with weak attenuation as the initial poles will reduce the number of iterations of the VF method. Such an effect of the chosen initial poles can be attributed to two reasons. First, if interconnect structures or other linear systems have low loss, then the time-domain response of this system will be weakly damped with small decay. This translates to small real parts of the poles of the systems in the Laplace domain. In this regard, for a highly lossy interconnect system, the choice of initial poles in Equation (2.67) may not be effective. Second, dominant poles of a linear system are located closely to the imaginary axis in the Laplace domain and only those poles significantly influence the time and frequency characteristics of a system. Initial poles following Equation (2.67) are located close to the imaginary axis in the Laplace domain and are good approximation to the actual dominant poles of a linear system.

2.4.4 Enhancement to the Original Vector Fitting Method

Over the years, many variations and improvement on the original VF method have been reported in the literature, for example, the time-domain VF method [24], and orthonormal VF by replacing the partial fractions in the original VF method with orthonormal rational functions [25]. Two pieces of simple but important enhancement to the original VF method are discussed in this section: (1) relaxed VF [26], and (2) fast implementation of the VF method [27].

2.4.4.1 Relaxed VF

The convergence performance of the original VF method, when fits noise contaminated frequency responses or the order of the model is less than that of the transfer function, is impeded due to the least-square normalization being set to one in Equation (2.34). This normalization affects the required pole relocation and may result in an increase in the least-quare error for Equation (2.36). Since the least-square solver produce a minimum error, the poles are relocated in small steps and the convergence may even stall. Another problem is that the least-square solution tends to produce a scaling function of Equation (2.34) with a small magnitude to scale down the fitting error and thus produces a smaller least-square error. The latter implies that the relocated poles will not minimize the original problem. For instance, the poles tend to shift too much toward low frequencies when fitting smooth functions.

A modification is introduced in [26] to alleviate the above-mentioned difficulties for the original VF method. It is achieved by replacing the high-frequency asymptotic requirement of the scaling function $\lambda(s)$ in Equation (2.34) with a more relaxed condition:

$$\lambda(s) = \tilde{c} + \sum_{k=1}^{Q} \frac{\tilde{r}_k}{s - \tilde{p}_k}, \qquad (2.68)$$

where $\tilde{c}$ is real. In order to avoid trivial solutions, one equation is added to the least-square problem in Equation (2.44):

$$\text{Re}\left\{ \sum_{m=1}^{M} \left(\tilde{c} + \sum_{k=1}^{Q} \frac{\tilde{r}_k}{s - \tilde{p}_k} \right) \right\} = M. \qquad (2.69)$$

Equation (2.69) is weighted in relation to the size of the transfer function in the least-square problem by

$$\text{weight} = \frac{\|w_e(s) \cdot H(s)\|_2}{M}, \qquad (2.70)$$

where $w_e(s)$ is an external weighting function. And since $\lambda(s)$ does not converge to unity at high frequencies during the iterations, Equation (2.53) should be modified to:

$$\Upsilon = G - \Lambda \cdot \tilde{c}^{-1} \cdot R^T. \qquad (2.71)$$

The zeros computed by Equation (2.71) is valid subject to a non-zero $\tilde{c}$. The solution is dropped and the least-square problem is solved again with a fixed value for $\tilde{c}$ in Equation (2.34): $\tilde{c} = \delta_{tol} \cdot \tilde{c}/|\tilde{c}|$, if the absolute value of $\tilde{c}$ is found to be smaller than a threshold, say, $\delta_{tol} = 1.0 \times 10^{-8}$.

Equation (2.69) enforces that the sum of the real part of $\lambda(s)$ over the given frequency samples equals a nonzero value. When the solution converges, $\lambda(s)$ will approach unity at all frequencies ($\tilde{r}_k = 0$, $\tilde{c} = 1$) as in the original VF formulation. The right-hand side in Equation (2.69) can take other numbers other than M since the resultant scaling of $\lambda(s)$ causes an identical scaling of $H(s)$ in Equation (2.35). In other words, Equation (2.69) does not impose any constraint on the original least-square problem other than preventing $\lambda(s)$ from becoming zero. Equation (2.69) alleviates the downscaling phenomenon by fixing the sum of the samples. It can be understood that a reduction in least-square error due to downscaling in some frequency range will be offset by a magnification of the least-square error at other frequencies. Moreover, the freed variable improves the pole relocating capability as the error magnification caused by a given pole relocation is reduced.

2.4.4.2 Fast Implementation of VF Method

For a linear system with a large number of ports, the broadband macromodeling with a common set of poles by the original VF method can consume large memory and take long computation time. A fast version of the VF method is proposed in Reference 27, which significantly reduces the computation time and memory usage by applying incomplete QR decomposition. A closer observation of Equations (2.43) and (2.57) reveals that only the residues $\tilde{r}_k$ for $\lambda(s)$ are used to find the poles by Equation (2.57), while all the residues $\hat{r}_k$ for $\theta(s)$ ($= \lambda(s)H(s)$) are calculated but never used. Much computational resources will be saved if the QR decomposition is applied to single-element least-square equations. Instead of formulating the least-square equations in Equation (2.61), QR decomposition is first operated on the least-square equations of each individual matrix element i:

$$\begin{bmatrix} A_a & A_b^i \end{bmatrix} = \begin{bmatrix} Q_i \end{bmatrix} \begin{bmatrix} R_i^{11} & R_i^{12} \\ 0 & R_i^{22} \end{bmatrix}, \quad i = 1, 2, \cdots, N. \tag{2.72}$$

Then all the decompositions in Equation (2.72) are assembled to make the following reduced set of least-square equations:

$$\begin{bmatrix} R_1^{22} \\ \vdots \\ R_N^{22} \end{bmatrix} \begin{bmatrix} \tilde{c}^M & \tilde{r}_1^M & \cdots & \tilde{r}_Q^M \end{bmatrix}^T = \begin{bmatrix} Q_1^T H_1 \\ \vdots \\ Q_N^T H_N \end{bmatrix}. \tag{2.73}$$

The resultant simplified equations in Equation (2.73) depend only on $\tilde{r}_k$ of $\lambda(s)$. The left-hand sides of Equations (2.72) and (2.73) are both significantly smaller than the sparse matrix in Equation (2.61), which reduce largely the memory usage and computational time. Once the poles of the system are identified, the subsequent identification of residues is performed as the eigenvalue solution to Equation (2.71). The computational complexity is thus reduced from $O((N^2 + 1)^2 Q^2 M \cdot N^2)$ to $O(Q^2 M \cdot N^2)$ for a network with N-input and N-output ports, M sampling frequencies, and Q common poles.

Note that the VF method, incorporated the above two pieces of latest development, is programmed by B. Gustavsen, and can be found online [28].

2.5 MACROMODEL SYNTHESIS

From the previous section, one finally obtains a macromodel with partial fractions of the transfer function $H(s)$ in Equation (2.21), either Y or S parameters, of a general N-port interconnect network by applying the VF method. The macromodel is expressed in the pole-residue form as follows:

$$H(s) = \begin{bmatrix} c^{1,1} + \sum_{k=1}^{Q} \dfrac{r_k^{1,1}}{s - p_k} & c^{1,2} + \sum_{k=1}^{Q} \dfrac{r_k^{1,2}}{s - p_k} & \cdots & c^{1,N} + \sum_{k=1}^{Q} \dfrac{r_k^{1,N}}{s - p_k} \\[2ex] c^{2,1} + \sum_{k=1}^{Q} \dfrac{r_k^{2,1}}{s - p_k} & c^{2,2} + \sum_{k=1}^{Q} \dfrac{r_k^{2,2}}{s - p_k} & \cdots & c^{2,N} + \sum_{k=1}^{Q} \dfrac{r_k^{2,N}}{s - p_k} \\[2ex] \vdots & \vdots & \ddots & \vdots \\[2ex] c^{N,1} + \sum_{k=1}^{Q} \dfrac{r_k^{N,1}}{s - p_k} & c^{N,2} + \sum_{k=1}^{Q} \dfrac{r_k^{N,2}}{s - p_k} & \cdots & c^{N,N} + \sum_{k=1}^{Q} \dfrac{r_k^{N,N}}{s - p_k} \end{bmatrix},$$

(2.74)

where p_k are the poles of the interconnect network, which are identical for all the entries in the transfer function. Q is the number of poles. $c^{i,j}$ and $r_k^{i,j}$ are the direct coupling constant and residues corresponding to the poles, respectively.

The derivation of partial differential equations from the macro-model in Equation (2.74) is referred to as macromodel synthesis [12]. In general, a set of first-order differential equations, which is also called state-space equations, can be formulated as

$$
\begin{cases}
\dfrac{d}{dt} x(t) = Ax(t) + Bu(t) \\[2mm]
\quad y(t) = Cx(t) + Du(t),
\end{cases}
\tag{2.75}
$$

where $A \in \mathbb{R}^{L \times L}$, $B \in \mathbb{R}^{L \times N}$, $C \in \mathbb{R}^{N \times L}$, and $D \in \mathbb{R}^{N \times N}$. N is the number of ports of the interconnect network. And L is the total number of states, which equals the product of the total number of poles and the total number of ports, that is, $L = Q \times N$. If the transfer function $H(s)$ is an admittance matrix, then the kth element of the input vector $u(t)$ and the output vector $y(t)$ corresponds to the voltage $v_k(t)$ and current $i_k(t)$ at port k, respectively. Whereas, if $H(s)$ is a scattering matrix, they represent the incident wave $a_k(t)$ and the reflected wave $b_k(t)$ at port k, respectively. The incident and reflected waves at the kth port are defined in terms of the port voltage $v_k(t)$ and the port current $i_k(t)$ with respect to an arbitrary reference impedance Z_0 at port k, by

$$
\begin{cases}
a_k(t) = \left[v_k(t) + z_0 i_k(t) \right] / \left(2\sqrt{z_0} \right) \\[2mm]
b_k(t) = \left[v_k(t) + z_0 i_k(t) \right] / \left(2\sqrt{z_0} \right).
\end{cases}
\tag{2.76}
$$

2.5.1 Jordan Canonical Method for Macromodel Synthesis

It is known that two equivalent state-space systems have the same matrix transfer function. Therefore, given a matrix-transfer function formulated in Equation (2.74), several forms of state-space realization can be obtained. In this chapter, the macromodel synthesis is realized by the Jordan-canonical method [12, 29].

Assuming that the submatrices A_r, B_r, and C_r for a general N-port network contain only real poles and their corresponding residues, and the submatrices A_c, B_c, and C_c comprise only complex conjugate poles and their corresponding residues, the Jordan-canonical realization of Equation (2.74) takes the following form:

$$
\begin{Bmatrix} \dot{x}_1 \\ \dot{x}_2 \\ \dot{x}_3 \end{Bmatrix} = \begin{bmatrix} A_r & 0 & 0 \\ 0 & A_c & 0 \\ 0 & 0 & A_c^* \end{bmatrix} \begin{Bmatrix} x_1 \\ x_2 \\ x_3 \end{Bmatrix} = \begin{bmatrix} B_r \\ B_c \\ B_c^* \end{bmatrix} u, \tag{2.77}
$$

$$
i = \begin{bmatrix} C_r & C_c & C_c^* \end{bmatrix} \begin{Bmatrix} x_1 \\ x_2 \\ x_3 \end{Bmatrix} + Du, \tag{2.78}
$$

where the asterisk denotes the complex conjugate. Since complex poles do not have a direct meaning in the time domain [12], the similarity transform is introduced by

$$
\tilde{x} = Tx, \tag{2.79}
$$

where $\tilde{x} = \{\tilde{x}_1 \quad \tilde{x}_2 \quad \tilde{x}_3\}^T$ and $x = \{x_1 \quad x_2 \quad x_3\}^T$ are the vectors containing state variables, and the transformation matrix is defined as [30]

$$
T = \begin{bmatrix} I & 0 & 0 \\ 0 & I & I \\ 0 & jI & -jI \end{bmatrix}, \tag{2.80}
$$

where I is an identity matrix, and j equals $\sqrt{-1}$.

It can be proved that the similarity transformation does not change the transfer function of the original system [29]. Consider the following two state-space systems,

$$
\begin{cases} \dot{x} = Ax + Bu \\ y = Cx + Du, \end{cases} \tag{2.81}
$$

$$
\begin{cases} \dot{\tilde{x}} = \tilde{A}\tilde{x} + \tilde{B}u \\ \tilde{y} = \tilde{C}\tilde{x} + \tilde{D}u. \end{cases} \tag{2.82}
$$

The state-space system in Equation (2.82) is obtained by the similarity transformation defined in Equation (2.79). Substituting

$x = T^{-1}\tilde{x}$, $\dot{x} = T^{-1}\dot{\tilde{x}}$ into Equation (2.81) and comparing it with Equation (2.82), one can easily obtain the following relation between the corresponding matrices of the two systems:

$$\begin{cases} \tilde{A} = TAT^{-1}, & \tilde{C} = CT^{-1} \\ \tilde{B} = TB, & \tilde{D} = D. \end{cases} \tag{2.83}$$

With Equation (2.83), one can prove that the transfer function of Equation (2.82) is the same as that of Equation (2.81),

$$\begin{aligned} \tilde{H}(s) &= \tilde{C}\left(sI - \tilde{A}\right)H - 1\tilde{B} + \tilde{D} \\ &= CT^{-1}\left(sI - TAT^{-1}\right)^{-1}TB + D \\ &= CT^{-1}\left[T\left(sT^{-1}T - A\right)T^{-1}\right]^{-1}TB + D \\ &= CT^{-1}T\left(sT^{-1}T - A\right)T^{-1}TB + D \\ &= C(s - A)B \\ &= H(s), \end{aligned} \tag{2.84}$$

where in the derivation of Equation (2.84), the orthogonality of the similarity transformation matrix T is used. Therefore, one can finally obtain the following state-space equations with all real-valued matrix entries:

$$\begin{bmatrix} \dot{\tilde{x}}_1 \\ \dot{\tilde{x}}_2 \\ \dot{\tilde{x}}_3 \end{bmatrix} = \begin{bmatrix} A_r & 0 & 0 \\ 0 & \text{Re}(A_c) & \text{Im}(A_c) \\ 0 & -\text{Im}(A_c) & \text{Re}(A_c) \end{bmatrix} \begin{bmatrix} \tilde{x}_1 \\ \tilde{x}_2 \\ \tilde{x}_3 \end{bmatrix} + \begin{bmatrix} B_r \\ 2\text{Re}(B_c) \\ -2\text{Im}(B_c) \end{bmatrix} u, \quad (2.85)$$

$$y = \begin{bmatrix} C_r & \text{Re}(C_c) & \text{Im}(C_c) \end{bmatrix} \begin{bmatrix} \tilde{x}_1 \\ \tilde{x}_2 \\ \tilde{x}_3 \end{bmatrix} + Du. \tag{2.86}$$

For a general N-port network characterized by $q1$ real poles and $q2$ complex conjugate pole pairs, the dimension of the matrix A is $(q1 + 2 * q2) * N$.

A two-port network containing two common poles is used as an example to illustrate the process of macromodel synthesis. Its transfer function of Y-matrix can be expressed as

$$
\begin{bmatrix}
c^{1,1} + \displaystyle\sum_{k=1}^{2} \dfrac{r_k^{1,1}}{s - p_k} & c^{1,2} + \displaystyle\sum_{k=1}^{2} \dfrac{r_k^{1,2}}{s - p_k} \\[4mm]
c^{2,1} + \displaystyle\sum_{k=1}^{2} \dfrac{r_k^{2,1}}{s - p_k} & c^{2,2} + \displaystyle\sum_{k=1}^{2} \dfrac{r_k^{2,2}}{s - p_k}
\end{bmatrix}
\begin{bmatrix} V_1 \\ V_2 \end{bmatrix}
=
\begin{bmatrix} I_1 \\ I_2 \end{bmatrix}.
\qquad (2.87)
$$

The Jordan-canonical method is used to realize the time-domain state-space equations. If all the poles are real, then only four state variables are needed and the final state-space equations are

$$
\begin{Bmatrix} \dot{x}_1 \\ \dot{x}_2 \\ \dot{x}_3 \\ \dot{x}_4 \end{Bmatrix}
=
\begin{bmatrix}
p_1 & 0 & 0 & 0 \\
0 & p_1 & 0 & 0 \\
0 & 0 & p_2 & 0 \\
0 & 0 & 0 & p_2
\end{bmatrix}
\begin{Bmatrix} x_1 \\ x_2 \\ x_3 \\ x_4 \end{Bmatrix}
+
\begin{bmatrix}
1 & 0 \\
0 & 1 \\
1 & 0 \\
0 & 1
\end{bmatrix}
\begin{bmatrix} v_1 \\ v_2 \end{bmatrix}
$$

$$
y = \begin{bmatrix} i_1 \\ i_2 \end{bmatrix}
=
\begin{bmatrix}
r_1^{1,1} & r_1^{1,2} & r_2^{1,1} & r_2^{1,2} \\
r_1^{2,1} & r_1^{2,2} & r_2^{2,1} & r_2^{2,2}
\end{bmatrix}
\begin{Bmatrix} x_1 \\ x_2 \\ x_3 \\ x_4 \end{Bmatrix}
+
\begin{bmatrix}
c^{1,1} & c^{1,2} \\
c^{2,1} & c^{2,2}
\end{bmatrix}
\begin{bmatrix} v_1 \\ v_2 \end{bmatrix}.
\qquad (2.88)
$$

If one pair of complex conjugate poles presents in Equation (2.87), which is denoted by $w' \pm jw''$ and its corresponding residues at different ports are $r'_{k,l} \pm jr''_{k,l}$ $(k,l = 1,2)$, the final state-space realization is

$$
\begin{Bmatrix} \dot{\tilde{x}}_1 \\ \dot{\tilde{x}}_2 \\ \dot{\tilde{x}}_3 \\ \dot{\tilde{x}}_4 \end{Bmatrix}
=
\begin{bmatrix}
w' & 0 & w'' & 0 \\
0 & w' & 0 & w'' \\
-w'' & 0 & w' & 0 \\
0 & -w'' & 0 & w'
\end{bmatrix}
\begin{Bmatrix} \tilde{x}_1 \\ \tilde{x}_2 \\ \tilde{x}_3 \\ \tilde{x}_4 \end{Bmatrix}
+
\begin{bmatrix}
2 & 0 \\
0 & 2 \\
0 & 0 \\
0 & 0
\end{bmatrix}
\begin{bmatrix} v_1 \\ v_2 \end{bmatrix}
$$

$$
y = \begin{bmatrix} i_1 \\ i_2 \end{bmatrix}
=
\begin{bmatrix}
r'_{1,1} & r'_{1,2} & r''_{1,1} & r''_{1,2} \\
r'_{2,1} & r'_{2,2} & r''_{2,1} & r''_{2,2}
\end{bmatrix}
\begin{Bmatrix} \tilde{x}_1 \\ \tilde{x}_2 \\ \tilde{x}_3 \\ \tilde{x}_4 \end{Bmatrix}.
\qquad (2.89)
$$

Finally, the state-space equations can be converted to equivalent circuits, which are compatible with the SPICE circuit simulator for time-domain analysis. This topic is examined in the following section.

2.5.2 Equivalent Circuits

The SPICE is a powerful general-purpose circuit simulation program for nonlinear DC, nonlinear transient and linear AC analyses, which is used to verify circuit designs and to predict the circuit behavior. It was originally developed at the Electronics Research Laboratory of the University of California, Berkeley in 1975. A few commercial versions of SPICE program are also widely used, such as PSpice [31] and HSPICE. For signal integrity analysis with a large number of nonlinear circuit elements, much effort can be saved by exploiting the many types of nonlinear circuit models embedded in the powerful SPICE circuit simulator.

However, the aforementioned SPICE simulators may not be able to accept directly the differential equations in Equation (2.75). In that case, the macromodel represented by Equation (2.75) can be converted to an equivalent circuit network consisting of passive elements and controlled voltage and current sources [10, 12].

2.5.2.1 Admittance Matrix-Based Equivalent Circuits

For the purpose of illustration, a simple case of a two-port network with two states variables characterized by admittance parameters is considered [12]:

$$
\begin{aligned}
\begin{Bmatrix} \dot{x}_1 \\ \dot{x}_2 \end{Bmatrix} &= \begin{bmatrix} a_{11} & a_{12} \\ a_{21} & a_{22} \end{bmatrix} \begin{Bmatrix} x_1 \\ x_2 \end{Bmatrix} + \begin{bmatrix} b_{11} & b_{12} \\ b_{21} & b_{22} \end{bmatrix} \begin{Bmatrix} v_1 \\ v_2 \end{Bmatrix} \\
\begin{Bmatrix} i_1 \\ i_2 \end{Bmatrix} &= \begin{bmatrix} c_{11} & c_{12} \\ c_{21} & c_{22} \end{bmatrix} \begin{Bmatrix} x_1 \\ x_2 \end{Bmatrix} + \begin{bmatrix} d_{11} & d_{12} \\ d_{21} & d_{22} \end{bmatrix} \begin{Bmatrix} v_1 \\ v_2 \end{Bmatrix}.
\end{aligned}
\tag{2.90}
$$

An equivalent-circuit network representing Equation (2.90) can be realized as shown in Figure 2.5. v_1, v_2 and i_1, i_2 are the port voltages and currents, respectively. State variables x_1 and x_2 are represented by the voltages of capacitors, whereas voltage-controlled current sources (VCCS) are used to replace the terms such as $c_{11}x_1$. Equivalent circuit

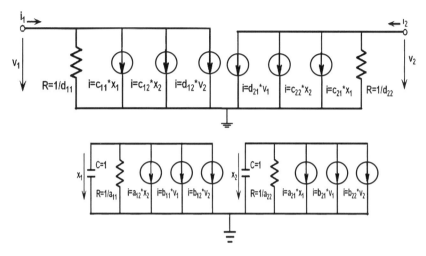

Figure 2.5 Illustration of the equivalent circuit realization of admittance matrix-based macromodel represented by Equation (2.90).

realization can be easily generalized for the case with more state variables or more ports.

2.5.2.2 Scattering Matrix-Based Equivalent Circuits

Similarly, a simple case of Equation (2.75) is considered, which represents a two-port network with two state variables characterized by scattering parameters [30]:

$$\begin{Bmatrix} \dot{x}_1 \\ \dot{x}_2 \end{Bmatrix} = \begin{bmatrix} a_{11} & a_{12} \\ a_{21} & a_{22} \end{bmatrix} \begin{Bmatrix} x_1 \\ x_2 \end{Bmatrix} + \begin{bmatrix} b_{11} & b_{12} \\ b_{21} & b_{22} \end{bmatrix} \begin{Bmatrix} a_1 \\ a_2 \end{Bmatrix}$$

$$\begin{Bmatrix} b_1 \\ b_2 \end{Bmatrix} = \begin{bmatrix} c_{11} & c_{12} \\ c_{21} & c_{22} \end{bmatrix} \begin{Bmatrix} x_1 \\ x_2 \end{Bmatrix} + \begin{bmatrix} d_{11} & d_{12} \\ d_{21} & d_{22} \end{bmatrix} \begin{Bmatrix} a_1 \\ a_2 \end{Bmatrix},$$

(2.91)

where a_k and b_k are the incident wave and the reflected wave at port k, respectively.

Additional equations relating the wave variables to the port voltages and currents need to be supplemented,

$$\begin{cases} i_k(t) = \dfrac{1}{\sqrt{Z_0}} \left[a_k(t) - b_k(t) \right] \\ v_k(t) = \sqrt{Z_0} \left[a_k(t) + b_k(t) \right], \end{cases}$$

(2.92)

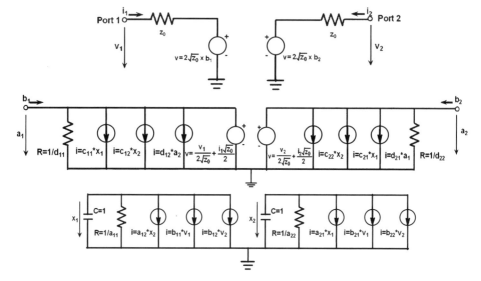

Figure 2.6 Illustration of the equivalent circuit realization of scattering matrix-based macromodel represented by Equations (2.91) and (2.92).

where Z_0 is the characteristic impedance of port k.

A similar equivalent circuit network representing Equations (2.91) and (2.92) can be realized as shown in Figure 2.6. It can be generalized to the cases with more state variables or ports.

2.6 STABILITY, CAUSALITY, AND PASSIVITY OF MACROMODEL

Before it can be used in time-domain simulation to produce stable results, the macromodel, which is obtained by the rational function approximation with least-square fitting approaches, must fulfill the requirements on stability, causality, and passivity, in addition to accuracy.

2.6.1 Stability

The linear system theory [29] presents the following points regarding the stability of a linear system. If all the real parts of the exponents

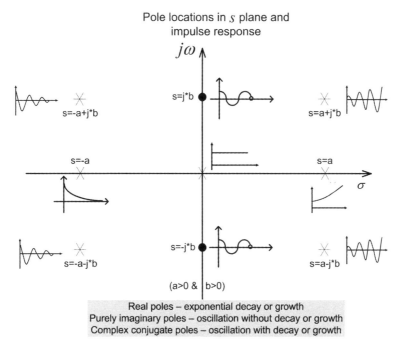

Figure 2.7   Locations of poles in the Laplace domain and their corresponding time-domain impulse responses.

representing a system in time domain are strictly negative, exponential stability occurs and the signals decay within an exponential envelope. If the real parts of the exponents are zero, the corresponding response of the system never decays or grows in amplitude, which is called marginal stability. If the real part of at least one of the exponents is positive, then its response grows without bound so as to cause the system to be unstable. Figure 2.7 illustrates the impulse response of a system subject to the different locations of the poles.

In the context of rational function approximation, the stability of a system in time domain correlates with the property of the poles. Therefore, in order for the fitting model to be useful for time-domain simulations, all the poles obtained from the rational function approximation must be stable. The condition to ensure the stability of the fitting model is that all the poles of the fitting model must lie in the left half of the complex plane, that is, $\text{Re}\{p_k\} \leq 0$.

However, some unstable poles may emerge during the VF process. The constraint on the fitting model is often enforced by some simple treatments, for example, directly deleting the unstable poles or flipping them around the imaginary axis to the left half plane [13] before computing their corresponding residues. Flipping an unstable pole p_k with $Re(p_k) > 0$ from the right half plane to the left half plane has the effect of multiplying the approximant by an all-pass function,

$$P(s) = \frac{s - p_k}{s - [-Re(p_k) + j\,Im(p_k)]},\tag{2.93}$$

and

$$|P(j\omega)| = \frac{\sqrt{|-Re(p_k)|^2 + |\omega - Im(p_k)|^2}}{\sqrt{|Re(p_k)|^2 + |\omega - Im(p_k)|^2}} = 1.\tag{2.94}$$

This reveals that the pole flipping does not change the amplitude of the system. The VF produces a multiport common-pole model with guaranteed stable poles.

2.6.2 Causality

Causality dictates that the effect cannot precede the cause, that is, the response of a causal system cannot happen before the input impulse. The time-domain response $h(t)$ of a causal system are vanishing for $t < 0$. For passive interconnect systems without internal sources, it means that the signal at the transmitting end always takes finite time or have finite delay before it reaches the receiving end. A system is said to be causal if it meets a set of necessary and sufficient conditions defined as Kronig–Kramers relations [32, 33], which show that the real part of the transfer function of a system can be expressed as an integral over its imaginary part, and vice versa:

$$H(j\omega) = \frac{1}{j\pi}\,PV\int_{-\infty}^{+\infty} \frac{H(j\omega')}{\omega - \omega'}\,d\omega'.\tag{2.95}$$

This integration can be split into real and imaginary parts

$$\text{Re}\{H(j\omega)\} = \frac{1}{\pi}\text{PV}\int_{-\infty}^{+\infty}\frac{\text{Im}\{H(j\omega')\}}{\omega-\omega'}d\omega'$$

$$\text{Im}\{H(j\omega)\} = -\frac{1}{\pi}\text{PV}\int_{-\infty}^{+\infty}\frac{\text{Re}\{H(j\omega')\}}{\omega-\omega'}d\omega',$$

(2.96)

where PV indicates the Cauchy principal value.

Causality violations in the original sampled data by measurement or simulation may seriously compromise the accuracy of the macromodel and the speed of convergence of the VF method. The reason is that the VF method, which minimizes by iterations the root-mean-square error between the macromodel and the given data, enforces all poles to appear in the left half plane. This enforcement implies both the stability and causality of the macromodel. However, if the original sampled data are not causal, the frequency response of a causal macromodel by construction will never match the original noncausal data [34]. Therefore, causality check must be done to the frequency-domain sampled data.

A simple and expedient causality check may be performed through the use of the Blumer index [15, 35]:

$$B = 1 - \left|\frac{\sum_{l=1}^{N}|\text{Re}\{H(\omega_i)\}|^2\Delta\omega_i - \sum_{l=1}^{N}|\text{Im}\{H(\omega_i)\}|^2\Delta\omega_i}{\sum_{l=1}^{N}|\text{Re}\{H(\omega_i)\}|^2\Delta\omega_i + \sum_{l=1}^{N}|\text{Im}\{H(\omega_i)\}|^2\Delta\omega_i}\right|.$$

(2.97)

The closer the Blumer index B of the given data is to one, the more trustworthy of the causality of the data.

However, robust causality check with Equation (2.96) is not a trivial task, because the available sampled data are band limited and the high frequency values of $H(j\omega)$ may not decay to zero. Therefore, the truncation error related to the limited band data, and the discretization error in numerical calculation of Equation (2.96) must be taken into account when one designs an approach for accurate causality check. We present a robust approach for causality check of sampled data—generalized Hilbert transform reported in Reference 34.

A given set of sampled data $H(j\omega)$ is causal if and only if the ideal reconstruction error

$$\Delta_n(j\omega) = H_n(j\omega) - H(j\omega)$$

(2.98)

approaches zero for all frequencies, where $H_n(j\omega)$ denotes the reconstructed transfer function. For practical numerical evaluation, the truncation error $E_n(j\omega)$ and discretization error $D_n(j\omega)$ must be considered, and a new error estimation is given by

$$\tilde{\Delta}_n(j\omega) = \tilde{H}_n(j\omega) - H(j\omega)$$
$$= \Delta_n(j\omega) + E_n(j\omega) + D_n(j\omega). \qquad (2.99)$$

For a given reconstructed transfer function, if

$$\left|\tilde{\Delta}_n(j\omega_k)\right| > E_n^{tot}(\omega_k), \qquad (2.100)$$

then $H(j\omega)$ is not causal; If

$$\left|\tilde{\Delta}_n(j\omega_k)\right| \leq E_n^{tot}(\omega_k), \qquad (2.101)$$

then any causality violation in the data is smaller than the numerical resolution, and not detectable. The numerical resolution is affected by the number of subtraction points n, but cannot be arbitrarily small due to the limited number of frequency sampled data. $E_n^{tot}(\omega)$ is the worst-case error given by

$$E_n^{tot}(\omega_k) = T_n(\omega) + \tilde{D}_n(\omega), \qquad (2.102)$$

where $T_n(\omega)$ is the upper bound of the truncation error $|E_n(j\omega)|$, and $\tilde{D}_n(\omega)$ is an estimate of the discretization error $|D_n(j\omega)|$.

The truncation error $E_n(j\omega)$ is defined as

$$E_n(j\omega) = \hat{H}_n(j\omega) - H_n(j\omega)$$
$$= \frac{\prod_{q=1}^n(\omega - \bar{\omega}_q)}{j\pi} \int_{\Omega_c} \frac{-H(j\omega')}{\prod_{q=1}^n(\omega - \bar{\omega}_q)} \frac{d\omega'}{\omega - \omega'}, \qquad (2.103)$$

where the truncation error is a function of number n and position of subtraction points $\bar{\omega}_q|_{q=1,\cdots,n}$. Note that Ω_c is the complement band of $\Omega = [-\omega_{max}, \omega_{max}]$, that is, $\Omega_c = \bar{\Omega}$. The increase in the number of subtractions n reduces the integrand of Equation (2.103) and thus a smaller truncation error $E_n(j\omega)$. $\hat{H}_n(j\omega)$, the reconstructed impulse response, is an approximation to the integration Equation (2.95) over a limited band $\Omega = [-\omega_{max}, \omega_{max}]$:

$$\hat{H}_n(j\omega) = \ell_H(j\omega) + \frac{\Pi_{q=1}^n(\omega-\bar{\omega}_q)}{j\pi} \cdot PV \int_\Omega \frac{H(j\omega')-\ell_H(j\omega')}{\Pi_{q=1}^n(\omega-\bar{\omega}_q)} \frac{d\omega'}{\omega-\omega'}$$
$$+ \frac{\Pi_{q=1}^n(\omega-\bar{\omega}_q)}{j\pi} \int_{\Omega_c} \frac{-\ell_H(j\omega')}{\Pi_{q=1}^n(\omega-\bar{\omega}_q)} \frac{d\omega'}{\omega-\omega'}, \tag{2.104}$$

and

$$\ell_H(j\omega) = \sum_{q=1}^n H(j\bar{\omega}_q) \prod_{\substack{p=1 \\ p\neq q}}^n \frac{\omega-\bar{\omega}_p}{\bar{\omega}_q-\bar{\omega}_p}. \tag{2.105}$$

It is proved in Reference 34 that

$$|E_n(j\omega)| \leq T_n(\omega), \tag{2.106}$$

under the assumption of $|H(jw)| \leq M|\omega^\alpha|$, for $\omega \in \Omega_c$ and $\alpha = 0,1,2\cdots$. And $T_n(\omega)$ is given by

$$T_n(\omega) = \frac{M}{n} \sum_{q=1}^n (\bar{\omega}_q)^\alpha \left\{ \begin{vmatrix} \ln \left| \dfrac{\omega_{max}-\bar{\omega}_q}{\omega_{max}-\omega} \right| \\ -(-1)^{\alpha+n} \left| \ln \dfrac{\omega_{max}+\bar{\omega}_q}{\omega_{max}+\omega} \right| \end{vmatrix} \right\} \cdot \prod_{\substack{p=1 \\ p\neq q}}^n \frac{|\omega-\bar{\omega}_p|}{\bar{\omega}_q-\bar{\omega}_p}. \tag{2.107}$$

Note that $T_n(\omega)$ for the scattering parameters satisfies $|H(j\omega)| = |S(j\omega)| \leq 1$, which corresponds to $\alpha = 0$ and $M = 0$. Equation (2.107) shows that the truncation error is bounded between any two consecutive subtraction points and decreases with the increase in the number of subtraction points n.

The discretization error $D_n(j\omega)$ due to the numerical error introduced by numerical integration is defined as

$$D_n(j\omega) = \tilde{H}_n(j\omega) - \hat{H}_n(j\omega), \tag{2.108}$$

where $\tilde{H}_n(j\omega)$ denotes the impulse response calculated from a given numerical quadrature rule of order v. Singularity extraction is used to evaluate Equation (2.95) before a quadrature rule is applied:

$$\text{PV}\int_{-\infty}^{+\infty} g(\omega')\frac{d\omega'}{\omega - \omega'} \simeq \int_{\Omega}[g(\omega') - g(\omega)]\frac{d\omega'}{\omega - \omega'} + g(\omega)\cdot\text{PV}\int_{\Omega}\frac{d\omega'}{\omega - \omega'}$$
$$= \int_{\Omega}\frac{g(\omega') - g(\omega)}{\omega - \omega'}d\omega + g(\omega)\cdot\ln\left|\frac{\omega_{\max} + \omega}{\omega_{\max} - \omega}\right|,$$

(2.109)

where the second term regarding the singularity is evaluated in a closed form. The first term, after the singularity extraction, is well defined and can be calculated with any quadrature rule. An estimate of the discretization error is given by Reference 34

$$\tilde{D}_n(\omega) = \left|\tilde{H}_n^{v1}(j\omega) - \tilde{H}_n^{v2}(j\omega)\right| \simeq \left|\tilde{D}_n(j\omega)\right|,$$

(2.110)

where $\tilde{H}_n^{v1}(j\omega)$ and $\tilde{H}_n^{v2}(j\omega)$ represent the transfer functions obtained by the integration in Equation (2.104) with two different quadrature rules of orders $v_1 < v_2$, for example, the Simpson's rule and the trapezoidal rule. Note that causality-constrained data interpolation is also developed in Reference 34 to generate consistent DC and low-frequency data to supplement the band-limited frequency-sampled data.

2.6.3 Passivity Assessment

Passive systems, such as interconnects, and other passive components, must be characterized by a passive macromodel. Otherwise, unexpected and unpredictable nonphysical results may happen in the time-domain simulation depending on the loads connected to the terminals/ports of a nonpassive macromodel. In the worst case, the time-domain simulation results may just become unstable. Therefore, passivity assessment or check is important and should be performed before the macromodel is used for time-domain simulation. Only scattering parameters are discussed in the following sections.

Several techniques can be used to check the passivity of a macromodel [36]. A straightforward approach is to test the passivity by frequency sweeping of the singular values of the transfer function. This approach is time-consuming for large models because of a large frequency band involved, and poor sampling may lead to erroneous results. To alleviate the shortcomings of the direct frequency sweeping method, purely algebraic conditions have been used to check the passivity. Those conditions include the linear matrix inequality (LMI), the algebraic Riccati equation (ARE), and the eigenvalues with a Hamiltonian matrix (HM).

A passive macromodel with S parameters in the Laplace domain can be written as

$$S(s) = C(sI - A)^{-1} B + D, \tag{2.111}$$

where A, B, C, and D are state-space matrices defined in Equations (2.75), (2.85), and (2.86). A set of sufficient and necessary conditions [37–39] must be complied with, that is, a scattering matrix ($S(s)$) represents a passive system if and only if

1. elements of $S(s)$ are analytic (no singularities) for $\Re\{s\} > 0$,
2. $S^*(s) = S(s^*)$,
3. $I - S^H(s)S(s)$ is a nonnegative-definite matrix for $\Re\{s\} > 0$,

where the superscript * denotes the complex conjugate, and H the transpose conjugate (Hermitian transpose). A complex Hermitian matrix $A = A^H$ is nonnegative definite if $x^H A x \geq 0$ for all complex nonzero vectors x. Condition 1 implying causality is fulfilled by all stable models. Condition 2 is automatically satisfied by construction, that is, if the poles and residues of a macromodel are generated such that they are either real or complex conjugate pairs. Condition 3, which can be assessed only for the imaginary axis $s = j\omega$, is consistent with the property Equation (2.15) in Section 2.2.3 for a passive S matrix.

Condition 3 shows that the model for a passive system has no energy gain, and its singular values $\sigma\{S(j\omega)\}$ are all bounded by one, that is,

$$\max_{i,\omega} \sigma_i(j\omega) \leq 1, \quad \text{for } \sigma_i(j\omega) \in \sigma(S(j\omega)), \tag{2.112}$$

where the singular values $\sigma(S(j\omega))$ are given by

$$\sigma(S(j\omega)) = \sqrt{eig(S^H(j\omega) \cdot S(j\omega))}. \tag{2.113}$$

This condition poses challenges to the passivity assessment because it requires all the scattering parameters and involves an infinite frequency spectrum from DC to infinity. Algebraic methods can be used for passivity assessment by Condition 3 without using frequency sweeping. Passivity assessment is facilitated by the means of an HM [36], which is motivated by the theories in the system and control engineering proposed in [40]. For scattering parameter representations, if A has no imaginary eigenvalues, $\gamma > 0$ is not a singular value of D, and, $\omega_0 \in \mathbb{R}$, then γ is a singular value of $H(j\omega)$ if and only if $j\omega_0$ is an eigenvalue of M_γ:

$$M_\gamma = \begin{pmatrix} A - B\left(D^T D - \gamma^2 I\right)^{-1} D^T C & -\gamma B\left(D^T D - \gamma^2 I\right)^{-1} B^T \\ \gamma C^T \left(DD^T - \gamma^2 I\right)^{-1} C & -A^T + C^T D\left(D^T D - \gamma^2 I\right)^{-1} B^T \end{pmatrix},$$

$$(2.114)$$

where A, B, C, and D are defined in Equations (2.75), (2.85), and (2.86). Because the singular values of the scattering parameters are real bounded, the threshold value of γ is equal to one in Equation (2.114). Imaginary eigenvalues of the HM corresponds to the frequency points where singular values of the scattering matrix $S(s)$ are equal to ones. The macromodel in Equation (2.75) is passive if the following HM has no imaginary eigenvalues:

$$M = \begin{pmatrix} A - B\left(D^T D - I\right)^{-1} D^T C & -B\left(D^T D - I\right)^{-1} B^T \\ C^T \left(DD^T - I\right)^{-1} C & -A^T + C^T D\left(D^T D - I\right)^{-1} B^T \end{pmatrix}. \quad (2.115)$$

Note that the HM is frequency independent. If no imaginary eigenvalues are found, the macromodel is passive. If imaginary eigenvalues appear, the macromodel is not passive and should be corrected by passivity enforcement schemes.

One difficulty in passivity assessment with the HM is the computational efficiency in solving for Hamiltonian eigenvalues. The size of the HM in Equation (2.115) is two times of the number of system states. And the computational cost to find Hamiltonian eigenvalue scales in a cubic power of the problem size and becomes excessive for large macromodel with large number of ports and macromodel orders. A multi-shift restarted Arnoldi iteration solver was used in Reference 41 to compute selectively purely imaginary eigenvalues of the HM. Further improvement was made in Reference 42 with an adaptive sampling technique to reduce the number of shifts used in the Arnoldi eigenvalue solver.

As discussed in Section 2.2, the passive networks with interconnects and packages are reciprocal and their scattering matrices are symmetrical. Therefore, a half-size test matrix approach was proposed in [43] to fully use this property to save computational cost in finding imaginary eigenvalues. The half-sized matrix is named in Reference 43 the passivity matrix P to differentiate with the HM. The passivity matrix P for a symmetrical scattering matrix is given by

$$P = \left(A - B(D-I)^{-1}C\right)\left(A - B(D+I)^{-1}C\right), \qquad (2.116)$$

and

$$eigs(M) = \pm\sqrt{eigs(P)}. \qquad (2.117)$$

The square root of the negative real eigenvalues of the passivity matrix P defines the frequencies $j\omega$ where the singular values of $S(j\omega)$ cross over the threshold of unity. Use of half-sized P reduces the computation time for the eigenvalues by eight times compared to use of M because the complexity of the computation is cubic. Note that if $(D - I)$ or $(D + I)$ are singular, then a modified half-size test matrix [44] has to be used:

$$\tilde{P} = \left(\tilde{A} - \tilde{B}(\tilde{D}-I)^{-1}\tilde{C}\right)\left(\tilde{A} - \tilde{B}(\tilde{D}+I)^{-1}\tilde{C}\right), \qquad (2.118)$$

where the following bijective mapping is used:

$$\tilde{A} = A^{-1}, \quad \tilde{B} = -\tilde{A}B, \quad \tilde{C} = C\tilde{A}, \quad \tilde{D} = D - C\tilde{A}B, \qquad (2.119)$$

and $\tilde{D} = S(0)$. If $S(0) - I$ or $S(0) + I$ are also singular, then the half-size test matrix is not applicable. Instead, the crossings have to be identified from the HM of the reciprocal system $(S(1/j\omega))$ by shifting the frequency axis to a new frequency point $j\tilde{\omega}$ where $S(j\tilde{\omega}) - I$ and $S(j\tilde{\omega}) + I$ are not singular [45].

2.6.4 Passivity Enforcement

The presence of imaginary eigenvalues for the HM in Equation (2.115) indicates that the macromodel is not passive. Passivity violations include in-band violations and out-of-band violations. In-band violations are usually small because of the use of robust and accurate fitting methods like the VF. Out-of-band violations are usually severe, and difficult to detect and correct. Passivity enforcement is a process to compensate those passivity violations and make the macromodel passive.

2.6.4.1 Overview of Approaches for Passivity Enforcement

Different schemes for passivity enforcement in the open literature can be roughly classified into two categories—*a priori* and *a posteriori* approaches. Comprehensive overview of them can be found in References 46 and 47, which is used in this section.

The *a priori* approach is an optimal approach performing on-the-fly passivity enforcement during the construction of macromodels. Guaranteed passive model can be constructed during the fitting process by a convex optimization algorithm [18]. In this approach, an initial model is first identified and the poles of the model are retained, that is, the state matrix A is fixed while residue matrices including C and D are recomputed, subject to the constraint of global positive realness for an admittance matrix, to reach an optimal fit to the original data. The advantage of this approach is that it is subject to the constraints of positive real lemma and is known to provide optimal solution. However, its implementation is very complicated and the cost of the approach is very high both in terms of CPU time and memory [18, 46].

The *a posteriori* approach is a suboptimal approach usually perturbing reside or pole matrices. Perturbation of poles by preserving residues was first presented in Reference 48 to enforce passivity. Later an approach was proposed in Reference 49 to enforce passivity by perturbing poles and preserving zeros. Analytical conditions are derived to guarantee that the maximum passivity violation monotonically decreases in each iteration step. Compared to residue perturbation approaches, the pole perturbation approaches perturb the model behavior to a larger extent.

Most of the work for passivity enforcement in the open literature is focusing on perturbing the residue matrices. Two suboptimal approaches with residue perturbation have been intensively reported—one is enforcing passivity at discrete frequencies by the linear or quadratic programming (QP), the other is enforcing the passivity by perturbing iteratively the imaginary eigenvalues of a HM (see section IV-C in Reference 50).

Perturbation of residue matrices for passivity enforcement was first introduced in References 51 and 52, for Y parameters and in Reference 39 for S parameters. Passivity is enforced at discrete frequencies by slightly correcting the model residues with minimal change to the admittance matrix in a predefined frequency band of interest. The disadvantage of such a method is its low computational efficiency due to the need to solve a large sparse QP problem, although sparse solvers can greatly improve the efficiency of the method. Similar to other suboptimal approaches, the iterations may sometimes diverge and fail to produce good results. A fast procedure for identifying regions of passivity violations and enforcing passivity was proposed in Reference 36 to obviate the need of frequency sweeping. It detects passivity violations by checking the presence of imaginary eigenvalues of the HM, and this method perturbs the model residues indirectly via the HM eigenvalues to give a minimal change to the model impulse energy. The resulting change to the model behavior is higher than the approach in Reference 51. The Hamiltonian perturbation approach is not able to differentiate between in-band and out-of band passivity violations, lacks least-squares weighting capability, often requires more iterations [46]. Choosing the residue matrix eigenvalues as free variables [53] obviates the need for sparse-matrix solvers and speeds up the enforcement of passivity by the residue perturbation method in Reference 51. Combing this approach with the passivity assessment via the HM and a robust iterative technique [54] for prevention of divergence, an efficient and reliable approach was achieved for admittance-based pole-residue models. Modal perturbation, introduced in Reference 55 for Y parameters, is to preserve the relative accuracy of the model. It is less useful for S parameters because the ratio between the largest and smallest eigenvalues of S is much smaller than that of Y. Perturbation of the eigenvalues of the residue matrix for Y-parameter models was proposed in Reference 53 and symmetrical S-parameter-based models in Reference 47. An approach for identifying the spectrum regions of passivity violations was introduced in Reference 40 by evaluating the eigenvalues of a HM associated with the model. The approach was later adopted by References 36 and 52 for the electrical macromodeling of interconnects and packages. The computational efficiency in dealing with large number of ports and model orders was further improved with the multishifts Restarted Arnoldi iteration solver [41] and fast sweep methods to reduce the number of shifts used with the Arnoldi eigenvalue solver [56, 42]. By making use of the properties of the scattering parameters of reciprocal systems, an approach with half-size test matrices was introduced to handle symmetrical models from Y parameters [57] and S-parameters [43, 47].

Passivity enforcement based on first-order perturbation techniques works iteratively with each iteration involving the following four main steps [56]:

1. evaluate the passivity of the original macromodel or correct the nonpassive macromodel in the previous iteration step;
2. identify the passivity violation regions;
3. determine the magnitude of maximum violation in each region; and
4. enforce the passivity with first-order perturbation of either the residue matrices, eigenvalues of residues matrices, or eigenvalues of Hamiltonian matrices.

2.6.4.2 Identifying Passivity Violating Regions

The procedure to identify the regions violating the passivity has been discussed in a few papers, such as References 36 and 52, which is summarized as follows.

First, we collect the purely imaginary eigenvalues of the HM in Equation (2.115) by discarding those with nonpositive imaginary parts, sort them in an ascending order $\omega_1 < \omega_2 < \cdots < \omega_T$, and arrange them into a vector $s_{\sigma1} = \{\omega_1, \omega_2, \cdots, \omega_T\}$, where T is the total number of the entries, and let $\omega_H = \omega_T$.

Due to the numerical error, it may not be easy to find the purely imaginary eigenvalues with zero real parts. One solution to this problem is given in Reference 52. Note that the eigenvalue spectrum of the HM is symmetric about both the real axis and the imaginary axis, that is,

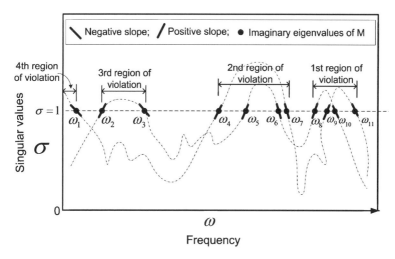

Figure 2.8 Regions of passivity violations are identified by checking the sign of the slopes of singular values crossing the threshold of unity.

four cyclic symmetrical eigenvalues come in a group. On the other hand, the imaginary eigenvalues are symmetric only with respect to the real axis. If the eigenvalues are symmetric only about the real axis, then they are purely imaginary.

Second, at the frequency corresponding to each of the above entries, we evaluate the slope (see Fig. 2.8) of the singular values of $S(j\omega)$ by [36]

$$\xi_i = \frac{j v_i^H J v_i}{v_i^H J M' v_i}, \tag{2.120}$$

where v_i is the right eigenvectors of the HM associated to $j\omega_i$. The first-order expansion matrix M' is also an HM and is defined as

$$M' =$$

$$\begin{pmatrix} -2B\left(D^T D - I\right)^{-2} D^T C & -B\left(2\left(D^T D - I\right)^{-2} + \left(D^T D - I\right)^{-1}\right)B^T \\ C^T\left(2\left(DD^T - I\right)^{-2} + \left(DD^T - I\right)^{-1}\right)C & 2C^T D\left(D^T D - I\right)^{-2} B^T \end{pmatrix}.$$

$$\tag{2.121}$$

The matrix J is real and skew symmetric, given by

$$J = \begin{pmatrix} 0 & I \\ -I & 0 \end{pmatrix}. \tag{2.122}$$

Note that $J^{-1} = J^T = -J$.

Third, we count the number of positive and negative slopes starting from ω_H. If the count of positive and negative slopes become equal, say at ω_k, then the first region of local passivity violation is established, that is, (ω_k, ω_H). For the example shown in Figure 2.8, the first region of passivity violation is (ω_8, ω_{11}).

Then we reset the count of slope to zero and continue the counting of slopes from ω_{k-1} to identify the next region of passivity violation. This process is repeated until all entries in the vector are exhausted. Figure 2.8 shows an example that the HM has 11 pairs of imaginary eigenvalues and their corresponding frequencies are from ω_1 to ω_{11}. Four regions of passivity violation are identified and marked in Figure 2.8. Note that the fourth region of passivity violation starts from zero frequency and the count of the slopes for this region is not zero, but one with a negative slope.

The above discussion assumes that the imaginary eigenvalues of the HM are simple. If the HM has repeated imaginary eigenvalues, then the well-established approach outlined in Reference 36 can be used. Note that the slope defined in Equation (2.120) is based on the first-order perturbation localized at a given imaginary eigenvalue $j\omega_i$, and it is not affected by the multiplicity of other eigenvalues distinct from $j\omega_i$.

1. *Nondefective Eigenspaces.* One imaginary eigenvalue $j\omega_i$ may have multiplicity $\mu_i > 1$ and a complete set of right eigenvectors for the associated eigenspace. Those eigenvectors are denoted as

$$\left\{ v_{i,v} \big|_{v=1,\cdots\mu_i} \right\}. \tag{2.123}$$

They form a basis of the eigensapce of dimension μ_i and present as a diagonal block in the Jordan canonical form

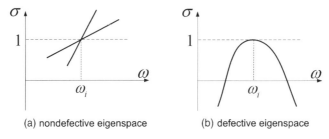

(a) nondefective eigenspace   (b) defective eigenspace

Figure 2.9 Multiplicity of the singular values at $j\omega_i$ with $\mu_i = 2$ [36].

associated to the Hamiltonian matix. Each of them has a slope defined as

$$\xi_{i,\upsilon} = \frac{jv_{i,\upsilon}^H J v_{i,\upsilon}}{v_{i,\upsilon}^H J M' v_{i,\upsilon}}, \quad \text{for } \upsilon = 1, \cdots \mu_i. \tag{2.124}$$

Each of these slopes is counted distinctively to identify the regions of passivity violation. Note that this situation occurs when two separate singular values crosses the threshold of unity at the same frequency point, as shown in Figure 2.9a.

2. *Defective Eigenspaces.* In this case the number of independent eigenvectors ρ_i is less than the multiplicity of the associated eigenvalue $j\omega_i$, that is, $\rho_i < \mu_i$. The situation could be very complicated. Thus, only the nonderogatory case with one eigenvector ($\rho = 1$) is considered, which is equivalent to the case of a single Jordan block corresponding to the imaginary eigenvalue $j\omega_i$. Figure 2.9b shows a situation that a maximum occurs at the threshold of unity. The singular value can be expressed as a quadratic expansion around ω_i:

$$\sigma(\omega) \approx 1 - a(\omega - \omega_i)^2, \quad a > 0, \tag{2.125}$$

and

$$\omega - \omega_i \approx \pm \sqrt{\frac{1 - \sigma(\omega)}{a}}. \tag{2.126}$$

In this situation, the singular value touches the threshold of unity without crossing it and should not be counted in finding regions of passivity violations. This case of defective HM may not occur in real applications [36], probably because of the numerical error.

Finally, we present an alternative way of identifying the regions of passivity violations without checking the slopes of the singular values at the crossing points [55]. It is simple and straightforward. First, a sorted list is created with all the purely imaginary eigenvalues of M, that is, $s_{\sigma 1} = \{\omega_1, \omega_2, \cdots, \omega_T\}$, and $\omega_1 < \omega_2 < \cdots < \omega_T$. Then, the singular values of $S(j\omega)$ at the midpoint of each interval, that is,

$$s = j\left\{\frac{\omega_1 + \omega_2}{2}, \cdots, \frac{\omega_{T-1} + \omega_T}{2}\right\},$$ are evaluated. If the singular value is greater than one at the sample $(\omega_1 + \omega_2)/2$, then the frequency band of $[\omega_1, \omega_2]$ defines a region of passivity violation. The above step is repeated till all the midpoint frequencies are exhausted. In addition, the singular values at $\omega_1/2$ and $2\omega_T$ should be checked to identify if either region at the two ends of the entire frequency interval, that is, $[0, \omega_1]$ and $[\omega_T, \infty)$, is violating passivity.

2.6.4.3 Determine the Magnitude of Maximum Violations

In order to compensate the regions of passivity violations, we need to know the amount of correction required to restore the passivity. Therefore, the magnitude of the local maximum in each passivity violation region has to be determined. This can be done by a frequency sweep with a bisection method [40]. The singular values returned by a general SVD algorithm are sorted according to their magnitude. Spurious maximum appears whenever two singular values cross. This problem is overcome by rearranging the sequence of the singular values together with columns U and V when moving from one frequency to the next. This is achieved by checking the change in the direction of the columns of U, using a switching-back procedure [47].

2.6.4.4 Passivity Enforcement Schemes

As discussed in Section 2.6.4.1, two suboptimal approaches exist for passivity enforcement by residue perturbation, that is the QP scheme—linear or QP for passivity enforcement at discrete frequencies, and the HM scheme—an HM approach with iterative perturbation of the imaginary eigenvalues. If x denotes an array containing all the residue perturbations, the above two schemes can be formulated as follows [50]:

$$\text{QP: } \begin{cases} \min x^T \Psi x \\ Rx \leq g, \end{cases} \text{ and HM: } \begin{cases} \min x^T \Psi x \\ Qx \leq r, \end{cases} \quad (2.127)$$

where the symmetric and positive definite matrix Ψ preserve the accuracy of the model for a perturbation with a suitable norm weighting. Ψ is related to a weighted controllability Gramian of the model, which will be formulated later. Note that a positive-definite matrix is analogous to a positive real number, and a symmetric matrix Ψ is positive definite if $x^T \Psi x > 0$ for all nonzero vectors x. The inequality constraint $Rx \leq g$ for the QP scheme is to restore the energy gain of the model below the unit threshold to achieve passivity. Conversely, the equality constraint $Qx = r$ for the HM scheme is to force the purely imaginary Hamiltonian eigenvalues to collapse and move off the imaginary axis, so as to achieve passivity. The QP scheme is more robust than the HM scheme for large passivity violations. The HM scheme is more effective for small and well-localized passivity violations because of less iteration steps required. A hybrid approach aiming to combine the strength of both schemes is recommended in Reference 50 for a robust passivity enforcement. The regions of passivity violations are first identified by a fast Hamiltonian eigenvalue solver, that is, multishift restarted Arnoldi iteration solver with adaptive sampling [42]. Two successive iteration loops follows: the QP scheme is first applied to remove large out-of-band passivity violations, succeeded by the HM scheme to remove any small residual passivity violations. A comparative study of two passivity enforcement schemes can be found in Reference 46. In the following sections, we will summarize the HM scheme reported in Reference 36 and the QP scheme reported in Reference 47 for passivity enforcement.

A. The HM Scheme. We present in this section the HM scheme, that is, passivity enforcement via perturbation of eigenvalues of the HM [36]. The HM scheme is effective for small passivity violations. Small passivity violations are loosely defined as those requiring a small perturbation to restore the passivity of the original system. In the HM scheme, the state matrix A with all the poles is not touched, the state matrix D with direct coupling at s $\rightarrow \infty$ is preserved, one of the transfer matrix B is retained, and only the other transfer matrix C is perturbed. Note that C, constructed from the residues, is also called the residue matrix. This approach is feasible with the guarantee of the underlying hypotheses of controllability and observability.

A.1 Criteria for Accurate Approximation. First, criteria for accurate approximation are derived to ensure that minimal changes are made to the original system response while restoring passivity. Let

$$\Delta_c = \tilde{C} - C, \tag{2.128}$$

where $\tilde{C}$ denotes the residue matrix obtained by applying the perturbation Δ_c. Then the response of the original system induced by the perturbation is changed by

$$dh(t) = L^{-1}\{dH(s)\} = \Delta_c \cdot e^{\{A \cdot t\}} B, \quad t \geq 0. \tag{2.129}$$

The new matrix $\tilde{C}$ is determined by minimizing the following functional with the cumulative energy of the perturbations to the impulse response

$$E = \int_0^\infty \|dh(t)\|_F^2 \, dt = \int_0^\infty tr\{dh(t) \cdot dh^T(t)\} dt, \tag{2.130}$$

where, tr denotes the trace of a matrix, that is, the sum of all the diagonal elements. Substituting Equation (2.129) into Equation (2.130) gives

$$E = tr(\Delta_c W \Delta_c^T), \tag{2.131}$$

where W is the controllability Gramian given by

$$W = \int_0^\infty e^{A \cdot t} BB^T e^{A^T \cdot t} dt. \tag{2.132}$$

It can be computed as the symmetric and positive definite solution of the Lyapunov equation

$$AW + WA^T = -BB^T. \tag{2.133}$$

The assumption of controllability entitles the Gramian to a Cholesky factorization

$$W = K^T K,$$ (2.134)

where K is a nonsingular upper triangular matrix. Therefore, the perturbation Δ_c in the new basis defined by the Cholesky factor is given by

$$\Delta_c^k = \Delta_c K^T.$$ (2.135)

Equation (2.131) is thus changed to

$$E = tr\left\{\Delta_c^k \left(\Delta_c^k\right)^T\right\} = \left\|\Delta_c^k\right\|_F^2 = \sum_{i,j}\left|\Delta_{c,ij}^k\right|^2.$$ (2.136)

It shows that minimizing the perturbation to the impulse response is equivalent to minimizing the perturbation to the residue matrix C with an appropriate coordinate system.

A.2 Perturbation and Passivity Enforcement. Second, after the creation of the perturbation criteria, an iterative algorithm is presented to compensate small passivity violations. In the previous sections, the regions of passivity violations are identified and the amount of passivity violations, that is, the largest singular value, is available for each region. First-order perturbation to the HM is performed to drive its imaginary eigenvalues off the imaginary axis, as shown in Figure 2.10. From the definition in Equation (2.115), the HM perturbed by Δ_M is given by

$$\tilde{M} = M + \Delta_M,$$ (2.137)

where the perturbation matrix Δ_M ignoring the second-order terms reads

$$\Delta_M = \begin{pmatrix} -B\left(D^T D - I\right)^{-1} D^T \Delta_C & 0 \\ C^T \left(DD^T - I\right)^{-1} \Delta_C + \Delta_C^T \left(DD^T - I\right)^{-1} C & \Delta_C^T D \left(D^T D - I\right)^{-1} B^T \end{pmatrix}.$$ (2.138)

Note that the imaginary eigenvalues of M corrected with a first-order perturbation still remain imaginary if they are subject to a

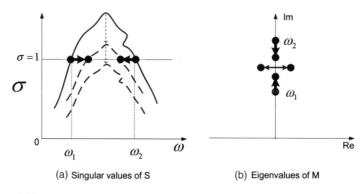

(a) Singular values of S   (b) Eigenvalues of M

Figure 2.10 Illustration of passivity enforcement of macromodel created from the scattering parameters. Note the one-to-one correspondence between the purely imaginary eigenvalues of the Hamiltonian matrix and the frequencies where the singular values of the transfer matrix cross or touch the threshold of unity. Perturbation of singular values corresponds to displace the eigenvalues off the imaginary axis.

sufficiently small perturbation. A quantity Δ_C should be determined to force the imaginary eigenvalue at $j\omega_i$ to a new location denoted as $j\tilde{\omega}_i$. The first-order approximation is given by

$$j\tilde{\omega}_i - j\omega_i \simeq \frac{v_i^H J \Delta_M v_i}{v_i^H J v_i}. \tag{2.139}$$

In order to express the numerator as a linear combination of the perturbation Δ_C, the eigenvector v_i is partitioned into $v_i = \left\{ v_{i1}^T, v_{i2}^T \right\}^T$ by using the induced block partition of the HM.

And

$$v_i^H J \Delta_M v_i = 2 \Re \left\{ z_i^H \Delta_C v_{i1} \right\}, \tag{2.140}$$

where

$$z_i^H = D\left(D^T D - I\right)^{-1} B^T v_{i2} + \left(DD^T - I\right)^{-1} C v_{i1}. \tag{2.141}$$

Substituting the new basis defined in Equation (2.134) leads to

$$v_i^H J \Delta_M v_i = 2 \Re \left\{ \left(v_{i1}^T K^{-1} \right) \otimes z_i^H \right\} vec\left(\Delta_C^K\right), \tag{2.142}$$

where the Kronecker product is defined as

$$A \otimes B = \begin{pmatrix} a_{11}B & \cdots & q_{1q}B \\ \vdots & \vdots & \vdots \\ a_{q1}B & \cdots & a_{qq}B \end{pmatrix}. \tag{2.143}$$

The following property holds

$$Y = BXA^T \Leftrightarrow vec(Y) = (A \otimes B) \cdot vec(X), \tag{2.144}$$

where $vec(X)$ denotes the vector storing the stacked columns of matrix X. This equivalence is used to restate in explicit form the linear constraint between the above two matrices Y and X.

In summary, the matrix perturbation Δ_C to displace the imaginary eigenvalue from $j\omega_i$ to $j\tilde{\omega}_i$ must fulfill the following condition:

$$2\Re\left\{\left(v_{i1}^T K^{-1}\right) \otimes z_i^H\right\} vec\left(\Delta_C^K\right) = -\Im\left\{v_i^H J v_i\right\}(\tilde{\omega}_i - \omega_i). \tag{2.145}$$

The above condition, derived by first-order approximation, is linear. All the constraints of Equation (2.145) for imaginary eigenvalues can be grouped into a compact matrix form, together with the condition of minimizing Equation (2.136):

$$Z \cdot vec\left(\Delta_C^K\right) = r, \quad \min\left\|vec\left(\Delta_C^K\right)\right\|_2. \tag{2.146}$$

Each row in matrix Z is comprised of the left-hand side of Equation (2.145). Note that Equation (2.146) is a standard least-square problem, which is usually underdetermined, because the number of unknowns is $n \times p$, compared with the number of linear constraints bounded by the dynamic order n. The condition at the right-hand side of Equation (2.146) ensures an accurate solution which minimizes the derivation from the original system response.

How to determine the location of the new eigenvalue is discussed now. As shown in Figure 2.10, reducing the singular value effectively moves each crossing point toward the direction of the slope; that is, eigenvalues with positive slopes are pushed toward higher frequencies, those with negative slopes toward lower frequencies. This observation gives only information on the correct sign that should be used for the

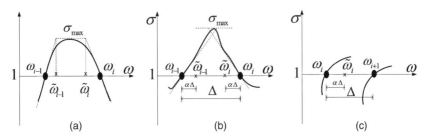

Figure 2.11   Determining the location of the perturbed eigenvalues in three scenarios. The initial eigenvalues are denoted by dots and the perturbed ones by crosses [36].

right-hand side of Equation (2.145). The amount of displacement for each eigenvalue has to be determined, and three scenarios are discussed below:

1. Figure 2.11a shows the first scenario which is a simple case of a localized passivity violation with only one singular value exceeding the critical value. The dashed lines denote the local slopes at the crossing points and an estimate of the local maximum σ_{max}. The new locations can be determined by the intersection between the upper bound and the two tangent lines at the crossing points:

$$(\tilde{\omega}_i - \omega_i) = \xi_i^{-1}(\sigma_{max} - 1). \tag{2.147}$$

The above scenario corresponds to a uniformly concave curve of the singular values over the violation interval.

2. Figure 2.11b shows another scenario with a concave curve of the singular values. An addition constraint is formulated to set an upper bound for the eigenvalue displacement:

$$\begin{aligned} (\tilde{\omega}_i - \omega_i) &\le \alpha(\omega_{i+1} - \omega_i), \quad \text{for } \xi_i > 0 \\ (\omega_i - \tilde{\omega}_i) &\le \alpha(\omega_i - \omega_{i-1}), \quad \text{for } \xi_i < 0. \end{aligned} \tag{2.148}$$

The scale factor α has empirical values of $0 < \alpha < 0.5$, and $\alpha = 0.3$ may be used. Smaller values of α guarantee the validity of the first-order approximation but increase the number of iteration steps.

3. Figure 2.11c shows another scenario when the slopes of two adjacent crossing points have equal sign. An upper bound similar to Equation (2.148) can be used to restrict the displacement of each eigenvalue.

More general cases may appear with any combination of the above three scenarios. Thus, the above procedure for displacing the eigenvalues must be iterated until all the imaginary eigenvalues are pushed off the imaginary axis. If q iterations are applied to displace all the imaginary eigenvalues off the imaginary axis, the residue matrix has gone though q first-order perturbations. The new residue matrix is given by

$$\tilde{C}_q = C + \sum_{m=1}^{q} \Delta_{c,m}. \tag{2.149}$$

The hypothesis of small passivity violation can be verified against the relative perturbation by checking the following condition:

$$\sum_{ij} \left| (\tilde{C}_q - C)_{ij} \right|^2 \ll \sum_{ij} |C_{ij}|^2. \tag{2.150}$$

A.3 Computation Cost. Last, a brief discussion is given on the complexity of the HM scheme for passivity enforcement. Computation of the imaginary eigenvalues of the HM requires $O(n^3)$ operations. The above computation must be repeated for each iteration step. The numerical solution of the Lyapunov equation Equation (2.133) and the Cholesky factorization Equation (2.134) requires also $O(n^3)$ operations, but they only need to be done once before the start of the iteration. Solving the least-square problem Equation (2.146) costs $n_i^2 np - n_i^3/3$, where n_i is the number of imaginary eigenvalues to be perturbed. This cost is small because usually $n_i \ll n$. Note that the dedicated fast eigenvalue solver discussed in Section 2.6.3 can speed up the computation of the imaginary eigenvalues of the HM.

B. The QP Scheme. The QP approach has been used for passivity enforcement through perturbation of the residue matrix, or perturbation of eigenvalues of the residue matrix. Compared to the residue matrix perturbation, the eigenvalue perturbation of the residue matrix has the advantage of obviating the need for a sparse solver and reducing the size of the QP problem to be solved, in particular for cases with a large number of ports. Only the method of perturbation of residue matrix eigenvalues [47] is presented in the following. Note that due to the reciprocal of the interconnect and package structures, the scattering matrix is symmetrical. Thus, the half-size test matrix [47] is used to identify the regions of passivity violations.

B.1 Perturbation. The singular values of the scattering matrix S can be obtained as the eigenvalues of the augmented matrix H as follows:

$$\begin{bmatrix} V & V \\ U & -U \end{bmatrix}^{-1} \begin{bmatrix} 0 & S^H \\ S & 0 \end{bmatrix} \begin{bmatrix} V & V \\ U & -U \end{bmatrix} = \begin{bmatrix} \Sigma & 0 \\ 0 & -\Sigma \end{bmatrix}, \qquad (2.151)$$

where

$$H = \begin{bmatrix} 0 & S^H \\ S & 0 \end{bmatrix}. \qquad (2.152)$$

An SVD on S is given by

$$S(s) = U(s) \sum (s) V(s). \qquad (2.153)$$

Applying matrix inverse to Equation (2.151) gives

$$\frac{1}{2}\begin{bmatrix} V^{-1} & V^{-1} \\ V^{-1} & -U^{-1} \end{bmatrix}^{-1} \begin{bmatrix} 0 & S^H \\ S & 0 \end{bmatrix} \begin{bmatrix} V & V \\ U & -U \end{bmatrix} = \begin{bmatrix} \Sigma & 0 \\ 0 & -\Sigma \end{bmatrix}, \qquad (2.154)$$

Retaining the partition of the matrices associated with the positive Σ block produces

$$\frac{1}{2}\begin{bmatrix} V^{-1} & U^{-1} \end{bmatrix} \begin{bmatrix} 0 & S^H \\ S & 0 \end{bmatrix} \begin{bmatrix} V \\ U \end{bmatrix} = \Sigma, \qquad (2.155)$$

which can be written in a compact form as

$$QHT = \Sigma. \qquad (2.156)$$

A perturbed singular value is related to the elements of H by applying a first-order perturbation to the truncated eigenvalue problem in Equation (2.156), which is a linear relation:

$$\Delta\sigma_i(s) = \Delta\lambda_i(H) = \frac{q_i^T \Delta H t_i}{q_i^T t_i}, \quad i = 1, \cdots n, \tag{2.157}$$

where q_i^T and t_i denote the ith row of Q and column of T, respectively.

If the eigenvectors are scaled to unit length, the denominator in Equation (2.157) is unity. Introducing the following partitioning

$$\Delta\sigma_i = \begin{bmatrix} q_{i,a}^T & q_{i,b}^T \end{bmatrix} \begin{bmatrix} 0 & \Delta S^H \\ \Delta S & 0 \end{bmatrix} \begin{bmatrix} t_{i,a} \\ t_{i,b} \end{bmatrix}, \tag{2.158}$$

that is,

$$\Delta\sigma_i = q_{i,b}^T \Delta S t_{i,a} + q_{i,a}^T \Delta S^H t_{i,b}. \tag{2.159}$$

Note that the two terms in Equation (2.159) are conjugate. Then Equation (2.159) gives

$$\Delta\sigma_i = 2 q_{i,b}^T \Delta S t_{i,a}, \quad i = 1, \cdots n. \tag{2.160}$$

B.2 Formulate the QP. Passivity enforcement is achieved by applying minimal perturbations ΔS to the model subject to a passivity constraint

$$\Delta S \simeq 0,$$
$$\sigma_i + \Delta\sigma_i < 1, \quad i = 1, \cdots n. \tag{2.161}$$

where

$$\Delta S = \sum_{m=1}^{N} \frac{\Delta R_m}{s - a_m} + \Delta D. \tag{2.162}$$

Equation (2.161) is formulated as a constrained least-square problem, solved by QP

$$\begin{cases} \min\limits_{\Delta x} \frac{1}{2}\left(\Delta x^T A_{sys}^T A_{sys} \Delta x\right) \\ B_{sys}\Delta x \leq g, \end{cases} \tag{2.163}$$

where Δx, which are free variables, contain the elements of $\{\Delta R_m\}$ and ΔD.

B.3 Formulate the Least-Squares Equation. First, the residue matrix, which is real and symmetric, is diagonalized

$$R_m = P_m \Lambda_{R_m} P_m^T, \tag{2.164}$$

where P_m is the eigenvector matrix, which is real with $P_m^{-1} = P_m^T$.

A first-order perturbation to the residue matrix gives

$$\frac{R_m + \Delta R_m}{s - a_m} \approx \frac{P_m\left(\Lambda_{R_m} + \Delta\Lambda_{R_m}\right)P_m^T}{s - a_m}. \tag{2.165}$$

The perturbed residue matrix eigenvalues $\Delta\Lambda_{R_m}$, not the residue matrix elements, are chosen as free variables in Equation (2.163) by means of

$$\Delta R_m = P_m \Delta\Lambda_{R_m} P_m^T. \tag{2.166}$$

In case of complex conjugate pairs of residues, the real and imaginary parts are diagonalized separately. Similarly, the eigenvalues of a nonzero D are used as free variables:

$$\Delta D = P_D \Delta\Lambda_D P_D^T. \tag{2.167}$$

Note that the eigenvector matrices P_m and P_D, which are known quantities, are obtained directly from R_m and D. All residue matrix elements are perturbed via a reduced set of variables during the passivity enforcement. The number of free variables is nN (number of ports times number of poles), compared to $n(n + 1)N/2$ when perturbing all the matrix elements by considering the symmetry of the matrix. The computation time for solving Equation (2.163) is greatly reduced because the QP requires basic operations in the order of cube with the problem size. After solving Equation (2.163), the changes $\{\Delta R_m\}$ and ΔD are recovered by using Equations (2.166) and (2.167), respectively.

Second, if it is a real pole-residue term, Equation (2.166) is expanded into

$$
\begin{aligned}
\frac{\Delta R}{s-a} &= \frac{1}{s-a}\sum_{j=1}^{n} p_j p_j^T \Delta\lambda_j \\
&= \frac{1}{s-a}\sum_{j=1}^{n} \Gamma_j \Delta\lambda_j \cong 0,
\end{aligned}
\tag{2.168}
$$

where p_j is jth column of P_m in Equation (2.166). The above equation can be written in a matrix-vector equation with $\Delta\lambda j$ in the vector of unknowns. An example of two ports ($n = 2$) is given by

$$
\frac{1}{s-a}\begin{bmatrix} \Gamma_1^{1,1} & \Gamma_2^{1,1} \\ 2\Gamma_1^{1,2} & \Gamma_2^{1,2} \\ \Gamma_1^{2,2} & \Gamma_2^{2,2} \end{bmatrix}\begin{bmatrix} \Delta\lambda_1 \\ \Delta\lambda_2 \end{bmatrix} \cong 0.
\tag{2.169}
$$

Note that enforcing symmetry leads to the factor of 2 in the second row of the Γ matrix.

Third, if it is a pole-residue term with one complex conjugate pair of poles, the pair is split into real and imaginary parts to ensure that the solution is conjugate:

$$
\begin{aligned}
\frac{\Delta R' + j\Delta R''}{s-(a'+ja'')} + \frac{\Delta R' - j\Delta R''}{s-(a'-ja'')} &= f(s)\Delta R' + g(s)\Delta R'' \\
f(s) &= \frac{1}{s-(a'+ja'')} + \frac{1}{s-(a'-ja'')} \\
g(s) &= \frac{j}{s-(a'+ja'')} + \frac{j}{s-(a'-ja'')}.
\end{aligned}
\tag{2.170}
$$

Substituting Equation (2.170) into Equation (2.168) gives

$$
f(s)\sum_{j=1}^{n} \Gamma_j'\Delta\lambda_j' + g(s)\sum_{j=1}^{n} \Gamma_j''\Delta\lambda_j'' \cong 0.
\tag{2.171}
$$

The above equation is rearranged into a system matrix A_{sys}, and again the real and imaginary parts are separated to make the free variables real quantities:

$$A_{sys}\Delta\tilde{x} = \begin{bmatrix} \Re A \\ \Im A \end{bmatrix}\Delta\tilde{x} = \begin{bmatrix} 0 \\ 0 \end{bmatrix}, \tag{2.172}$$

where A_{sys} has $\bar{M}n(n+1)$ M rows and Nn columns. $\bar{M}$ is the number of frequency samples, n the number of ports, and N the order of the model. To avoid that A_{sys} becomes excessive for large number of ports, the equations $A_{sys}^T A_{sys}$ in Equation (2.163) is directly expressed as a sum of outer products:

$$A_{sys}^T A_{sys} = \sum_{row} A_{sys,row}^T A_{sys,row}. \tag{2.173}$$

B.4 Derive the Constraint Equation. For the case of a real pole-residue term, the relation between a perturbation of a free variable $\Delta\lambda(j)$ and the singular value $\Delta\sigma$ can be obtained by replacing ΔS in Equation (2.160) with Equation (2.168):

$$\Delta\sigma_i = \frac{2\sum_{j=1}^{n}\left(q_{i,b}^T p_j\right)\left(p_j^T t_{i,a}\right)\Delta\lambda_j}{s-a}. \tag{2.174}$$

Note that the computation cost is low because of only inner products involved.

For the case of a complex conjugate pair, the contribution from the jth eigenvalue is

$$\Delta\sigma_i = \begin{bmatrix} k_1(s) & k_2(s) \end{bmatrix}\begin{bmatrix} \Delta\lambda_j' \\ \Delta\lambda_j'' \end{bmatrix}$$
$$k_1(s) = 2f(s)\left(q_{i,b}^T p_j'\right)\left(p_j'^T t_{i,a}\right) \tag{2.175}$$
$$k_2(s) = 2g(s)\left(q_{i,b}^T p_j''\right)\left(p_j''^T t_{i,a}\right),$$

where the prime and double prime denote the contribution from $\Delta R'$ and $\Delta R''$, respectively.

The proportional term D can be dealt with as in Equations (2.169) and (2.174), but with the denominator $(s - \alpha)$ replaced by unity. The final constraint matrix B_{sys} in Equation (2.163) is replaced with its real parts.

B.5 Other Issues. The conditioning of Equation (2.163) is improved by scaling the columns of A_{sys}. These columns cannot be scaled to unit

Euclidian length, because A_{sys} is not expressed explicitly in Section B.4. It is given in Reference 47 that the scaling is $||s - a||_2$ for a real pole, and $||1/f(s)||_2$ and $||1/g(s)||_2$ for a complex pair. The final solution to Equation (2.163) is recovered by multiplication with the same scaling factors.

The frequency samples for the least-squares part in Equation (2.163) can just use those for identifying the pole-residue macromodel. In order to further improve the conditioning of Equation (2.163), additional frequency samples are added at frequencies which correspond to the location of out-of-band poles. These samples are given a low weighting in the least-squares problem to avoid possible impact on in-band results. Frequency samples for the constraint part in Equation (2.163) are chosen as those corresponding to the maximum of the singular values in the regions of passivity violation. The number of rows in B_{sys} is thus reduced, and the computational efficiency is thus increased.

Each row in A_{sys} corresponds to one element of S at one frequency sample. The quality of the perturbed model can be improved by appropriate row weighting, for example, weighting with the inverse of the magnitude of the elements to achieve relative error control.

The passivity enforcement must be done repeatedly due to the nonlinearity of Equation (2.163), because a correction to the model may cause new passivity violations. A_{sys} is formed during the first iteration and reused in the subsequent iteration, thereby the computation cost is reduced. Therefore, the first iteration is computationally more expensive than the rest.

A summary of the passivity enforcement of scattering matrix-based macromodel with the perturbation of residue matrix eigenvalues is given by Reference 47:

1. Identify the regions of passivity violations by using a half-size test matrix in Equation (2.116);
2. Calculate S at frequencies of maximum violations;
3. Apply singular value decomposition: $[U, \Sigma, V] = \mathrm{svd}(S)$;
4. Form $q_{i,b}^T$ and $t_{i,a}$ with the first-order perturbation Equation (2.160);
5. Diagonalize matrices $\{\Delta R_m\}$ and ΔD by Equations (2.166) and (2.167);

6. Compute the reduced size perturbation Equations (2.174) and (2.175);

7. Form the constraint Equation (2.163);

8. In the first iteration, calculate the reduced size cost function Equation (2.171) at frequencies defining the fitting band;

9. In the first iteration, create the cost Equation (2.163);

10. Solve Equation (2.163) using the QP;

11. Update the pole-residue macromodel.

12. Repeat the above steps till all the passivity violation regions are compensated.

2.6.5 Other Issues

A few other issues regarding the macromodeling are briefly discussed in this section.

Macromodel Order. *A priori* model-order selection helps generate a macromodel with a minimum size for efficient simulations with accuracy control. A heuristic approach was used in Reference 58 to estimate the order of the macromodel by applying experimental observation of the frequency response in frequency-sampled data via the smith chart. More approaches may be found in References 59–61.

Frequency Band. Sampled data covering a finite range of the spectrum can be contaminated by measurement error or simulation error. Obtaining DC response of the structure and including it in the development of the macromodel is essential for correctly predicting the steady state response in subsequent system-level simulation [15]. The *S*-parameter data are usually truncated toward DC at a low, but finite frequency. Low-frequency data obtained through a full-wave electromagnetic field solver may not be accurate, unless special care is exercised to ensure the robustness of the field solver at low frequencies, for example, References 62 and 63. Extrapolation can be used to extend the measured or numerically obtained *S*-parameter data to DC. To ensure the physical consistency of the extrapolated data, causality constraints are often imposed [15]. The highest frequency for the scattering parameters should be chosen according to the guideline of $\omega_H = 2\pi \, 0.35/t_r$, where t_r is the rise time of the excitation signal.

Macromodel Ports. Ports provide interface between the macro-model and the external circuitry. Similar to physical experiment requiring proper de-embedding, electromagnetic simulation also needs numerical de-embedding of the ports to get accurate results. Possible interaction among the ports may also affect the accuracy of the simulation results.

2.7 MACROMODELING APPLIED TO HIGH-SPEED INTERCONNECTS AND CIRCUITS

Numerical examples are presented to demonstrate the macromodeling method for simulation of high-speed interconnects and passive circuits. The sampled data used in these examples are all scattering parameters.

2.7.1 A Lumped Circuit with Nonlinear Components

A lumped circuit with nonlinear components is shown in Figure 2.12a. The nonlinear component, that is, the CMOS inverter used in Figure

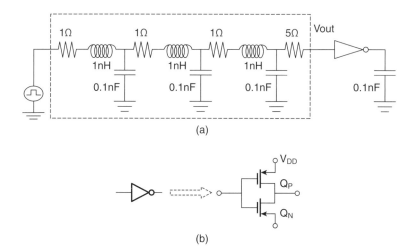

(a)

(b)

Figure 2.12   (a) a lumped circuit with nonlinear components; (b) the inverter realized by two MOSFET [30].

2.12a, is realized by two MOSFET transistors, as indicated in Figure 2.12b.

The scattering parameters for the two-port network enclosed in the dashed-line box in Figure 2.12a can be obtained by using analytical formulae, and the reference impedance Z_R used for each port is 30 Ω. The analytical formulae for the scattering parameters are obtained based on the definition of scattering parameters, and the Kirchhoff's current law (KCL) and Kirchhoff's voltage law (KVL) for circuits:

$$S_{11} = \frac{b_1}{a_1}\bigg|_{a_2=0} = \frac{V_1/I_1 + Z_R}{V_1/I_1 + Z_R},$$

$$V_1/I_1 = \frac{\beta Z_2}{\beta + Z_2} + Z_1, \quad \beta = \frac{\alpha Z_2}{\alpha + Z_2} + Z_1, \quad \alpha = \frac{(Z_R+5)Z_2}{(Z_R+5)+Z_2} + Z_1,$$

$$Z_1 = 1 + j\omega \cdot (1 \times 10^{-9}), \quad Z_2 = \frac{1}{j\omega \cdot (0.1 \times 10^{-9})}$$

$$(2.176)$$

$$S_{22} = \frac{b_2}{a_2}\bigg|_{a_1=0} = \frac{V_2/I_2 + Z_R}{V_2/I_2 + Z_R},$$

$$V_2/I_2 = \frac{\beta Z_2}{\beta + Z_2} + 5, \quad \beta = \frac{\alpha'' Z_2}{\alpha'' + Z_2} + Z_1, \quad \alpha = \frac{(Z_R+Z_1)Z_2}{(Z_R+Z_1)+Z_2} + Z_1,$$

$$(2.177)$$

$$S_{21} = \frac{b_2}{a_1}\bigg|_{a_2=0} - \frac{2}{\kappa + \dfrac{Z_R}{Z_1}(\kappa - \zeta)},$$

$$\kappa = Z_1\left(\frac{\gamma' - \alpha'}{Z_1} + \frac{\gamma'}{Z_2} + \frac{\zeta}{Z_2}\right) + \zeta, \quad \gamma' = \alpha'\beta' - \frac{Z_1}{5}, \quad (2.178)$$

$$\beta' = 1 + \frac{Z_1}{Z_2} + \frac{Z_1}{5}, \quad \alpha' = \frac{5}{Z_R} + 1, \quad \zeta = \left(2 + \frac{Z_1}{Z_2}\right)\gamma' - \alpha',$$

where the excitation port is denoted as port 1, and the output port is port 2. Z_1 denotes the impedance of the serial branch of the resistor and inductor, and Z_2 denotes the impedance of the parallel branch of capacitors. Note that the last resistor of the circuit in Figure 2.12a is 5 Ohms.

The VF method is used to generate the pole-residue model of the network characterized by the scattering matrix. A macromodel with six poles including two real and four complex poles (see Table 2.1 and Fig. 2.13), is built to represent the original transfer function of the two-port network. Good agreement can be observed between the analytical results and the results by the macromodel with the VF method (see Fig. 2.14). Note that interested readers may compare the poles, which are

Table 2.1
Two Real Poles, Two Pairs of Complex Conjugate Poles, and the Corresponding Residues Identified by Vector Fitting Method [30]

Poles	Residues (S11)	Residues (S12 and S21)	Residues (S22)
−0.2029	0.21	0.1853	0.1636
−30.6775	−60.6235	−6.57e−06	2.43e−07
−0.6520+3.1517j	0.1561−0.0504j	−0.1388+0.0302j	0.1222−0.0142j
−0.6520−3.1517j	0.1561+0.0504j	−0.1388−0.0302j	0.1222+0.0142j
−0.5506+5.4617j	0.0507−0.0189j	0.0462−0.0088j	0.0409−0.0007j
−0.5506−5.4617j	0.0507+0.0189j	0.0462+0.0088j	0.0409+0.0007j

All the values are normalized by 1.0e9.

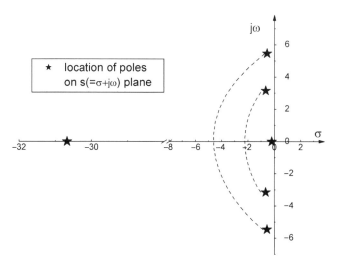

Figure 2.13   Distribution of the poles, obtained by the vector fitting method, in the s-plane [30].

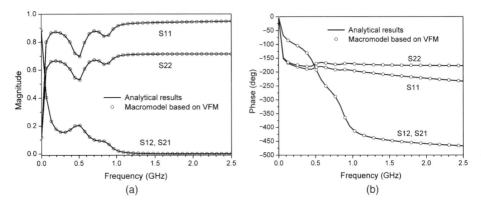

Figure 2.14 Comparison of the scattering parameters by simulation and from the macromodel [30].

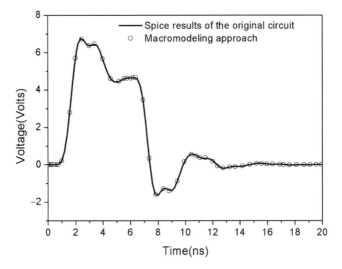

Figure 2.15 Transient voltage waveform V_{out} at the output port of the circuit [30].

listed in Table 2.1 and obtained by the VF methods, with the exact solution of the poles obtained by setting the numerators in Equation (2.176) to Equation (2.178) to zeros.

The transient simulation results are plotted in Figure 2.15. The results with the macromodel are compared with those produced by direct SPICE simulation, and good agreements can be observed. The

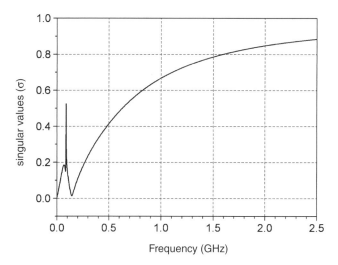

Figure 2.16   Singular values of the scattering parameters of the macromodel, which are all below the threshold of unity.

excitation source used for the transient simulation is a pulse with a rise/fall time of 0.5 ns and a pulse width of 5 ns.

Note that the singular values of the scattering parameters of the macromodel are all below the threshold of unity (see Fig. 2.16), which shows that the above macromodel is passive, and thus ensures stable time-domain simulation results.

2.7.2 Vertically Natural Capacitors (VNCAPs)

On-chip decoupling capacitors, including metal-oxide-semiconductor (MOS) capacitors, metal-insulator-metal (MIM) capacitors and lateral flux capacitors play a vital role in sustaining a reliable and low noise power supply to the on-chip devices. On-chip VNCAPs, which belong to lateral flux capacitors, is used as an important type of decoupling capacitors for RF integrated circuits and packaging, especially after the semiconductor industry ushers in the 65-nm technology era. A simple three-layer VNCAP is shown in Figure 2.17.

The scattering parameters of the above VNCAP are simulated by a commercial finite-element solver. The VF method is used to fit a pole-residue macromodel of the two-port scattering matrix. Figures

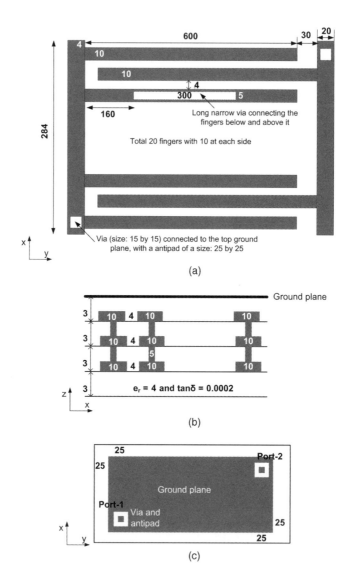

Figure 2.17 Configuration and dimension of a three-layer vertically natural capacitor (VNCAP): (a) the detailed configuration of one layer of the VNCAP; (b) cross-section view of the three-layer VNCAP; (c) top view of the VNCAP—a solid ground plane on top of a substrate, and two ports are also indicated. Note that all units are in micro-meters.

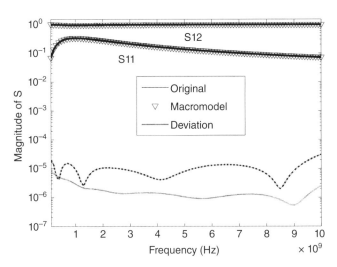

Figure 2.18 The magnitude of the scattering parameters—original versus the fitted macromodel, and the deviation between them.

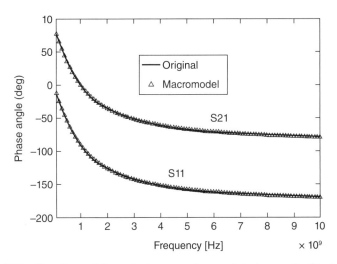

Figure 2.19 The phases of the scattering parameters—original versus the fitted macromodel.

2.18 and 2.19 show that the magnitude and the phases of the fitted model agree well with those of the original scattering parameters. The order of the macromodel is six, and the root-mean-square error for the fitted model is 6.7003e-006. Figure 2.20 reveals that the initial macromodel is passive except for a narrow low-frequency band. A small

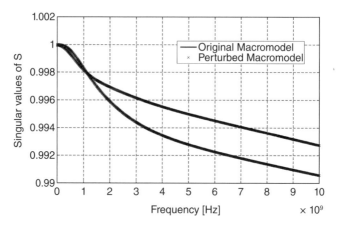

Figure 2.20   The singular values of the macromodel before and after the passivity enforcement.

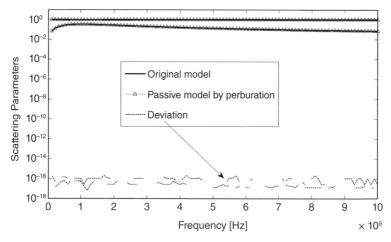

Figure 2.21   The original macromodel is compensated for the passivity violation by perturbation of residue matrix eigenvalues.

perturbation restores the passivity of the macromodel (see Fig. 2.21). The passivity violation is compensated by the perturbation of residue matrix eigenvalues [47]. Figure 2.21 shows that the perturbation to the original macromodel is very small (in the order of 10^{-15}) to restore its passivity.

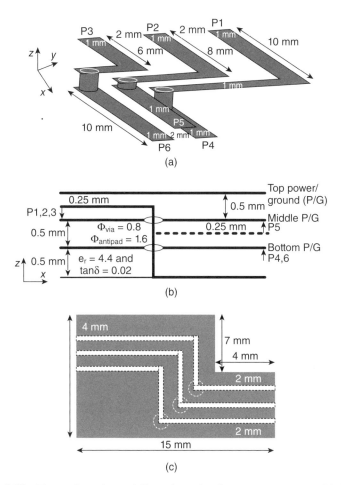

Figure 2.22   The configuration and dimensions of an interconnect structure: (a) the interconnect structure including four striplines, two microstrip lines, and three vias, with all the substrates and power/ground planes removed; (b) the cross-section view of the interconnect structure also including three power/ground planes, and three substrates; (c) top view of the power/ground plane with traces—the three planes are identical but with irregular shapes.

2.7.3 Stripline-to-Microstrip Line Transition with Vias

An interconnect structure is shown in Figure 2.22, which has three irregularly shaped power/ground planes, three striplines transitioned by vias to two microstrip lines at the bottom of the structure and a stripline

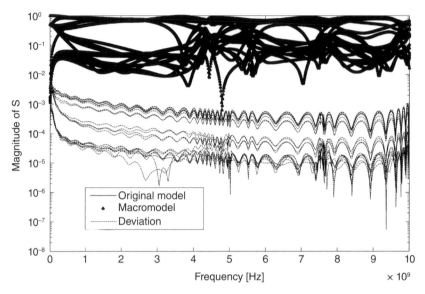

Figure 2.23 The magnitude of the scattering parameters—original versus the fitted macromodel with 80th order, and the deviation between them.

at the middle substrate. The scattering parameters of the six-port inter-connect network are fitted by the VF method. The results are shown in Figure 2.23. The order of the macromodel used to produce the results in Figure 2.23 is 80, and the root-mean-square error is 5.46e-04. However, the accuracy of the fitted model is low at both ends of the frequency band of (0, 10) GHz.

The passivity of the macromodel obtained by the VF method is examined by the HM approach. The singular values of the scattering parameters of the macromodel are plotted in Figure 2.24. Although the scattering parameters of the structure are extrapolated to DC by a commercial three-dimensional (3D) electromagnetic solver, and Figure 2.24 reveals that the singular values exceed the threshold of unity only at a narrow band at the low-frequency end, initial attempt to enforce passivity by perturbation of residue matrix eigenvalues is not successful due to the large out-of-band violations shown in Figure 2.24. Table 2.2 shows that increasing the order of the macromodel reduces the root-

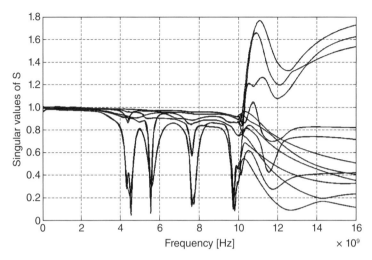

Figure 2.24   The singular values of the macromodel with 80th order before passivity enforcement—large out-of-band violation beyond 10 GHz.

Table 2.2
The Root-Mean-Square-Error, and Maximum Passivity Violation Due to Different Model Orders Used in the Vector Fitting Method

Model order	Root-mean-square error	Maximum violation
30	0.0024	0.005,146,7
80	5.46e−04	0.7619
180	4.34e−04	146.5103
300	3.96e−04	1734.2963

mean-square error, but it may cause larger out-of-band passivity violations.

Finally, we choose the order of the model to be 36. The VF method thus gives a root-mean-square of 9.5072e-004. This low-order model does not give large out-of-band passivity violations (see Fig. 2.25). The violations are compensated in a few iterations and a passive macromodel is obtained (see Fig. 2.26).

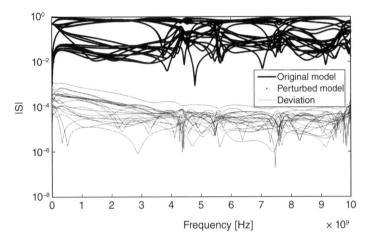

Figure 2.25 Singular values of the macromodel with 36th order before and after passivity enforcement.

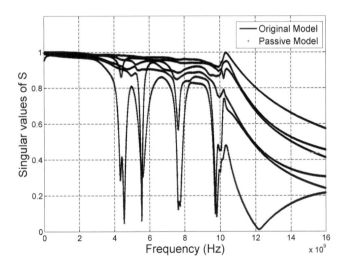

Figure 2.26 The passive macromodel obtained after passivity enforcement.

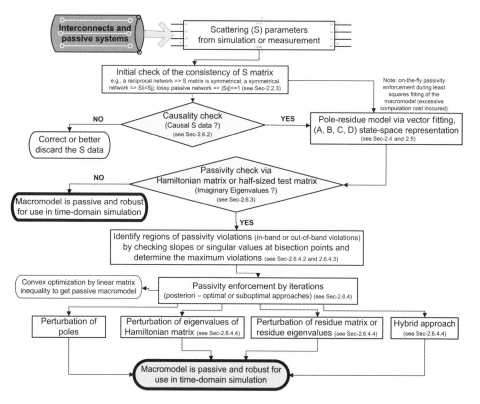

Figure 2.27   Summary of the entire flow of macromodeling with scattering parameters.

2.8 CONCLUSION

First, the entire flow for the macromodeling of interconnects and passive circuits, which are characterized by scattering matrices, is summarized in Figure 2.27.

Second, the latest development, which is not captured in this chapter, as well as the future direction in the area of macromodeling is briefly discussed. A complete procedure together with key techniques for macromodeling has been developed as presented in this chapter. The macromodeling technique is either written as separate codes [28] or simulation tools, or incorporated into commercial software. Research efforts in macromodeling is recently channeled to a few new areas:

1. ***Delay-Aware Macromodeling***: If the interconnect is electrically long compared to the operation wavelength, the propagation delay must be extracted explicitly to facilitate efficient macromodeling. New passivity enforcement schemes are being developed for this type of delay-based macromodeling [64].

2. ***Parametric macromodeling***: The conventional macromodeling approach uses the frequency as the only variable in the model. The macromodeling must be performed again if the scattering parameters are changed because of the changes in geometries and material properties. Parametric macromodeling is being proposed to overcome the above limit [65]. Parametric macromodeling is useful for design and optimization, for which the macromodel is repeatedly updated.

3. ***Sensitivity analysis with macromodeling***: Design sensitivity information is crucial in engineering problems such as optimization, statistical, yield, and tolerance analysis [66]. Macromodeling may be applied to sensitivity analysis. In addition, other topics regarding macromodeling are also interesting to study, such as new passivity enforcement schemes, and automatic model order estimation.

REFERENCES

[1] E. CHIPROUT and M. NAKHLA, *Asymptotic Waveform Evaluation and Moment Matching for Interconnect Analysis*, Kluwer, Boston, MA, 1993.

[2] R. ACHAR and M. S. NAKHLA, Efficient transient simulation of embedded sub-networks characterized by S-parameters in the presence of nonlinear elements, *IEEE Trans. Microw. Theory Tech.*, vol. 46, pp. 2356–2363, 1998.

[3] I. IERDIN, M. NAKHLA, and R. ACHAR, Circuit analysis of electromagnetic radiations and field coupling effects for networks with embedded full-wave modules, *IEEE Trans. Electromagn. Compat.*, vol. 42, pp. 449–460, 2000.

[4] L. M. SILVEIRA, I. M. ELFADEL, J. K. WHITE, M. CHILUKURI, and K. S. KUNDERT, Efficient frequency-domain modeling and circuit simulation of transmission lines, *IEEE Trans. Comp. Packag. Manuf. Technol. B*, vol. 17, pp. 505–513, 1994.

[5] E. C. CHANG and S.-M. KANG, Transient simulation of lossy coupled transmission lines using iterative linear least square fitting and piecewise recursive convolution, *IEEE Trans. Circuits Syst. I*, vol. 43, no. 11, pp. 923–932, 1996.

[6] W. T. BEYENE and J. E. SCHUTT-AINE, Efficient transient simulation of high-speed interconnects characterized by sampled data, *IEEE Trans. Comp. Packag. Manuf Technol. B*, vol. 21, no. 1, pp. 105–114, 1998.

[7] T. MANGOLD and P. RUSSER, Full-wave modeling and automatic equivalent-circuit generation of millimeter-wave planar and multilayer structures, *IEEE Trans. Microw. Theory Tech.*, vol. 47, pp. 851–858, 1999.

[8] M. ELZINGA, K. L. VIRGA, and J. L. PRINCE, Improved global rational approximation macromodeling algorithm for networks characterized by frequency-sampled data, *IEEE Trans. Microw. Theory Tech.*, vol. 48, pp. 1461–1468, 2000.

[9] T. WATANABE and H. ASAI, Synthesis of time-domain models for interconnects having 3-D structure based on FDTD method, *IEEE Trans. Circuits Syst. II*, vol. 47, pp. 302–305, 2000.

[10] R. NEUMAYER, F. HASLINGER, A. STELZER, and R. WIEGEL, Synthesis of SPICE-compatible broadband electrical models from N-port scattering parameter data, in *Proc. IEEE Symp. Electromagn. Compat.*, Minnesota, August 2002, pp. 469–474.

[11] S. LIN and E. S. KUH, Transient simulation of lossy interconnects based on the recursive convolution formulation, *IEEE Trans. Circuits Syst. I*, vol. 39, pp. 879–892, 1992.

[12] R. ACHAR and M. S. NAKHLA, Simulation of high-speed interconnects, *Proc. IEEE*, vol. 89, pp. 693–728, 2001.

[13] B. GUSTAVSEN and A. SEMLYEN, Rational approximation of frequency domain responses by vector fitting, *IEEE Trans. Power Deliv.*, vol. 14, pp. 1052–1061, 1999.

[14] W. PINELLO, J. MORSEY, and A. C. CANGELARIS, Synthesis of SPICE-compatible broadband electrical models for pins and vias, in *Proc. 51st Electronic. Components and Technology Conf.*, Orlando, FL, May 2001, pp. 518–522.

[15] E.-P. LI, X. WEI, A. C. CANGELLARIS, E.-X. LIU, Y. ZHANG, et al., Progress review of electromagnetic compatibility analysis technologies for packages, printed circuit boards, and novel interconnects, *IEEE Trans. Electromagn. Compat.*, vol. 52, no. 2, pp. 248–265, 2010 (invited).

[16] J. VLACH and K. SINGHAL, *Computer Methods for Circuit Analysis and Design*, Van Nostrand, New York, 1983.

[17] R. PINTELON, P. GUILLAUME, Y. ROLAIN, J. SCHOUKENS, and H. V. HAMME, Parametric identification of transfer functions in the frequency domain—A survey, *IEEE Trans. Autom. Control*, vol. 39, no. 11, pp. 2245–2260, 1994.

[18] C. P. COELHO, J. PHILLIPS, and L. M. SILVEIRA, A convex programming approach for generating guaranteed passive approximations to tabulated frequency-data, *IEEE Trans. Comput. Aided Des. Integr. Circuits Syst.*, vol. 23, no. 2, pp. 293–301, 2004.

[19] C. K. SANATHANAN and J. KOERNER, Transfer function synthesis as a ratio of two complex polynomials, *IEEE Trans. Autom. Control*, vol. AC-8, no. 1, pp. 56–58, 1963.

[20] D. DESCHRIJVER, B. GUSTAVSEN, and T. DHAENE, Advancements in iterative methods for rational approximation in the frequency domain, *IEEE Trans. Power Deliv.*, vol. 22, no. 3, pp. 1633–1642, 2007.

[21] G. H. GOLUB and C. F. Van LOAN, *Matrix Computations*, Johns Hopkins University Press, Baltimore, MD, 1983.

[22] C. D. MEYER, *Matrix Analysis and Applied Linear Algebra*, Society for Industrial and Applied Mathematics, Philadelphia, 2000.

[23] L. ELSNER, A remark on simultaneous inclusions of the zeros of a polynomial by Gershgorin theorem, *Numer. Math.*, vol. 21, pp. 425–427, 1973.

[24] S. GRIVET-TALOCIA, Package macromodeling via time-domain vector fitting, *IEEE Microw. Wireless Compon. Lett.*, vol. 13, no. 11, pp. 472–474, 2003.

[25] D. DESCHRIJVER, B. HAEGEMAN, and T. DHAENE, Orthonormal vector fitting: A robust macromodeling tool for rational approximation of frequency domain responses, *IEEE Trans. Adv. Packag.*, vol. 30, no. 2, pp. 216–225, 2007.

[26] B. GUSTAVSEN, Improving the pole relocating properties of vector fitting, *IEEE Trans. Power Deliv.*, vol. 21, no. 3, pp. 1587–1592, 2006.

[27] D. DESCHRIJVER, M. MROZOWSKI, T. DHAENE, and D. De ZUTTER, Macromodeling of multiport systems using a fast implementation of the vector fitting method, *IEEE Microw. Wireless Compon. Lett.*, vol. 18, no. 6, pp. 383–385, 2008.

[28] The vector fitting website, http://www.energy.sintef.no/Produkt/VECTFIT/index.asp.

[29] C. T. CHEN, *Linear System Theory and Design*, Oxford University Press, New York, 1998.

[30] E.-P. LI, E.-X. LIU, L.-W. LI, and M.-S. LEONG, A coupled efficient and systematic full-wave time-domain macromodeling and circuit simulation method for signal integrity analysis of high-speed interconnects, *IEEE Trans. Adv. Packag.*, vol. 27, no. 1, pp. 213–223, 2004.

[31] A. VLADIMIRESCU, *The PSpice Book*, John Wiley & Sons, New York, 1994.

[32] S. RAMO, J. R. WHINNERY, and T. VANDUZER, *Fields and Waves in Communication Electronics*, John Wiley & Sons, New York, 1965.

[33] G. B. ARFKEN and H. J. WEBER, *Mathematical Methods for Physicists*, 6th ed., Academic Press, San Diego, 2005.

[34] P. TRIVERIO and S. GRIVET-TALOCIA, Robust causality characterization via generalized dispersion relations, *IEEE Trans. Adv. Packag.*, vol. 31, no. 3, pp. 579–593, 2008.

[35] P. TRIVERIO and S. GRIVET-TALOCIA, A robust causality verification tool for tabulated frequency data, in *Proc. 10th IEEE Workshop Signal Propag. Interconnects*, Berlin, Germany, May 2006, pp. 65–68.

[36] S. GRIVET-TALOCIA, Passivity enforcement via perturbation of Hamiltonian matrices, *IEEE Trans. Circuits Syst. I*, vol. 51, no. 9, pp. 1755–1769, 2004.

[37] M. R. WOHLERS, *Lumped and Distributed Passive Networks*, Academic, New York, 1969.

[38] P. TRIVERIO, S. GRIVET-TALOCIA, M. S. NAKHLA, F. G. CANAVERO, and R. ACHAR, Stability, causality, and passivity in electrical interconnect models, *IEEE Trans. Adv. Packag.*, vol. 30, no. 4, pp. 795–808, 2007.

[39] D. SARASWAT, R. ACHAR, and M. NAKHLA, On passivity enforcement for macromodels of S-parameter based tabulated subnetworks, in *Proc. IEEE Int. Symp. Circuits Syst.*, May 23–26 2005, pp. 3777–3780.

[40] S. BOYD, V. BALAKRISHNAN, and P. KABAMBA, A bisection method for computing the H_∞ norm of a transfer matrix and related problems, *Math. Control Signals Syst.*, vol. 2, no. 3, pp. 207–219, 1989.

[41] S. GRIVET-TALOCIA and A. UBOLLI, On the generation of large passive macro-models for complex interconnect structures, *IEEE Trans. Adv. Packag.*, vol. 29, no. 1, pp. 39–54, 2006.

[42] S. GRIVET-TALOCIA, An adaptive sampling technique for passivity characterization and enforcement of large interconnect macromodels, *IEEE Trans. Adv. Packag.*, vol. 30, no. 2, pp. 226–237, 2007.

[43] B. GUSTAVSEN and A. SEMLYEN, Fast passivity assessment for S-parameter rational models via a half-size test matrix, *IEEE Trans. Microw. Theory Tech.*, vol. 56, no. 12, pp. 2701–2708, 2008.

[44] D. DESCHRIJVER and T. DHAENE, Modified half-size test matrix for robust passivity assessment of S-parameter macromodels, *IEEE Microw. Wireless Compon. Lett.*, vol. 19, no. 5, pp. 263–265, 2009.

[45] R. N. SHORTEN, P. CURRAN, K. WULFF, and E. ZEHEB, A note on spectral conditions for positive realness of transfer function matrices, *IEEE Trans. Autom. Control*, vol. 53, no. 5, pp. 1258–1261, 2008.

[46] S. GRIVET-TALOCIA and A. UBOLLI, A comparative study of passivity enforcement schemes for linear lumped macromodels, *IEEE Trans. Adv. Packag.*, vol. 31, no. 4, pp. 673–683, 2008.

[47] B. GUSTAVSEN, Fast passivity enforcement for S-parameter models by perturbation of residue matrix eigenvalues, *IEEE Trans. Adv. Packag.*, vol. 33, no. 1, pp. 257–265, 2010.

[48] A. LAMECKI and M. MROZOWSKI, Equivalent SPICE circuits with guaranteed passivity from nonpassive models, *IEEE Trans. Microw. Theory Tech.*, vol. 55, no. 3, pp. 526–532, 2007.

[49] D. DESCHRIJVER and T. DHAENE, Fast passivity enforcement of S-parameter macromodels by pole perturbation, *IEEE Trans. Microw. Theory Tech.*, vol. 57, no. 3, pp. 620–626, 2009.

[50] M. SWAMINATHAN, C. DAEHYUN, S. GRIVET-TALOCIA, K. BHARATH, V. LADDHA, and X. JIANYONG, Designing and modeling for power integrity, *IEEE Trans. Electromagn. Compat.*, vol. 52, no. 2, pp. 288–310, 2010.

[51] B. GUSTAVSEN and A. SEMLYEN, Enforcing passivity for admittance matrices approximated by rational functions, *IEEE Trans. Power Syst.*, vol. 16, no. 1, pp. 97–104, 2001.

[52] D. SARASWAT, R. ACHAR, and M. S. NAKHLA, Global passivity enforcement algorithm for macromodels of interconnect subnetworks characterized by tabulated data, *IEEE Trans. VLSI Syst*, vol. 13, no. 7, pp. 819–832, 2005.

[53] B. GUSTAVSEN, Fast passivity enforcement for pole-residue models by perturbation of residue matrix eigenvalues, *IEEE Trans. Power Deliv.*, vol. 23, no. 4, pp. 2278–2285, 2008.

[54] B. GUSTAVSEN, Computer code for passivity enforcement of rational macromodels by residue perturbation, *IEEE Trans. Adv. Packag.*, vol. 30, no. 2, pp. 209–215, 2007.

[55] B. GUSTAVSEN, Passivity enforcement of rational models via modal perturbation, *IEEE Trans. Power Deliv.*, vol. 23, no. 2, pp. 768–775, 2008.

[56] D. Saraswat, R. Achar, and M. S. Nakhla, Fast passivity verification and enforcement via reciprocal systems for interconnects with large order macromodels, *IEEE Trans. VLSI Syst.*, vol. 15, no. 1, pp. 48–59, 2007.

[57] A. Semlyen and B. Gustavsen, A half-size singularity test matrix for fast and reliable passivity assessment of rational models, *IEEE Trans. Power Deliv.*, vol. 24, no. 1, pp. 345–351, 2009.

[58] N. Stevens, D. Deschrijver, and T. Dhaene, Fast automatic order estimation of rational macromodels for signal integrity analysis, in *Proc. IEEE Workshop on Signal Propagation on Interconnects*, May 2007, pp. 89–92.

[59] Y. Rolain, J. Schoukens, and R. Pintelon, Order estimation for linear time-invariant systems using frequency domain identification methods, *IEEE Trans. Autom. Control*, vol. 42, no. 10, pp. 1408–1417, 1997.

[60] S.-H. Min, H. Lee, E. Song, Y.-S. Choi, T.-J. Cho, et al., Model-order estimation and reduction of distributed interconnects via improved vector fitting, in *IEEE 14th Topical Meeting on Electrical Performance of Electronic Packaging*, October 2005, pp. 43–46.

[61] S. Grivet-Talocia, M. Bandinu, and F. G. Canavero, An automatic algorithm for equivalent circuit extraction from noisy frequency responses, in *International Symposium on Electromagnetic Compatibility*, 2005, pp. 163–168.

[62] S.-H. Lee, K. Mao, and J.-M. Jin, A complete finite-element analysis of multilayer anisotropic transmission lines from DC to terahertz frequencies, *IEEE Trans. Adv. Packag.*, vol. 31, no. 2, pp. 326–338, 2008.

[63] F. P. Andriulli, H. Bagci, F. Vipiana, G. Vecchi, and E. Michielssen, Analysis and regularization of the TD-EFIE low-frequency breakdown, *IEEE Trans. Antenn. Propag.*, vol. 57, no. 7, pp. 2034–2046, 2009.

[64] A. Charest, M. Nakhla, and R. Achar, Scattering domain passivity verification and enforcement of delayed rational functions, *IEEE Microw. Wireless Compon. Lett.*, vol. 19, no. 10, pp. 605–607, 2009.

[65] P. Triverio, S. Grivet-Talocia, M. Bandinu, and F. G. Canavero, Geometrically parameterized circuit models of printed circuit board traces inclusive of antenna coupling, *IEEE Trans. Electromagn. Compat.*, vol. 52, no. 2, pp. 471–478, 2010.

[66] N. K. Nikolova, J. W. Bandler, and M. H. Bakr, Adjoint techniques for sensitivity analysis in high-frequency structure CAD, *IEEE Trans. Microw. Theory Tech.*, vol. 52, no. 1, pp. 403–419, 2004.

CHAPTER 3

2.5D Simulation Method for 3D Integrated Systems

3.1 INTRODUCTION

Vias are usually employed in the electronic packages with the shape of circular cylinders. Thus, the theory of multiple scattering among many parallel conducting cylinders [1] can be used to model them efficiently. The theory of scattering by conducting cylinders (vias) in the presence of two perfect electric conductors (PECs) [2] has been applied to study the problem of vias in three-dimensional (3D) integrated structures [3, 4]. In this research, instead of using the Green's function approach in References 3 and 4 to obtain the corresponding formulae, we will directly apply the parallel-plate waveguide (PPWG) theory, which is a relatively simple and straightforward way to tackle the problem of scattering by cylinders in the presence of two or more PEC planes. Without loss of generality, we assume that the power-ground (P-G) planes in an electronic package are made of PECs, which may be of finite thickness, and the vias are circular PEC cylinders.

In this chapter, the semi-analytical scattering matrix method (SMM) based on the N-body scattering theory is presented for modeling of 3D

Electrical Modeling and Design for 3D System Integration: 3D Integrated Circuits and Packaging, Signal Integrity, Power Integrity and EMC, First Edition. Er-Ping Li.

electronic package integration and printed circuit boards (PCB) with multiple vias. Using the modal expansion of fields in a PPWG, the formula derivation of the SMM is presented in detail. In the conventional SMM, the P-G planes are assumed to be infinitely large so it cannot capture the resonant behavior of the real-world packages. In particular, the SMM method has been extended to solve the finite domain of P-G planes in coupling with a novel boundary modeling method proposed by the author's group. This method has demonstrated its unique features which is capable to efficiently handle the complex real-world package and PCB structures.

In the latter part of the chapter, numerical examples are presented for validation of the implemented SMM algorithm with the proposed frequency-dependent cylinder layer (FDCL). The extended method is not only capable to simulate the finite-sized P-G planes, but it can also simulate the arbitrarily shaped boundary of the planes. This is one prominent feature of the FDCL modeling method.

3.2 MULTIPLE SCATTERING METHOD FOR ELECTRONIC PACKAGE MODELING WITH OPEN BOUNDARY PROBLEMS

3.2.1 Modal Expansion of Fields in a Parallel-Plate Waveguide (PPWG)

The source-free Maxwell's equations are given as

$$\nabla \times \mathbf{E} = -j\omega\mu\mathbf{H}, \tag{3.1}$$

$$\nabla \times \mathbf{H} = j\omega\varepsilon\mathbf{E}, \tag{3.2}$$

$$\nabla \cdot \mathbf{E} = 0, \tag{3.3}$$

$$\nabla \cdot \mathbf{H} = 0. \tag{3.4}$$

Two adjacent conductor planes either power or ground can be considered as a PPWG. Assuming that the z-axis is normal to the surface of the P-G planes and the electromagnetic fields have $e^{-j\beta z}$ dependence where β is the propagation wavenumber along the guiding direction z. For the PPWG structure, two independent solutions of the above Maxwell's equations in cylindrical coordinate are given as

$$E_z(\rho, \phi, z) = \sum_{n=-\infty}^{\infty} \sum_{m=0}^{\infty} \left[a_{mn}^E J_n(k_\rho \rho) + b_{mn}^E H_n^{(2)}(k_\rho \rho) \right] C_m e^{jn\phi} \quad (3.5)$$

for *TM* waves,

$$H_z(\rho, \phi, z) = \sum_{n=-\infty}^{\infty} \sum_{m=1}^{\infty} \left[a_{mn}^H J_n(k_\rho \rho) + b_{mn}^H H_n^{(2)}(k_\rho \rho) \right] S_m e^{jn\phi} \quad (3.6)$$

for *TE* waves,

where a_{mn}^E and b_{mn}^E are the expansion coefficients of the incoming and outgoing transverse magnetic (TM) waves, a_{mn}^H and b_{mn}^H are the expansion coefficients of the incoming and outgoing transverse electric (TE) waves, respectively. $k^2 = \omega^2 \mu \varepsilon = k_\rho^2 + \beta_m^2$, $\beta_m = k_z = m\pi/d$, where d is the spacing of the adjacent P-G planes, and μ and ε represent the permeability and permittivity of the dielectric sandwiched between the P-G planes. The terms C_m and S_m stands for $C_m = \cos(\beta_m z)$ and $S_m = \sin(\beta_m z)$, respectively. An $e^{j\omega t}$ time dependence is assumed throughout the formulation.

Other components of **E** and **H** related to E_z and H_z are calculated by

$$\begin{bmatrix} \mathbf{E}_s \\ \mathbf{H}_s \end{bmatrix} = \frac{1}{k_\rho^2} \begin{bmatrix} \dfrac{\partial}{\partial z} & j\omega\mu\hat{z}\times \\ -j\omega\varepsilon\hat{z}\times & \dfrac{\partial}{\partial z} \end{bmatrix} \begin{bmatrix} \nabla_s E_z \\ \nabla_s H_z \end{bmatrix}. \quad (3.7)$$

The operator ∇_s represents the gradient in transverse direction and in cylindrical coordinates, and it can be written as

$$\nabla_s \equiv \hat{\rho}\frac{\partial}{\partial \rho} + \hat{\phi}\frac{1}{\rho}\frac{\partial}{\partial \phi}. \quad (3.8)$$

Then, using modal expansion approach, the E_z and H_z components of an incident wave are expressed as:

$$\begin{aligned} E_z^{inc} &= \sum_m \sum_n a_{mn}^E \cos(\beta_m z) J_n(k_\rho \rho) e^{jn\phi} \quad \text{for TM}^z \text{ mode,} \\ H_z^{inc} &= \sum_m \sum_n a_{mn}^H \sin(\beta_m z) J_n(k_\rho \rho) e^{jn\phi} \quad \text{for TE}^z \text{ mode.} \end{aligned} \quad (3.9)$$

The modal expansion of the scattered fields E_z^{scat} and H_z^{scat} can be expressed similar to those in Equation (3.9) by using b_{mn}^E and b_{mn}^H as the unknown expansion coefficients. Substituting Equation (3.9) into Equation (3.7), we can obtain all other components of the electromagnetic fields corresponding to TM^z and TE^z modes.

Since the total field is a summation of the incident and scattered fields, we can finally obtain the following expressions for the total tangential electromagnetic fields in cylindrical coordinates, normal to $\hat{\rho}$ in the ith PPWG formed by pair of P-G planes.

$$
\mathbf{E}_t^{(i)} = \sum_{n=-\infty}^{\infty} \left\{ \sum_{m=0}^{\infty} \left[a_{mn}^{E(i)} J_{mn}^{(i)} + b_{mn}^{E(i)} H_{mn}^{(i)} \right] \mathbf{e}_{tmn}^{E(i)} \right.
$$
$$
\left. + \sum_{m=1}^{\infty} \left[a_{mn}^{H(i)} J_{mn}'^{(i)} + b_{mn}^{H(i)} H_{mn}'^{(i)} \right] \mathbf{e}_{tmn}^{H(i)} \right\} e^{jn\phi}, \tag{3.10}
$$

$$
\mathbf{H}_t^{(i)} = \sum_{n=-\infty}^{\infty} \left\{ \sum_{m=0}^{\infty} \left[a_{mn}^{E(i)} J_{mn}'^{(i)} + b_{mn}^{E(i)} H_{mn}'^{(i)} \right] \mathbf{h}_{tmn}^{E(i)} \right.
$$
$$
\left. + \sum_{m=1}^{\infty} \left[a_{mn}^{H(i)} J_{mn}^{(i)} + b_{mn}^{H(i)} H_{mn}^{(i)} \right] \mathbf{h}_{tmn}^{H(i)} \right\} e^{jn\phi}, \tag{3.11}
$$

where the eigenvectors are defined as

$$
\mathbf{e}_{tmn}^{E(i)} = C_m^{(i)} \hat{z} - \frac{jn\beta_m^{(i)}}{k_\rho^{2(i)} \rho} S_m^{(i)} \hat{\varphi}
$$
$$
\mathbf{h}_{tmn}^{E(i)} = -\frac{j\omega\varepsilon}{k_\rho^{(i)}} C_m^{(i)} \hat{\varphi}, \tag{3.12}
$$

for mn^{th} TM mode, and

$$
\mathbf{e}_{tmn}^{H(i)} = \frac{j\omega\mu}{k_\rho^{(i)}} S_m^{(i)} \hat{\varphi}
$$
$$
\mathbf{h}_{tmn}^{E(i)} = S_m^{(i)} \hat{z} + \frac{jn\beta_m^{(i)}}{k_\rho^{2(i)} \rho} C_m^{(i)} \hat{\varphi}, \tag{3.13}
$$

for mn^{th} TE mode. The terms $C_m^{(i)}$ and $S_m^{(i)}$ are defined as $C_m^{(i)} = \cos\left(\beta_m^{(i)}(z - z_i)\right)$ and $S_m^{(i)} = \sin\left(\beta_m^{(i)}(z - z_i)\right)$, respectively. And

$z \in [z_i, \ z_i + h_i]$; h_i is the height of the waveguide. Symbols $J_{mn}^{(i)}$, $J_{mn}'^{(i)}$, $H_{mn}^{(i)}$, and $H_{mn}'^{(i)}$ represent the following Bessel and Hankel functions:

$$
\begin{aligned}
J_{mn}^{(i)} &= J_n\left(k_m^{(i)}\rho\right), & J_{mn}'^{(i)} &= J_n'\left(k_m^{(i)}\rho\right), \\
H_{mn}^{(i)} &= H_n^{(2)}\left(k_m^{(i)}\rho\right), & H_{mn}'^{(i)} &= H_n^{(2)\prime}\left(k_m^{(i)}\rho\right),
\end{aligned}
\tag{3.14}
$$

where $k_m^{2(i)} = k_\rho^{2(i)} = k^2 - \beta_m^{2(i)}$.

3.2.2 Multiple Scattering Coefficients among Cylindrical PEC and Perfect Magnetic Conductor (PMC) Vias

The boundary condition for the PMC is given as $\hat{n} \times \mathbf{H} = 0$. The total magnetic field on the surface of qth PMC cylinder with radius r_q in the ith parallel-plate layer is given by

$$
\begin{aligned}
\mathbf{H}_t^{(i)}\left(r_q, \phi, z\right) = \sum_{n=-\infty}^{\infty} &\left\{ \sum_{m=0}^{\infty}\left[a_{mn}^{E(i)}J_n'\left(k_\rho r_q\right) + b_{mn}^{E(i)}H_n^{(2)\prime}\left(k_\rho r_q\right)\right]\frac{-j\omega\varepsilon}{k_\rho}C_m\hat{\varphi} \right. \\
&+ \sum_{m=1}^{\infty}\left[a_{mn}^{H(i)}J_n\left(k_\rho r_q\right) + b_{mn}^{H(i)}H_n^{(2)}\left(k_\rho r_q\right)\right] \\
&\left. \left(\frac{jn\beta_m}{k_\rho^2 r_q}C_m\hat{\varphi} + S_m\hat{z}\right) \right\}e^{jn\phi} = 0,
\end{aligned}
\tag{3.15}
$$

for any value of $z \in [0,d]$. Then,

$$
b_{mn(q)}^{H(i)} = T_{mn(q)}^{E(i)}a_{mn(q)}^{H(i)},
\tag{3.16}
$$

$$
b_{mn(q)}^{E(i)} = T_{mn(q)}^{H(i)}a_{mn(q)}^{E(i)},
\tag{3.17}
$$

where

$$
T_{mn(q)}^{E(i)} = -\frac{J_n\left(k_\rho r_q\right)}{H_n^{(2)}\left(k_\rho r_q\right)},
\tag{3.18}
$$

$$
T_{mn(q)}^{H(i)} = -\frac{J_n'\left(k_\rho r_q\right)}{H_n^{(2)\prime}\left(k_\rho r_q\right)} = -\frac{J_{n+1}\left(k_\rho r_q\right) - J_{n-1}\left(k_\rho r_q\right)}{H_{n+1}^{(2)}\left(k_\rho r_q\right) - H_{n-1}^{(2)}\left(k_\rho r_q\right)},
\tag{3.19}
$$

with $k_\rho = k_m = \sqrt{k^2 - \beta_m^2}$. The equations can be written in matrix form as

$$\begin{bmatrix} b_{mn(q)}^{E(i)} \\ b_{mn(q)}^{H(i)} \end{bmatrix} = \begin{bmatrix} T_{mn(q)}^{H(i)} & 0 \\ 0 & T_{mn(q)}^{E(i)} \end{bmatrix} \begin{bmatrix} a_{mn(q)}^{E(i)} \\ a_{mn(q)}^{H(i)} \end{bmatrix}. \tag{3.20}$$

The boundary condition for the PEC is given as $\hat{n} \times \mathbf{E} = 0$. The total magnetic field on the surface of qth PEC cylinder with radius r_q in the ith parallel-plate layer is given by

$$\mathbf{E}_t^{(i)}(r_q, \phi, z) = \sum_{n=-\infty}^{\infty} \left\{ \sum_{m=0}^{\infty} \left[a_{mn}^{E(i)} J_n(k_\rho r_q) + b_{mn}^{E(i)} H_n^{(2)}(k_\rho r_q) \right] \right.$$

$$\left(\frac{-jn\beta_m}{k_\rho^2 r_q} S_m \hat{\phi} + C_m \hat{z} \right) + \sum_{m=1}^{\infty} \left[a_{mn}^{H(i)} J_n'(k_\rho r_q) + b_{mn}^{H(i)} H_n^{(2)\prime}(k_\rho r_q) \right]$$

$$\left. \frac{j\omega\mu}{k_\rho} S_m \hat{\phi} \right\} e^{jn\phi} = 0, \tag{3.21}$$

for any value of $z \in [0,d]$. Then,

$$b_{mn(q)}^{E(i)} = T_{mn(q)}^{E(i)} a_{mn(q)}^{E(i)}, \tag{3.22}$$

$$b_{mn(q)}^{H(i)} = T_{mn(q)}^{H(i)} a_{mn(q)}^{H(i)}. \tag{3.23}$$

The equations can be written in matrix form as

$$\begin{bmatrix} b_{mn(q)}^{E(i)} \\ b_{mn(q)}^{H(i)} \end{bmatrix} = \begin{bmatrix} T_{mn(q)}^{E(i)} & 0 \\ 0 & T_{mn(q)}^{H(i)} \end{bmatrix} \begin{bmatrix} a_{mn(q)}^{E(i)} \\ a_{mn(q)}^{H(i)} \end{bmatrix}. \tag{3.24}$$

For the scattering analysis from the PMC and PEC cylinders, the different z-direction modes (related to different index m) are decoupled, and different ϕ-direction modes (related to different index n) are decoupled, and then the TM (E-) and TE (H-) modes for the cases of PMC and PEC cylinders are considered as decoupled.

The following short discussion proves that TE and TM modes generated by PEC cylinder are decoupled in the PPWG.

The boundary condition at the surface of the PEC cylinder is: $E_t|_{\rho=a} = 0$, that is,

$$\sum_{m=0}^{\infty}\left\{\left[a_{mn}^{E(i)}J_{mn}^{(i)} + b_{mn}^{E(i)}H_{mn}^{(i)}\right]\mathbf{e}_{t,mn}^{E(i)} + \left[a_{mn}^{H(i)}J_{mn}^{\prime(i)} + b_{mn}^{H(i)}H_{mn}^{\prime(i)}\right]\mathbf{e}_{t,mn}^{H(i)}\right\} = 0. \quad (3.25)$$

Substituting the corresponding equations in Equations (3.12) and (3.13) into Equation (3.25), we have

$$\sum_{m=0}^{\infty}\left\{\left[a_{mn}^{E(i)}J_{mn}^{(i)} + b_{mn}^{E(i)}H_{mn}^{(i)}\right]\left[\cos\left(\beta_m^{(i)}(z-z_i)\right)\hat{z} - \frac{jn\beta_m^{(i)}}{(k_m^{(i)})^2\rho}\sin\left(\beta_m^{(i)}(z-z_i)\right)\hat{\varphi}\right]\right.$$
$$\left. + \left[a_{mn}^{H(i)}J_{mn}^{\prime(i)} + b_{mn}^{H(i)}H_{mn}^{\prime(i)}\right]\frac{j\omega\mu}{k_m^{(i)}}\sin\left(\beta_m^{(i)}(z-z_i)\right)\hat{\varphi}\right\} = 0. \quad (3.26)$$

Grouping all the terms in Equation (3.26) w.r.t $\hat{z}$ and $\hat{\varphi}$ components, we get

$$\sum_{m=0}^{\infty}\left[a_{mn}^{E(i)}J_{mn}^{(i)} + b_{mn}^{E(i)}H_{mn}^{(i)}\right]\cos\left(\beta_m^{(i)}(z-z_i)\right)\hat{z}$$
$$+ \sum_{m=0}^{\infty}\left\{\begin{array}{l}\left[a_{mn}^{E(i)}J_{mn}^{(i)} + b_{mn}^{E(i)}H_{mn}^{(i)}\right]\dfrac{-jn\beta_m^{(i)}}{(k_m^{(i)})^2\rho}\\[2ex] + \left[a_{mn}^{H(i)}J_{mn}^{\prime(i)} + b_{mn}^{H(i)}H_{mn}^{\prime(i)}\right]\dfrac{j\omega\mu}{k_m^{(i)}}\end{array}\right\}\sin\left(\beta_m^{(i)}(z-z_i)\right)\hat{\varphi} = 0.$$

$$(3.27)$$

The sine and cosine functions $\sin\left(\beta_m^{(i)}(z-z_i)\right)$ and $\cos\left(\beta_m^{(i)}(z-z_i)\right)$ in Equation (3.27) are not always zero, so we have

$$a_{mn}^{E(i)}J_{mn}^{(i)} + b_{mn}^{E(i)}H_{mn}^{(i)} = 0, \quad (3.28)$$

and

$$\left[a_{mn}^{E(i)}J_{mn}^{(i)} + b_{mn}^{E(i)}H_{mn}^{(i)}\right]\frac{-jn\beta_m^{(i)}}{(k_m^{(i)})^2\rho} + \left[a_{mn}^{H(i)}J_{mn}^{\prime(i)} + b_{mn}^{H(i)}H_{mn}^{\prime(i)}\right]\frac{j\omega\mu}{k_m^{(i)}} = 0. \quad (3.29)$$

Because of Equation (3.28), the expansion coefficients in Equation (3.29) for TE and TM modes become independent, that is, TE and TM

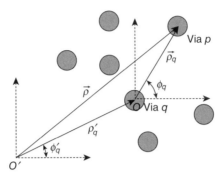

Figure 3.1   A set of random cylindrical vias (2D view).

modes for the PEC cylinders are totally decoupled; and different modes n are also decoupled. Finally, we have

$$a_{mn}^{E(i)} J_{mn}^{(i)} + b_{mn}^{E(i)} H_{mn}^{(i)} = 0, \tag{3.30}$$

$$a_{mn}^{H(i)} J_{mn}'^{(i)} + b_{mn}^{H(i)} H_{mn}'^{(i)} = 0, \tag{3.31}$$

or

$$b_{mn}^{E(i)} = -\frac{J_{mn}^{(i)}}{H_{mn}^{(i)}} a_{mn}^{E(i)}, \quad b_{mn}^{H(i)} = -\frac{J_{mn}'^{(i)}}{H_{mn}'^{(i)}} a_{mn}^{E(i)}. \tag{3.32}$$

Consider a set of randomly distributed cylindrical vias as shown in Figure 3.1, where the vias can have different radius and may be present in different layers of an electronic package.

By taking into account the multiple scattering among N_c cylindrical vias, the scattered field at an observation point p can be expressed as

$$E^{scat}(\vec{\rho}) = \sum_{q=1}^{N_c} \sum_{m=0}^{M_q} \sum_{n=-N_q}^{N_q} b_{qmn} H_n^{(2)}(k_m \rho_q) e^{jn\phi_q} \cos(\beta_m z), \tag{3.33}$$

where (ρ_q, ϕ_q) are the local coordinates with $\rho_q = |\vec{\rho} - \vec{\rho}_q|$ and $\phi_q = \arg(\vec{\rho} - \vec{\rho}_q)$. $M_q + 1$ represents the truncation number of modes in the PPWG structure, and $2N_q + 1$ is that of the Hankel functions used to express the scattered waves of the qth via. b_{qmn} denotes the unknown expansion coefficients for the scattered field.

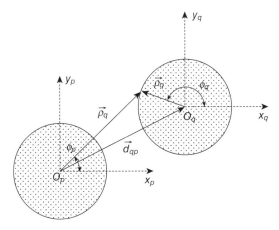

Figure 3.2   A schematic of cylindrical coordinates for translational addition theorem.

The addition theorem of the Bessel functions for the translation of cylindrical coordinates from cylinder p to cylinder q is given as

$$H_m^{(2)}(k_\rho\rho_p)e^{jm\phi_p} = \sum_{m=-\infty}^{\infty}\left[H_{n-m}^{(2)}(k_\rho d_{qp})e^{-j(n-m)\theta_{qp}}\right]J_n(k_\rho\rho_q)e^{jn\phi_q}, \quad (3.34)$$

where $\rho_q < d_{pq}$; $[\rho_p, \rho_q, \phi_p, \phi_q, d_{qp}, \theta_{qp}] \in$ Real; $k_\rho \in$ Complex; $k_\rho \neq 0$, and the terms here are expressed in global coordinate system (Fig. 3.2).

According to Equation (3.5), we define the following incoming and outgoing modes for TM case:

$$E_{zmn}^{(a)E} = J_n(k_\rho\rho)C_m e^{jn\phi}, \quad (3.35)$$

$$E_{zmn}^{(b)E} = H_n^{(2)}(k_\rho\rho)C_m e^{jn\phi}. \quad (3.36)$$

Substitute Equations (3.35) and (3.36) into Equation (3.7), we get the tangential modes for TM case.

$$\mathbf{E}_{smn}^{(a)E} = \frac{1}{k_\rho^2}\frac{\partial}{\partial z}\nabla_S E_{zmn}^{(a)E}, \quad (3.37)$$

$$\mathbf{H}_{smn}^{(a)E} = \frac{-j\omega\varepsilon}{k_\rho^2}\hat{z} \times \nabla_S E_{zmn}^{(a)E}, \tag{3.38}$$

for incoming wave, and

$$\mathbf{E}_{smn}^{(b)E} = \frac{1}{k_\rho^2}\frac{\partial}{\partial z}\nabla_S E_{zmn}^{(b)E}, \tag{3.39}$$

$$\mathbf{H}_{smn}^{(b)E} = -\frac{j\omega\varepsilon}{k_\rho^2}\hat{z} \times \nabla_S E_{zmn}^{(b)E}, \tag{3.40}$$

for outgoing wave.

According to Equation (3.6), we define the following incoming and outgoing modes for TE case.

$$H_{zmn}^{(a)H} = J_n(k_\rho\rho)S_m e^{jn\phi}, \tag{3.41}$$

$$H_{zmn}^{(b)H} = H_n^{(2)}(k_\rho\rho)S_m e^{jn\phi}. \tag{3.42}$$

Substitute Equations (3.41) and (3.42) into Equation (3.7), we get the tangential modes for TE case.

$$\mathbf{E}_{smn}^{(a)H} = \frac{j\omega\mu}{k_\rho^2}\hat{z} \times \nabla_S H_{zmn}^{(a)H}, \tag{3.43}$$

$$\mathbf{H}_{smn}^{(a)H} = \frac{1}{k_\rho^2}\frac{\partial}{\partial z}\nabla_S H_{zmn}^{(a)H}, \tag{3.44}$$

for incoming wave, and

$$\mathbf{E}_{smn}^{(b)H} = \frac{j\omega\mu}{k_\rho^2}\hat{z} \times \nabla_S H_{zmn}^{(b)H}, \tag{3.45}$$

$$\mathbf{H}_{smn}^{(b)H} = \frac{1}{k_\rho^2}\frac{\partial}{\partial z}\nabla_S H_{zmn}^{(b)H}, \tag{3.46}$$

for outgoing wave.

The outgoing *TM* wave from the pth cylinder can be written as

$$
\begin{aligned}
E_{zm(p)}^{(b)E} &= \sum_{n_p=-N_p}^{N_p} b_{mn_p(p)}^E E_{zmn_p(p)}^{(b)E} \\
&= \sum_{n_q=-N_q}^{N_q} \left[\sum_{n_p=-N_p}^{N_p} H_{n_q-n_p}^{(2)}\left(k_\rho d_{qp}\right) e^{-j(n_q-n_p)\theta_{qp}} b_{mn_p(p)}^E \right] J_{n_q}\left(k_\rho \rho_q\right) C_m e^{jn_q\phi_q} \\
&= \sum_{n_q=-N_q}^{N_q} a_{mn_q(q)}^E E_{zmn_q(q)}^{(a)E} \\
&= E_{zm(q)}^{(a)E}.
\end{aligned}
\tag{3.47}
$$

Since $\rho_q \in q^{th}$ cylinder's boundary, so $\rho_q < d_{qp}$. The incoming wave coefficient for the qth cylinder is then given as

$$
a_{mn_q(q)}^E = \sum_{n_p=-N_p}^{N_p} H_{n_q-n_p}^{(2)}\left(k_\rho d_{qp}\right) e^{-j(n_q-n_p)\theta_{qp}} b_{mn_p(p)}^E,
\tag{3.48}
$$

and $a_{mn_q(q)}^E$ is independent of the terms ρ_p, ρ_q, ϕ_p, and ϕ_q.

For considering the different coordinates, the value of ∇_s should not change, which means for any function $f(\boldsymbol{\rho})$,

$$
\nabla_s^{(p)} f(\boldsymbol{\rho}) = \nabla_s^{(q)} f(\boldsymbol{\rho}).
\tag{3.49}
$$

Substitute Equation (3.47) into Equation (3.49), we get

$$
\begin{aligned}
\mathbf{E}_{sm(p)}^{(b)E} &= \frac{1}{k_\rho^2} \frac{\partial}{\partial z} \nabla_s^{(p)} E_{zm(p)}^{(b)E} \\
&= \frac{1}{k_\rho^2} \frac{\partial}{\partial z} \nabla_s^{(p)} \sum_{n_q=-N_q}^{N_q} a_{mn_q(q)}^E E_{zmn_q(q)}^{(a)E} \\
&= \sum_{n_q=-N_q}^{N_q} a_{mn_q(q)}^E \frac{1}{k_\rho^2} \frac{\partial}{\partial z} \nabla_s^{(q)} E_{zmn_q(q)}^{(a)E} \\
&= \sum_{n_q=-N_q}^{N_q} a_{mn_q(q)}^E \mathbf{E}_{smn(q)}^{(a)E} \\
&= \mathbf{E}_{sm(q)}^{(a)E}.
\end{aligned}
\tag{3.50}
$$

Similarly,

$$\mathbf{H}_{sm(p)}^{(b)E} = \sum_{n_q=-N_q}^{N_q} a_{mn_q(q)}^{E} \mathbf{H}_{smn(q)}^{(a)E} = \mathbf{H}_{sm(q)}^{(a)E}. \tag{3.51}$$

The outgoing waves away from the pth cylinder are translated into the incoming waves of the pth cylinder. The relationship between these coefficients is derived in Equation (3.48).

Similarly, we can get the exact same relationship between the translational coefficients for TE case:

$$a_{mn_q(q)}^{H} = \sum_{n_p=-N_p}^{N_p} H_{n_q-n_p}^{(2)} \left(k_p d_{qp} \right) e^{-j(n_q-n_p)\theta_{qp}} b_{mn_p(p)}^{H}. \tag{3.52}$$

For PEC cylinders,

$$b_{m(q)}^{E} = \overline{T}_{m(q)}^{E} \left[a_{m(q)}^{Einc} + \sum_{i,i\neq j} \overline{\alpha}_{m}^{(qp)} b_{m(p)}^{E} \right], \tag{3.53}$$

$$b_{m(q)}^{H} = \overline{T}_{m(q)}^{H} \left[a_{m(q)}^{Hinc} + \sum_{i,i\neq j} \overline{\alpha}_{m}^{(qp)} b_{m(p)}^{H} \right]. \tag{3.54}$$

For PMC cylinders,

$$b_{m(q)}^{E} = \overline{T}_{m(q)}^{H} \left[a_{m(q)}^{Einc} + \sum_{i,i\neq j} \overline{\alpha}_{m}^{(qp)} b_{m(p)}^{E} \right], \tag{3.55}$$

$$b_{m(q)}^{H} = \overline{T}_{m(q)}^{E} \left[a_{m(q)}^{Hinc} + \sum_{i,i\neq j} \overline{\alpha}_{m}^{(qp)} b_{m(p)}^{H} \right], \tag{3.56}$$

where $\overline{T}_{m(q)}^{E/H}$ is a diagonal matrix with its elements $T_{mn(q)}^{E/H}$ given in Equations (3.18) and (3.19).

Finally, the unknown coefficient vector b_q is summarized in the following equation:

$$b_q = \overline{T}_q \left(a_q + \sum_{p=1;p\neq q}^{N_c} \overline{\alpha}_{qp} b_p \right), \tag{3.57}$$

where $\overline{T}_q$ stands for the T-matrix of the qth via; a_q denotes the expansion coefficients of the wave incident on the qth via. Matrix $\overline{\alpha}_{qp}$ is the translation matrix representing the wave scattered by the pth via incident onto the qth via. The matrix elements of $\overline{\alpha}_{qp}$ can be obtained as

$$\alpha_{qp}(n_p, n_q) = H^{(2)}_{n_q - n_p}\left(k_\rho d_{qp}\right)e^{-j(n_q - n_p)\theta_{qp}}. \qquad (3.58)$$

Consolidating Equation (3.57) for all the vias yields the following equation for multiple scattering of cylinders:

$$(\overline{I} - \overline{T}\,\overline{S})b = \overline{T}a, \qquad (3.59)$$

where $\overline{I}$ is the unit matrix and $\overline{T}$ is the block diagonal matrix consisting of the $\overline{T}_q$ matrices for all cylinders ($q = 1, \cdots, N_c$). $b = [b_1, b_2, \cdots, b_{N_c}]^T$ stands for the unknown expansion vector of scattered waves and $a = [a_1, a_2, \cdots, a_{N_c}]^T$ is the expansion vector of incident waves on all of the vias. The matrix $\overline{S}$ is the combined translation matrix written as

$$\overline{S} = \begin{bmatrix} 0 & \overline{\alpha}_{12} & \cdots & \overline{\alpha}_{1N_c} \\ \overline{\alpha}_{21} & 0 & \cdots & \overline{\alpha}_{2N_c} \\ \vdots & \vdots & \ddots & \vdots \\ \overline{\alpha}_{N_c 1} & \overline{\alpha}_{N_c 2} & \cdots & 0 \end{bmatrix}. \qquad (3.60)$$

The dimension of matrix $\overline{S}$ is $N = \sum_{p=1}^{N_c}(M_p + 1)(2N_p + 1)$.

We can obtain the unknown coefficients vector b by solving Equation (3.59). The boundary of the package is modeled with a proposed novel boundary modeling technique. The detailed discussion for the modeling technique will be presented in Section 3.3.

3.2.3 Excitation Source and Network Parameter Extraction

A signal trace passing through three P-G planes is shown in Figure 3.3a. The central part of the trace is a through via which generates electromagnetic waves propagating in the PPWG formed by the three P-G planes. This is the most common excitation source for the P-G plane structure.

For using the formulation in the previous section, the equivalence principle is applied to replace the annular via antipad with PECs in the

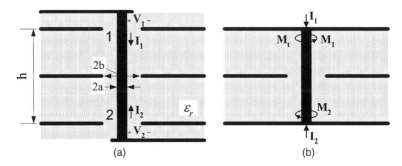

Figure 3.3   (a) Signal via passing through three P-G planes; (b) equivalent model for the source via for calculation of entries in the admittance matrix: Port 1 and Port 2 are excited at the antipad region with an equivalent magnetic current source alternatively.

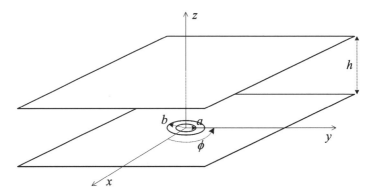

Figure 3.4   A magnetic frill current on the bottom PEC plane of an infinite parallel plate waveguide.

P-G planes and an equivalent magnetic source is added at the original via antipad region (see Fig. 3.3b). The structure shown in Figure 3.3 can be considered as a two-port network and the top and bottom via antipad regions are designated as Port 1 and Port 2, respectively. In order to facilitate the subsequent signal and power integrity analysis, we need to evaluate the admittance parameters of the two-port network shown in Figure 3.3a.

The magnetic current source considered here is an angular magnetic current ring source, which is also called a magnetic frill current. The magnetic frill current is placed on a perfectly conducting ground plane (PEC) (see Fig. 3.4). For modeling of the packaging problem in this research, the magnetic frill current is an equivalent source which is due to the electric field at the aperture on the bottom PEC plane. The field at the aperture is assumed to be of the transverse electromagnetic (TEM) mode:

$$\mathbf{E}(\rho, z = 0) = \frac{V_0}{\rho \ln(b/a)} \hat{\rho} \quad \text{for } a \leq \rho \leq b, \tag{3.61}$$

where V_0 is the modal voltage at the aperture.

The equivalent current, which is a magnetic frill current, is given by

$$\mathbf{M}(\rho, z) = \mathbf{E} \times \hat{n} = M_\phi \hat{\varphi} \quad \text{for } a \leq \rho \leq b, \tag{3.62}$$

and

$$M_\phi(\rho, z) = -\frac{V_0}{\rho \ln(b/a)} \delta(z). \tag{3.63}$$

Thus, the equivalent magnetic current at Port 1(2) can be written as

$$\mathbf{M}_{1(2)} = \mathbf{E} \times \hat{z} = -\frac{V_{1(2)}}{\rho \ln(b/a)} \hat{\varphi}. \tag{3.64}$$

Since the electric field in Equation (3.61) is independent of ϕ, the magnetic frill current M_ϕ is also absent of angular variation, that is, the structure is of rotational symmetry. Then, we have only three electromagnetic components—H_ϕ, E_ρ, and E_z, and the other three components are zero—$E_\phi = H_\rho = H_z = 0$ (see Reference 5, p. 266).

H_ϕ can be found from the following differential equation:

$$\frac{\partial}{\partial \rho}\left[\frac{1}{\rho}\frac{\partial}{\partial \rho}(\rho H_\phi)\right] + k^2 H_\phi + \frac{\partial^2 H_\phi}{\partial z^2} = j\omega\varepsilon M_\phi. \tag{3.65}$$

The associated electric field components are

$$E_\rho = -\frac{1}{j\omega\varepsilon}\frac{\partial H_\phi}{\partial z}, \tag{3.66}$$

$$E_\phi = \frac{1}{j\omega\varepsilon}\frac{1}{\rho}\frac{\partial}{\partial \rho}(\rho H_\phi). \tag{3.67}$$

The solution of Equation (3.65) can be written in terms of a magnetic Green's function as

$$H_\phi(\rho, z) = -j\omega\varepsilon \int_a^b M_\phi(\rho', z') G^H(\rho, z; \rho', z') d\rho. \qquad (3.68)$$

The magnetic Green's function represents the magnetic field due to a unit magnetic current loop. It can be derived by the method of separation of variables. Here only the final expression of the Green's function is given as

$$
\begin{aligned}
G^H(\rho, z; \rho', z') \\
= -\frac{j\pi\rho'}{2h} \sum_{m=0}^{\infty} \frac{1}{2-\delta_{m0}} J_1(k_\rho\rho_<) H_1^{(2)}(k_\rho\rho_>) \cos(k_z z) \cos(k_z z'), \qquad (3.69)
\end{aligned}
$$

where

$$
\begin{aligned}
k &= \omega\mu\varepsilon \\
k_z &= \beta_m = \frac{m\pi}{h} \qquad\qquad (3.70) \\
k_\rho &= k_m = \sqrt{k^2 - k_z^2}, \quad \mathrm{Im}(k_\rho) \le 0.
\end{aligned}
$$

Substituting Equation (3.69) into Equation (3.68), we can obtain the following expressions for H_ϕ.

$$
\begin{aligned}
H_\phi(\rho, z) = \frac{-\pi V_0}{2h\ln(b/a)} \frac{k^2}{\eta} \sum_{m=0}^{\infty} \frac{1}{2-\delta_{m0}} \frac{1}{k_\rho} \cos(k_z z) \\
\cdot \left\{
\begin{array}{l}
H_1^{(2)}(k_\rho\rho)[J_0(k_\rho b) - J_0(k_\rho a)], \quad \text{for } \rho \ge b \\[2mm]
j\dfrac{2}{\pi k_\rho\rho} + J_1(k_\rho\rho)H_0^{(2)}(k_\rho b) - H_1^{(2)}(k_\rho\rho)J_0(k_\rho a), \\[2mm]
\qquad \text{for } a \le \rho \le b \\[2mm]
J_1(k_\rho\rho)[H_0^{(2)}(k_\rho b) - H_0^{(2)}(k_\rho a)], \quad \text{for } \rho \le a,
\end{array}
\right.
\end{aligned}
\qquad (3.71)
$$

where $\eta = \sqrt{\mu/\varepsilon}$.

Now we derive the coefficients in the T-matrix method for the incident wave due to the magnetic frill current.

Recall that for the TM^z mode, we have the following expression for the E_z component as

$$E_z = \sum_{m=0}^{\infty} \sum_{n=-\infty}^{\infty} a_{mn}^E Z_n(k_\rho\rho)\cos(k_z z)e^{jn\phi}. \tag{3.72}$$

To find the coefficient of the incident wave, regarding with the magnetic frill current, we use Equation (3.71) and the following expansion

$$H_\phi = \sum_{m=0}^{\infty} \sum_{n=-\infty}^{\infty} a_{mn}^E \frac{-j\omega\varepsilon}{k_\rho} J_n'(k_\rho\rho)\cos(k_z z)e^{jn\phi}. \tag{3.73}$$

As mentioned in the preceding section, the magnetic frill current in a PPWG will only excite the TM^z modes. We have $a_{mn}^H = 0$ for all the TE^z modes.

Assume that a magnetic frill current is at via "s," then the H_ϕ incident wave on the via "s" due to the current using Equation (3.71) (for the case of $\rho \leq a$) is

$$H_\phi(\rho, z) = \frac{-\pi V_0 k^2}{2h\ln(b/a)\eta}$$

$$\sum_{m=0}^{\infty} \frac{1}{(2-\delta_{m0})k_\rho} \cos(k_z z) J_1(k_\rho\rho)\left[H_0^{(2)}(k_\rho b) - H_0^{(2)}(k_\rho a)\right].$$

$$\tag{3.74}$$

Using the following recurrence relation for a cylindrical Bessel function B:

$$\frac{d}{dz}\left[z^p B_p(z)\right] = z^{p-1} B_{p-1}(z), \tag{3.75}$$

if the order p is zero,

$$\frac{dB_0(z)}{dz}\left[z^p\right] = B_{-1}(z) = -B_1(z), \tag{3.76}$$

then, we can rewrite Equation (3.74) in the following form by using Equation (3.76).

$$
\begin{aligned}
H_\phi(\rho, z) &= \frac{-\pi V_0 k^2}{2h\ln(b/a)\eta} \sum_{m=0}^{\infty} \frac{\left[H_0^{(2)}(k_\rho b) - H_0^{(2)}(k_\rho a)\right]}{(2-\delta_{m0})k_\rho} \cos(k_z z) J_1(k_\rho \rho_s) \\
&= \frac{-\pi V_0 k^2}{2h\ln(b/a)\eta} \sum_{m=0}^{\infty} \frac{\left[H_0^{(2)}(k_\rho b) - H_0^{(2)}(k_\rho a)\right]}{(2-\delta_{m0})k_\rho} \cos(k_z z)\left[-J_0'(k_\rho \rho_s)\right] \\
&= \sum_{n=0}^{\infty}\sum_{m=0}^{\infty} \delta_{n0} \frac{\pi V_0 k^2}{h\ln(b/a)\eta} \frac{\left[H_0^{(2)}(k_\rho b) - H_0^{(2)}(k_\rho a)\right]}{k_\rho(1+\delta_{m0})} \\
&\quad \left[-J_0'(k_\rho \rho_s)\right]\cos(k_z z)e^{jn\phi}.
\end{aligned} \tag{3.77}
$$

Comparing Equation (3.77) with Equation (3.73), we obtain the coefficient of the incident wave for the via "*s*" as

$$
a_{mn}^E = \delta_{n0} \frac{j\pi V_s k}{h\ln(b/a)} \frac{\left[H_0^{(2)}(k_\rho b) - H_0^{(2)}(k_\rho a)\right]}{(1+\delta_{m0})}. \tag{3.78}
$$

Similarly, we can derive the coefficient of the incident wave on via "*q*" ($q \neq s$) due to the magnetic frill current at via "*s*" as follows:

$$
\begin{aligned}
H_\phi(\rho, z) &= \frac{-\pi V_s k^2}{2h\ln(b/a)\eta} \sum_{m=0}^{\infty} \frac{1}{(2-\delta_{m0})k_\rho} \cos(k_z z) H_1^{(2)}(k_\rho \rho) \\
&\quad \left[J_0(k_\rho b) - J_0(k_\rho a)\right] \\
&= \frac{-\pi V_s k^2}{2h\ln(b/a)\eta} \sum_{m=0}^{\infty} \frac{\left[J_0(k_\rho b) - J_0(k_\rho a)\right]}{(2-\delta_{m0})k_\rho} \cos(k_z z) \\
&\quad \left[-H_0^{(2)'}(k_\rho |\boldsymbol{\rho} - \boldsymbol{\rho}_s|)\right].
\end{aligned} \tag{3.79}
$$

Referring to Equation (3.34) with Figure 3.2, the addition theorem for translation of the coordinates is given as

$$
H_m^{(2)}(k_\rho \rho_p)e^{jm\phi_p} = \sum_{m=-\infty}^{\infty} \left[H_{n-m}^{(2)}(k_\rho d_{qp})e^{-j(n-m)\theta_{qp}}\right] J_n(k_\rho \rho_q)e^{jn\phi_q}.
$$

Then, the last term in Equation (3.79) can be expanded by using cylindrical harmonics as

$$
H_0^{(2)'}(k_\rho \rho_s) = \sum_{n=-\infty}^{\infty} J_n'(k_\rho \rho_q)e^{jn\phi_q} H_n^{(2)}(k_\rho \rho_{sq})e^{-jn\phi_{sq}}. \tag{3.80}
$$

Substituting Equation (3.80) into Equation (3.79) and comparing it to Equation (3.73), we obtain the coefficient of the incident wave for the via "q."

$$a_{mn}^E = \frac{j\pi V_s k}{h\ln(b/a)} \frac{[J_0(k_\rho b) - J_0(k_\rho a)]}{(1+\delta_{m0})} H_n^{(2)}(k_\rho \rho_{sq}) e^{-jn\phi_{sq}}. \quad (3.81)$$

The coefficient of the incident wave due to a magnetic frill current in the PPWG is summarized as

$$a_{mn}^E = \begin{cases} \delta_{n0} \dfrac{j\pi V_s k}{h\ln(b/a)} \dfrac{\left[H_0^{(2)}(k_\rho b) - H_0^{(2)}(k_\rho a)\right]}{(1+\delta_{m0})} & \text{for via } s \\[4mm] \dfrac{j\pi V_s k}{h\ln(b/a)} \dfrac{\left[J_0(k_\rho b) - J_0(k_\rho a)\right]}{(1+\delta_{m0})} H_n^{(2)}(k_\rho \rho_{sq}) e^{-jn\phi_{sq}} & \text{for via } q \neq s \end{cases}$$

$$(3.82)$$

and $a_{mn}^H = 0$.

The alternative derivation for the incident wave coefficients is presented as follows. Recalling Equation (3.64), the equivalent magnetic current at Port 1(2) is given by

$$\mathbf{M}_{1(2)} = \mathbf{E} \times \hat{z} = -\frac{V_{1(2)}}{\rho\ln(b/a)} \hat{\varphi}.$$

In order to derive I_1, a "testing" ring of unity amplitude magnetic current $\mathbf{M}_t$ is applied around the signal via at the antipad region. The magnetic frill current is equivalent to a delta-gap source and the electric field $\mathbf{E}^{src}$ of the delta-gap source at $\rho = a$ is expressed as

$$\mathbf{E}^{src} = \begin{cases} -\dfrac{1}{\Delta}, & -\Delta < z < 0 \\[3mm] 0, & \text{otherwise.} \end{cases} \quad (3.83)$$

The field $\mathbf{E}^{inc}$ in exterior region $a < \rho < b$ can be expressed in cylindrical waves as

$$E_z^{src} = \sum_{m=0}^M C_m H_0^{(2)}(k_m\rho)\cos(\beta_m z), \quad (3.84)$$

where C_m stands for the wave coefficient for each mode with the expression of

$$C_m = \frac{-2}{h(1+\delta_{m0})H_0^{(2)}(k_m a)}\operatorname{sinc}\left(\frac{m\pi\Delta}{h}\right), \qquad (3.85)$$

and $\operatorname{sinc}(x) = \sin(x)/x$.

By applying the $\mathbf{E}^{inc}$ expression for the signal (source) via "q," the incident field at any P-G via "p" can be expressed as

$$\mathbf{E}_z^{inc} = \sum_{m=0}^{M} C_m H_0^{(2)}\left(k_m |\boldsymbol{\rho}_p - \boldsymbol{\rho}_q|\right)\cos\left(\beta_m z\right)\hat{z}. \qquad (3.86)$$

Using the translational addition theorem,

$$H_0^{(2)}(k_m \rho_p) = \sum_{n=-\infty}^{\infty} H_n^{(2)}(k_m \rho_{qp})e^{-jn\phi_{qp}} J_m(k_m \rho_q)e^{jm\phi_q}, \qquad (3.87)$$

the incident wave coefficients $a_{mn}^{E/H}$ are obtained as

$$a_{mn}^E = C_m H_n^{(2)}(k_m \rho_{qp})e^{-jn\phi_{qp}}, \quad \text{and} \quad a_{mn}^H = 0. \qquad (3.88)$$

The incident coefficient for TE mode is considered as zero since the excitation field $\mathbf{E}^{inc}$ from the delta-gap source is expressed in TM mode only.

From the above equation, we obtain the coefficient for the incident wave, which is used to calculate the scattered waves as outlined in the previous section. The Y-matrix elements are calculated as

$$Y_{11} = \frac{I_1}{V_1} = \frac{1}{\ln(b/a)}\int_s \frac{1}{\rho}\hat{\varphi}\cdot\mathbf{H}_t(z=d)dS$$
$$Y_{21} = \frac{I_2}{V_1} = \frac{1}{\ln(b/a)}\int_s \frac{1}{\rho}\hat{\varphi}\cdot\mathbf{H}_t(z=0)dS. \qquad (3.89)$$

The other two entries of the admittance parameters of the two-port network can be obtained by repeating the same procedures.

The calculation for mutual admittance $\mathbf{Y}$ between the pth and q^{th} P-G vias can be performed in the following procedure.

For p^{th} PEC via,

$$\oint_l \mathbf{H}^{(q)} \cdot d\hat{l} = -I_q(z) + j\omega\varepsilon \oint_S \mathbf{E}^{(q)} \cdot d\hat{s} = -I_q(z), \quad \text{where } \mathbf{E}^{(q)}\big|_{\rho=r_q} = 0.$$

(3.90)

For $p \neq q$,

$$Y_{qp} = \frac{I_q(d)}{V_p} = I_q(d) = -\oint_l \mathbf{H}^{(q)}\big|_{z=d} \cdot d\hat{l} = -r_q \int_0^{2\pi} H_\phi^{(q)}\big|_{z=d} d\phi. \quad (3.91)$$

Finally, we get

$$
\begin{aligned}
Y_{qp} &= -2\pi r_q \sum_{m=0}^{M} \left[a_{m0}^{E(q)} J_0'\left(k_\rho r_q\right) + b_{m0}^{E(q)} H_0'^{(2)}\left(k_\rho r_q\right) \right] \frac{-j\omega\varepsilon}{k_\rho}(-1)^m \\
&= j\omega\varepsilon 2\pi r_q \sum_{m=0}^{M} \left[a_{m0}^{E(q)} J_0'\left(k_\rho r_q\right) + b_{m0}^{E(q)} H_0'^{(2)}\left(k_\rho r_q\right) \right] \frac{(-1)^m}{k_\rho} \\
&= j\omega\varepsilon 2\pi r_q \sum_{m=0}^{M} b_{m0}^{E(q)} \left[\frac{J_0'\left(k_\rho r_q\right)}{T_{m0}^{E(q)}} + H_0'^{(2)}\left(k_\rho r_q\right) \right] \frac{(-1)^m}{k_\rho} \\
&= 4\omega\varepsilon \sum_{m=0}^{M} \frac{(-1)^m}{k_\rho^2 J_0\left(k_\rho r_q\right)} b_{m0}^{E(q)}.
\end{aligned}
$$

(3.92)

For $p = q$,

$$
\begin{aligned}
Y_{pp} &= -\oint_l \left(\mathbf{H}^{(p)} + \mathbf{H}^{inc}\big|_{z=d} \right) \cdot d\hat{l} \\
&= 4\omega\varepsilon \sum_{m=0}^{M} \frac{(-1)^m}{k_\rho^2 J_0\left(k_\rho r_p\right)} b_{m0}^{E(p)} - \frac{j\omega\varepsilon 2\pi r_p}{d} \sum_{m=0}^{M} \frac{\delta_m H_1^{(2)}\left(k_\rho r_p\right)}{k_\rho H_0^{(2)}\left(k_\rho r_p\right)}.
\end{aligned}
$$

(3.93)

3.2.4 Implementation of Effective Matrix-Vector Multiplication (MVM) in Linear Equations

To solve the matrix equation for multiple scattering of the vias given in Equation (3.59), the effective MVM for linear equation system is implemented in this section. For the number of vias N_c, the unknown wave coefficient vector for the qth via can be rewritten in the following form:

$$b_{m(q)}^{E} = \bar{T}_{m(q)} \cdot \sum_{p=1;p\neq q \& q\neq s}^{N_c} \bar{\alpha}_{m}^{(qp)} \cdot \left(b_{m(p)}^{E} + b_{m(s)}^{Einc} \right), \quad \text{for } m = 0, 1, \cdots, M, \quad (3.94)$$

where m stands the mode number of the incident or scattered fields, and $b_{m(q)}^{Einc}$ represents the excitation coefficient vector for the source via "s." Equation (3.94) can be solved for each mode m. The diagonal matrix $\bar{T}_{m(q)}$ is expressed, if the qth via is a PEC cylinder, as

$$\bar{T}_{m(q)} = \begin{bmatrix} T_{m(-N_q)}^{E} & 0 & \cdots & 0 \\ 0 & \ddots & 0 & \vdots \\ \vdots & 0 & \ddots & 0 \\ 0 & \cdots & 0 & T_{mN_q}^{E} \end{bmatrix}, \quad (3.95)$$

and, if the qth via is a PMC cylinder, as

$$\bar{T}_{m(q)} = \begin{bmatrix} T_{m(-N_q)}^{H} & 0 & \cdots & 0 \\ 0 & \ddots & 0 & \vdots \\ \vdots & 0 & \ddots & 0 \\ 0 & \cdots & 0 & T_{mN_q}^{H} \end{bmatrix}. \quad (3.96)$$

The wave translation matrix $\bar{\alpha}_{m}^{(qp)}$ can be divided as

$$\bar{\alpha}_{m}^{(qp)}\left(\rho_{qp}, \phi_{qp} \right) = {}_{N}^{-N}\left[H_{n_q-n_p}^{(2)}\left(k_m \rho_{qp} \right) e^{-j(n_q-n_p)\phi_{qp}} \right]$$
$$= \bar{U}^{(qp)*} \cdot \bar{H}_{m}^{(qp)} \cdot \bar{U}^{(qp)}, \quad (3.97)$$

where

$$\bar{U}^{(qp)} = \begin{bmatrix} e^{-jN\phi_{qp}} & 0 & 0 \\ 0 & \ddots & 0 \\ 0 & 0 & e^{jN\phi_{qp}} \end{bmatrix}, \quad (3.98)$$

$$
\bar{\boldsymbol{H}}_m^{(qp)} =
\begin{bmatrix}
H_0^{(2)}(\cdot) & -H_1^{(2)}(\cdot) & H_2^{(2)}(\cdot) & \cdots & -H_{2N-1}^{(2)}(\cdot) & H_{2N}^{(2)}(\cdot) \\
H_1^{(2)}(\cdot) & H_0^{(2)}(\cdot) & -H_1^{(2)}(\cdot) & H_2^{(2)}(\cdot) & \cdots & -H_{2N-1}^{(2)}(\cdot) \\
H_2^{(2)}(\cdot) & H_1^{(2)}(\cdot) & \ddots & \ddots & \ddots & \vdots \\
\vdots & H_2^{(2)}(\cdot) & \ddots & \ddots & \ddots & H_2^{(2)}(\cdot) \\
H_{2N-1}^{(2)}(\cdot) & \cdots & \ddots & \ddots & \ddots & -H_1^{(2)}(\cdot) \\
H_{2N}^{(2)}(\cdot) & H_{2N-1}^{(2)}(\cdot) & \cdots & H_2^{(2)}(\cdot) & H_1^{(2)}(\cdot) & H_0^{(2)}(\cdot)
\end{bmatrix},
$$

$$(3.99)$$

that the argument $(\cdot)$ for the Hankel function of second kind is given as $(k_m \rho_{qp})$.

The MVM of the matrix $\bar{\boldsymbol{H}}_m^{(qp)}$ is

$$
\bar{\boldsymbol{H}}_m^{(qp)} \cdot \boldsymbol{x} = H_0^{(2)}
\begin{bmatrix} x_1 \\ x_2 \\ \vdots \\ x_{2N} \\ x_{2N+1} \end{bmatrix}
+ H_1^{(2)}
\left(
\begin{bmatrix} x_0 \\ x_1 \\ \vdots \\ x_{2N-1} \\ x_{2N} \end{bmatrix}
-
\begin{bmatrix} x_2 \\ x_3 \\ \vdots \\ x_{2N+1} \\ 0 \end{bmatrix}
\right)
$$

$$
+ H_2^{(2)}
\left(
\begin{bmatrix} 0 \\ 0 \\ x_1 \\ \vdots \\ x_{2N-1} \end{bmatrix}
+
\begin{bmatrix} x_3 \\ \vdots \\ x_{2N+1} \\ 0 \\ 0 \end{bmatrix}
\right)
+ \cdots + H_2^{(2)}
\left(
\begin{bmatrix} 0 \\ \vdots \\ 0 \\ x_1 \\ \vdots \\ x_{N+1} \end{bmatrix}
+ (-1)^N
\begin{bmatrix} x_{N+1} \\ \vdots \\ x_{2N} \\ x_{2N+1} \\ \vdots \\ 0 \end{bmatrix}
\right)
$$

$$
+ \cdots + H_{2N}^{(2)}
\left(
\begin{bmatrix} 0 \\ \vdots \\ \vdots \\ 0 \\ x_1 \end{bmatrix}
+
\begin{bmatrix} x_{2N+1} \\ 0 \\ \vdots \\ \vdots \\ 0 \end{bmatrix}
\right).
$$

$$(3.100)$$

The computing cost of the above MVM is $3N^2 + 2N$ instead of the original MVM $(2N + 1)^2$. Since $\phi_{pq} = \phi_{qp} + \pi$, hence

$$\bar{\alpha}_m^{(pq)} = \left[H_{n_p - n_q}^{(2)} \left(k_m \rho_{qp} \right) e^{-j(n_p - n_q)\phi_{pq}} \right]$$
$$= \bar{V} \cdot \bar{\alpha}_m^{(qp)} \cdot \bar{V} \tag{3.101}$$
$$= \bar{V} \cdot \bar{U}^{(qp)*} \cdot \bar{H}_m^{(qp)} \cdot \bar{U}^{(qp)} \cdot \bar{V},$$

where, $\bar{V}$ is a constant diagonal matrix and given as

$$\bar{V} = \begin{bmatrix} (-1)^{-N} & 0 & 0 \\ 0 & \ddots & 0 \\ 0 & 0 & (-1)^N \end{bmatrix}. \tag{3.102}$$

The whole matrix can be obtained as

$$\begin{bmatrix} b_{m(1)}^E \\ b_{m(2)}^E \\ \vdots \\ b_{m(N_c)}^E \end{bmatrix} = \begin{bmatrix} \bar{T}_{m(1)} & 0 & \cdots & 0 \\ 0 & \bar{T}_{m(2)} & \ddots & \vdots \\ \vdots & \ddots & \ddots & 0 \\ 0 & \cdots & 0 & \bar{T}_{m(N_c)} \end{bmatrix}$$

$$\cdot \begin{bmatrix} 0 & \bar{\alpha}_m^{(12)} & \cdots & \bar{\alpha}_m^{(1N_c)} \\ \bar{\alpha}_m^{(21)} & 0 & \bar{\alpha}_m^{(23)} & \vdots \\ \vdots & \bar{\alpha}_m^{(32)} & \ddots & \bar{\alpha}_m^{(N_c-1N_c)} \\ \bar{\alpha}_m^{(N_c 1)} & \cdots & \bar{\alpha}_m^{(N_c N_c-1)} & 0 \end{bmatrix}$$

$$\cdot \left(\begin{bmatrix} b_{m(1)}^E \\ b_{m(2)}^E \\ \vdots \\ b_{m(N_c)}^E \end{bmatrix} + \begin{bmatrix} 0 \\ b_{m(s)}^E \\ 0 \\ 0 \end{bmatrix} \right), \tag{3.103}$$

$$b_m^E = \bar{T}_m \cdot \bar{\alpha}_m \cdot \left(b_m^E + b_m^{E_{inc}} \right), \tag{3.104}$$

where $\bar{\alpha}_m$ can be written as the sum of a lower and upper triangular matrices as $\bar{\alpha}_m = \bar{\alpha}_m^L + \bar{\alpha}_m^U$, and the wave coefficient vector $b_{m(s)}^{E_{inc}}$ of the source via "s" is given as

$$b_{m(s)}^{E_{inc}} = \begin{bmatrix} b_{m0}^{E_{inc}} \\ 0 \\ 0 \\ 0 \end{bmatrix}. \tag{3.105}$$

3.2.5 Numerical Examples for Single-Layer Power-Ground Planes

3.2.5.1 Validation of the SMM Algorithm

Figure 3.5 shows an example to verify the SMM algorithm for modeling P-G planes with multiple vias where a signal trace passes through a pair of conductor planes with 12 shorting vias. The conductor planes are assumed to be infinitely large.

The simulation results are checked against those from the Ansoft HFSS. The E_z field distribution is plotted in Figure 3.6. The normalized

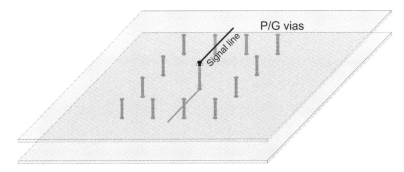

Figure 3.5   Vias of a signal trace passing through two conductor planes. The signal via is enclosed by 12 shorting vias connecting the two planes [11].

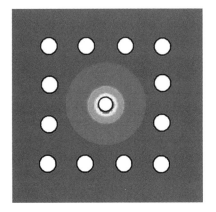 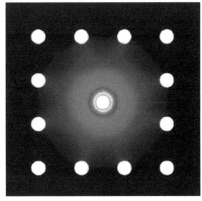

Figure 3.6   Comparison of the E_z field distribution at 1 GHz: our simulation results (left) versus Ansoft HFSS simulation results (right) [11]. The vias are drawn as white dots. (See color insert.)

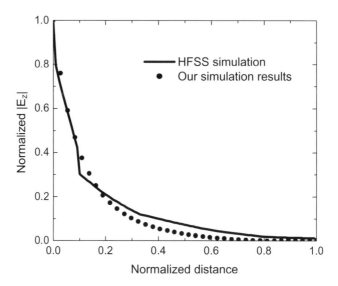

Figure 3.7 Validation of the simulated results for E_z with those from the Ansoft HFSS [11].

value of the E_z component along a horizontal line from the edge of the central via is given in Figure 3.7. In both cases the results agree quite well. The admittance parameter (Y11) of the structure is also calculated up to 5 GHz. The results again match well with the HFSS solutions as shown in Figure 3.8.

To simulate the coupling effect from the multiple vias, one example with a large number of shorting vias is presented in Figure 3.9. The vias are formed in the center block of 256 vias and the outer ring of 348 vias and the signal via is located between them. The SMM algorithm is used to calculate the E-field distribution and the computing is done within few minutes. In Figure 3.9, the field distribution is plotted to show the effect of multiple scattering from the P-G vias.

3.2.5.2 Co-simulation Examples

The signal traces in the top/bottom domain of a package can be modeled as transmission lines. Two problems need to be solved to extract the resistance, inductance, capacitance, and conductance (*RLCG*) parameters: an electrostatic problem for capacitance and conductance parameters, and a magneto-static problem for inductance and resis-

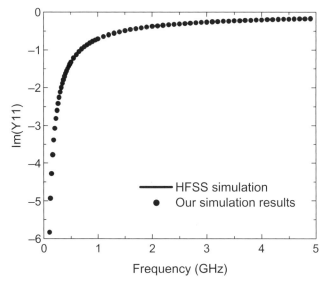

Figure 3.8 Y11 for the two-port network formed by the plate-through via [11].

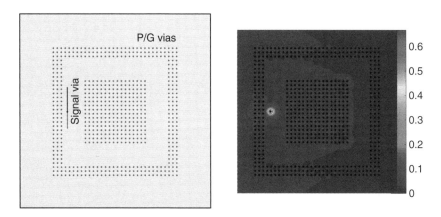

Figure 3.9 Field distribution of multiple scattering among the shorting vias. (See color insert.)

tance matrices. We apply the integral equation method solved by the method of moments (MoM) [6, 7] to obtain those *RLCG* parameters. For signal traces with discontinuity and vertical via bends, an analytical formula is applied to obtain their equivalent circuits. This approach finally leads to an equivalent circuit for the top/bottom domain. The

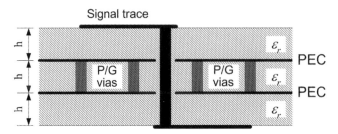

Figure 3.10 Schematic diagram of a signal trace packaging through two PEC planes [11].

admittance matrix (Y) of the inner domain can be converted to an equivalent circuit as mentioned in the earlier section. Combining the equivalent circuits for the top/bottom domain and the inner domain makes it possible to perform system-level signal integrity analysis of an electronic package.

A numerical example is shown here to demonstrate the system-level modeling approach. A schematic diagram for the example is shown in Figure 3.10 and its 3D view is shown in Figure 3.5. The following parameters are used: the width and length of the signal trace at the top and bottom layers are both 0.25 mm and 1.5 mm, respectively; the thickness of the substrate h = 600 μm; the radius of the through via and all the shorting vias are 150 μm and the dielectric constant of the substrate is 4.2.

The p.u.l. (per unit length) parameters of the signal trace are calculated from the 2D integral equation solved by the MoM. The simulated p.u.l. capacitance and inductance are 20.52 pF/m and 610.35 nH/m, respectively. The two-port admittance parameters of the through via have been computed in the previous subsection. These admittance parameters have taken account of the multiple coupling of the through via with the 12 shorting vias. The computational time for it is about 5 minutes on a laptop (CPU: 1.3 GHz, memory: 512 MB) compared to about 30 minutes used by the commercial HFSS at a workstation with similar CPU speed but much larger memory.

Then, the equivalent network for the signal trace is constructed with the p.u.l. parameters L,C and the admittance matrix [Y], and a source (a pulse with 1 ns rise/fall time) and some lumped circuit elements (R_s, C, R_L) are also added to the equivalent network for signal

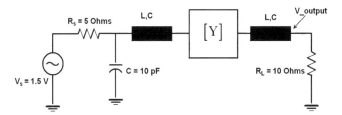

Figure 3.11 Schematic diagram of the equivalent circuit used for SPICE simulation [11].

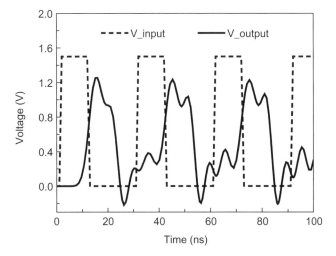

Figure 3.12 Voltage response at the far end of the signal trace in the presence of 12 shorting vias and two PEC planes [11].

integrity analysis as shown in Figure 3.11. Then, we perform the SI simulation using a Simulation Program with Integrated Circuit Emphasis (SPICE) simulator. The voltage response at the far end of the signal trace is shown in Figure 3.12, which demonstrates the system-level electrical co-simulation capability.

3.2.5.3 Simulation for P-G Planes Decoupling

The following example from Reference 8 is considered to demonstrate the capability of the method for modeling of the P-G planes decoupling. The schematic diagram of the test setup is shown in Figure 3.13.

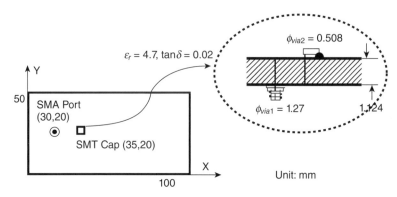

Figure 3.13 Schematic diagram of a pair of power-ground planes with an SMT decoupling capacitor close to an input SMA port [11].

An SMT decoupling capacitor is located near the input SMA port. The capacitance of the SMT capacitor is 8.14 nF, and the equivalent serial resistance and inductance are 666 mΩ and 1.57 nH, respectively.

The structure in Figure 3.13 is modeled as a two-port network: Port 1 is the input of the SMA connector, and Port 2 is at the end of the via attached to the SMT capacitor. The simulation result of the input impedance looking into the SMA port is shown in Figure 3.14. Results are compared with the measurements in Reference 8. The simulation results agree well with the measurements up to 700 MHz. The two results still follow the same trend beyond 700 MHz. But the resonant frequency predicted by our method is smaller than what was measured. The reason is that the dielectric constant of the substrate shows some frequency dependence and becomes lower than what was assumed in our simulation. Since the resonant frequency of the P-G planes is inversely proportional to the square root of the dielectric constant, the frequency shift was expected. Assuming a fixed dielectric constant of $\varepsilon = 4.7$ in our simulation, the predicted resonant frequency in the high frequency band should be lower than the measurement result. This argument is further supported when we compare the frequencies associated with the peaks (shown as long vertical dashed lines across the graph in Fig. 3.14) for the simulation result and the analytical result in Figure 3.14. They correlate to each other because both the simulation and analytical results are based on the same fixed dielectric constant of $\varepsilon = 4.7$. The difference in magnitude between the two results is probably due to the

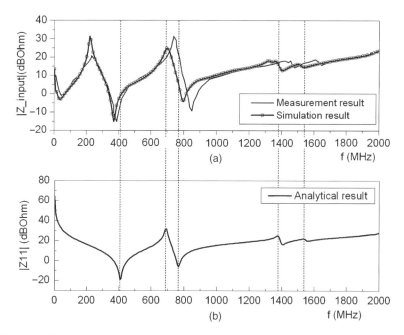

Figure 3.14   Comparison of the results of the input impedance: simulation, measurement, and analytical results [11].

variation of the loss tangent over frequency. Because both the frequency-dependent data for the dielectric constant and the loss tangent are not available, a thorough comparison between both results is not possible. The discrete SMT capacitor is a local decoupling capacitor for the SMA input port. Figure 3.14 reveals that it is only effective at low frequencies. As the frequency goes up, the P-G planes resonance will take over, and we may need more decoupling capacitors to maintain the low impedance over the operating frequency band.

3.3 NOVEL BOUNDARY MODELING METHOD FOR SIMULATION OF FINITE-DOMAIN POWER-GROUND PLANES

The multiple scattering theory of cylinders (vias) has been discussed in Section 3.2.2 and the conventional SMM algorithm for analysis of single-layer P-G planes is implemented and validated with numerical simulation results. However, the P-G planes are assumed to be infinitely large, which cannot capture the resonant behavior of real-world electronic package structure; in other words, the wave reflections from the package's boundary are not taken into account. In this section, a novel boundary modeling method is proposed to overcome this limitation. We propose a novel boundary modeling method: FDCL and then, using the proposed FDCL, the extended SMM algorithm is implemented to simulate the finite-domain package structures.

3.3.1 Perfect Magnetic Conductor (PMC) Boundary

The thickness of each layer in a parallel-plate structure like an electronic package is usually much smaller than the operating wavelength. Furthermore, the substrate material is usually homogeneous. Therefore, the fields inside such a layer in a parallel-plate structure satisfy a simplified two-dimensional (2D) Helmholtz wave equation in terms of E_z. By simple derivation we know that along the periphery of each layer, the magnetic field lines intersect the periphery normally [9]; in other words, such an open boundary constitutes a "perfect magnetic conductor (PMC) wall." Therefore, the original open domain at the periphery of a package can be accurately truncated by a PMC boundary.

However, the abovementioned modal expansion method is cumbersome in directly modeling arbitrarily shaped boundary of a finite-sized parallel-plate structure. The reason is that the cylindrical mode expansion associated with radial waveguides and cylindrical vias cannot be directly used for noncylindrical structures like the periphery of a package. Innovative boundary modeling methods are highly demanded because one cannot obtain accurate simulation results without correctly modeling the boundary of finite-sized parallel-plate structures.

3.3.2 Frequency-Dependent Cylinder Layer (FDCL)

A novel boundary modeling method, named the FDCL is proposed to resolve the issue outlined in the previous section. The FDCL is devised such that virtual PMC cylinders are postulated to replace the original continuous boundary of each layer of a parallel-plate structure as shown in Figure 3.15. The PMC cylinders are arranged in series at the position of the original finite boundary, and their radii vary with different simulation frequencies. All the PMC cylinders in the FDCL collectively

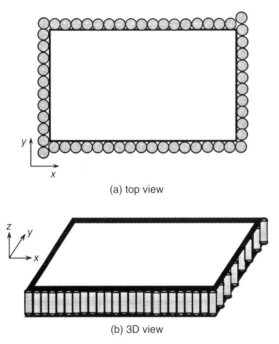

(a) top view

(b) 3D view

Figure 3.15   Illustration of the implementation of the FDCL (frequency-dependent cylinder layer) boundary, where a series of virtual PMC (perfect magnetic conductor) cylinders (shaded) is placed at the periphery to replace the original continuous finite boundary of a package structure [14].

function as a PMC boundary. Thus, the FDCL is a simple yet powerful boundary modeling method.

The superiority of the proposed boundary modeling method is that PMC cylinders are compatible with the modal expansion with T-matrix method (SMM) which has demonstrated its significance and efficiency in package analysis in References 10 and 11, and they can be straight-forwardly incorporated into the linear system of equations in Equation (3.59). The original linear system of equations need to be slightly modified and augmented by incorporating the T-matrix for PMC cylinders into the translation matrix ($\overline{T}$) in Equation (3.59).

Now we derive the T-matrix for the PMC cylinders in the FDCL. The boundary condition for the PMC cylinders shall comply $\hat{n} \times \mathbf{H} = 0$ on their surfaces. Applying the boundary condition to the PMC cylinders in the FDCL for the ith layer (PPWG) and using Equations (3.11–3.13), we obtain the following:

$$\hat{n} \times \mathbf{H} = \mathbf{H}_t^{(i)}(r_b, \phi, z)$$

$$= \sum_{n=-\infty}^{\infty} \left\{ \sum_{m=0}^{\infty} \left[a_{mn}^{E(i)} J_n'(k_m^{(i)} r_b) + b_{mn}^{E(i)} H_n'^{(2)}(k_m^{(i)} r_b) \right] \right.$$

$$\frac{-j\omega\varepsilon\cos\left(\beta_m^{(i)}(z-z_i)\right)}{k_m^{(i)}} \hat{\varphi} + \sum_{m=1}^{\infty} \left[a_{mn}^{H(i)} J_n(k_m^{(i)} r_b) + b_{mn}^{H(i)} H_n^{(2)}(k_m^{(i)} r_b) \right] \cdot$$

$$\left. \left[\sin\left(\beta_m^{(i)}(z-z_i)\right) \hat{z} + \frac{jn\beta_m^{(i)}\cos\left(\beta_m^{(i)}(z-z_i)\right)}{\left(k_m^{(i)}\right)^2 r_b} \hat{\varphi} \right] \right\} e^{jn\phi}$$

$$= 0,$$

$$(3.106)$$

where r_b is the radius of PMC cylinders. The following T-matrix for the PMC cylinders in the FDCL are derived as

$$\overline{T}_{mn}^{(i)} = \begin{bmatrix} T_{mn}^{E(i)} & 0 \\ 0 & T_{mn}^{H(i)} \end{bmatrix} = \begin{bmatrix} -\dfrac{J_n'(k_m^{(i)} r_b)}{H_n^{(2)\prime}(k_m^{(i)} r_b)} & 0 \\ 0 & -\dfrac{J_n(k_m^{(i)} r_b)}{H_n^{(2)}(k_m^{(i)} r_b)} \end{bmatrix}. \quad (3.107)$$

3.3.2.1 Considering the Radii of PMC Cylinders for Boundary

To apply the virtual PMC cylinders at the periphery of the package, the radii of the PMC cylinders need to be carefully chosen. The dominant mode for each layer of the package parallel-plate structure is the TM^z mode. The polarization of electric fields is along the z-direction, while the polarization of magnetic fields is in the x–y plane. Since the axes of the PMC cylinders in the FDCL are along the z-direction, the orientation of the PMC cylinders is perpendicular to the polarization of the magnetic fields of the domain mode. Therefore, if the radii of the PMC cylinders are too small compared to the operating wavelength, the electromagnetic wave will penetrate the PMC cylinder layer without reflection. On the other hand, if the radii of the PMC cylinders are too large compared to the operating wavelength, most of the electromagnetic wave will be trapped in the large gap enclosed by the PMC cylinders and the periphery of the package structure, instead of being reflected by the PMC cylinders. In short, the PMC cylinders for both cases fail to function as an ideal PMC boundary. Based on the intensive analysis of simulation, *the radius of the PMC cylinders in the FDCL must vary in accordance with operating frequency.* Therefore, for each operating frequency "*f*," we developed the following empirical formula to governing the radii of the PMC cylinders in the FDCL as

$$k_g \cdot r_b = \zeta, \tag{3.108}$$

where r_b stands for the radius of the PMC cylinders in the FDCL; $k_g = 2\pi/\lambda_g$ with λ_g is the guided wavelength determined by the electrical property of the substrate material. ζ is a predefined constant to adjust the frequency-dependent radii of the PMC cylinders in the FDCL for accuracy. We will discuss how to choose ζ in the next section.

3.3.3 Validations of FDCL

As mentioned in the previous section, the radii of the PMC cylinders in the FDCL for the boundary modeling are determined by the constant (ζ) specified in Equation (3.108). We use one example of test board with dimension of 75×50 mm, thickness of 1.2 mm, relative dielectric constant ε_r of 4.2, and loss tangent $\tan \delta$ of 0.015 to discuss the choice of ζ. In Figure 3.16, the results are compared with the reference solution [9]. By using the constant radius for all frequencies, it fails to provide all the resonant frequencies. On the other hand, the results of the frequency-dependent radii show very good agreement with reference solution and it accurately predicts all the resonant frequencies. The analyzed simulation results are presented in Figure 3.17 to choose the optimal value of parameter ζ. For the FDCL with cylinders of frequency-dependent radii, different values of ζ also affect the accuracy of the simulation results. Intensive numerical experiments show that an optimal value of ζ is within the range of 0.4–0.5 for accurate simulation results.

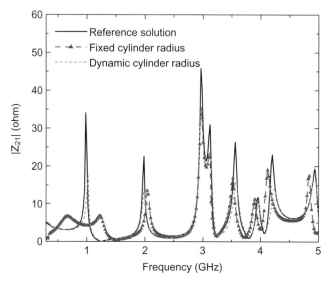

Figure 3.16 Comparison of the extended SMM results with fixed and dynamic radius of the PMC cylinders in the FDCL and the reference solution [14]. (See color insert.)

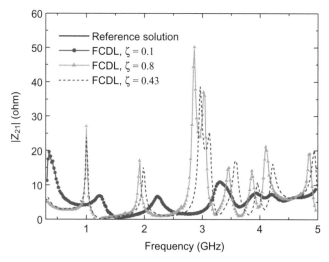

Figure 3.17 Effects of the different values of ζ on the accuracy of the simulation results by the FDCL boundary modeling method [14]. (See color insert.)

3.4 NUMERICAL SIMULATIONS FOR FINITE STRUCTURES

3.4.1 Extended Scattering Matrix Method (SMM) Algorithm for Finite Structure Simulation

For validation of the extended SMM with proposed FDCL, we fabricate several test boards as shown in Figures 3.18 and 3.22. The first test board has dimension of 156×106 mm, thickness of 1.2 mm, relative dielectric constant of 4.1, and loss tangent of 0.015. The locations of the ports are given as Port 1 (46,26), Port 2 (122,53), and Port 3 (46,76), all in millimeters as shown in Figure 3.18. The S-parameters between the ports are simulated using the SMM algorithm with the FDCL approach, and the results are compared with measurement data in Figures 3.19–3.21. The simulated results of the proposed method are agreed well with the measurement data for the S-parameters. The resonant frequency points are matched very well, but the magnitudes of the simulation are slightly higher than ones of the measurement. The reason is that the dielectric constant of the substrate shows some frequency dependence, and the loss tangent becomes larger than what it is assumed in the simulation. Since the dielectric loss in the dielectric substrate layer is related to the loss tangent, the magnitude variation in the S-parameters was expected. Assuming a fixed dielectric constant ε_r of 4.1 with loss tangent $\tan\delta = 0.015$ in our simulation, the slight

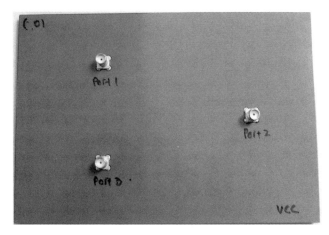

Figure 3.18 Test printed circuit board (PCB) for measurement. (See color insert.)

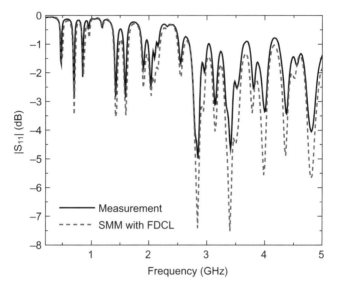

Figure 3.19   The scattering (S_{11}) parameter for Port 1 of the test board in Figure 3.18. The simulated result by the SMM with FDCL method is compared against the measurement data.

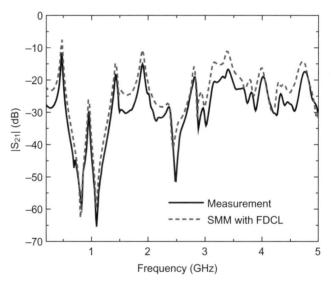

Figure 3.20   The scattering (S_{21}) parameter for Port 1 and 2 of the test board in Figure 3.18. The simulated result by the SMM with FDCL method is compared against the measurement data.

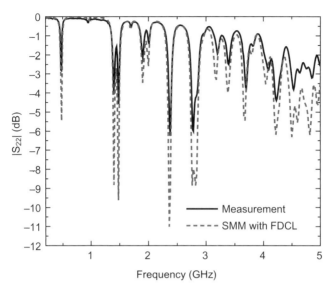

Figure 3.21   The scattering (S_{22}) parameter for Port 2 of the test board in Figure 3.18. The simulated result by the SMM with FDCL method is compared against the measurement data.

magnitude variation in the high frequency band should be found in comparison of the results.

Other examples of test PCBs are considered for the analysis of P-G vias between the planes to reduce the coupling effect between the signal vias or consider the low transfer impedance in the package for stable power distribution over frequency band. The test PCBs for the analysis are given in Figure 3.22. The test PCB1 has two signal vias (P1 and P2) only, and no P-G (shorting) vias as shown in Figure 3.22a. In the test PCB2 and PCB3, the shorting vias are placed with different configurations in order to reduce the coupling effect between the two signal vias as shown in Figure 3.22b,c. The features of the boards are dimension of $75 \times 50\,mm$, thickness of $1.1\,mm$, relative dielectric constant of 4.1, and loss tangent of 0.015. The port locations P1 and P2 are shown as in the figures. The simulated results for the S-parameters are presented with the measurement data. The S_{11} at Port 1 and the S_{21} between Ports 1 and 2 for both boards are compared with measured data in Figures 3.23 and 3.24, respectively. The results are shown good agreement. The top and bottom conductor planes and the PMC boundary at

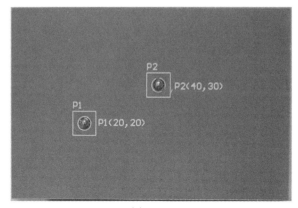

(a) PCB1

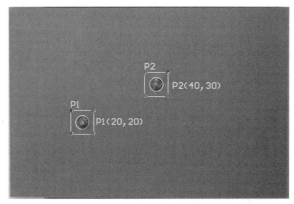

(b) PCB2

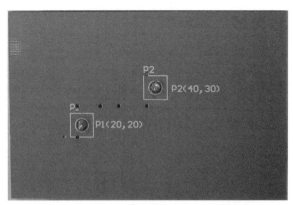

(c) PCB3

Figure 3.22 Test printed circuit boards (PCBs) for analysis of the coupling effect between the signal vias.

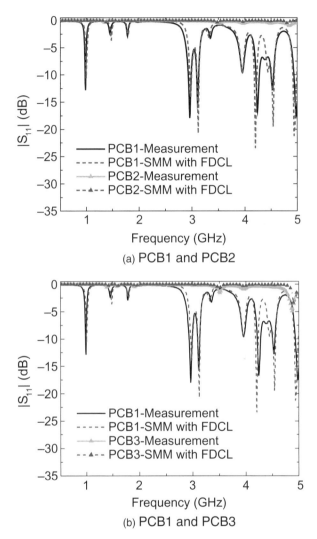

(a) PCB1 and PCB2

(b) PCB1 and PCB3

Figure 3.23 Comparison of the S_{11} parameters between the simulated results and measurement data for the test boards in Figure 3.22.

the periphery of the test board form a cavity, whose resonant frequencies are accurately captured by the FDCL modeling method. The slight difference in the magnitude of the S-parameters between the simulation and the measurement is probably due to the variation of the loss tangent over frequencies.

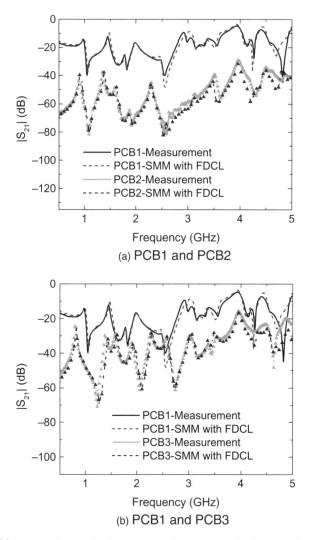

(a) PCB1 and PCB2

(b) PCB1 and PCB3

Figure 3.24   Comparison of the S_{21} parameters between the simulated results and measurement data for the test boards in Figure 3.22.

3.4.1.1 Experimental Setup for S-Parameter Measurements

Figure 3.25 shows the experimental setup for S-parameter measurement of the test PCBs in Figures 3.18 and 3.22. We use a vector network analyzer HP 8510C and S-parameter test set HP from Agilent to measure

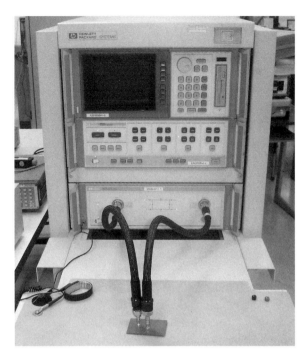

Figure 3.25   Experimental setup using Agilent HP 8510C Vector Network Analyzer and HP 8517B S-parameter Test Set to measure the test vehicles in Figures 3.18 and 3.22.

the S-parameters of the test vehicles in a frequency range from 45 MHz to 50 GHz with 2.4 mm coaxial cable.

3.4.2 Modeling of Arbitrarily Shaped Boundary Structures

One prominent feature of the FDCL modeling method is that it is versatile in handling arbitrarily shaped boundary of P-G planes and cutout structures in the planes. We develop a special formation procedure of the PMC cylinders for the arbitrarily shaped periphery of the packages and plane cutout structures.

An L-shaped test board with cutout is used as an example to demonstrate the flexibility of the FDCL for modeling irregularly shaped boundary. The example, taken from Reference 12, is shown in Figure 3.26. On the power plane there is a rectangular cutout whose diagonal points are (35,25) and (50,10), all in millimeters, respectively. The

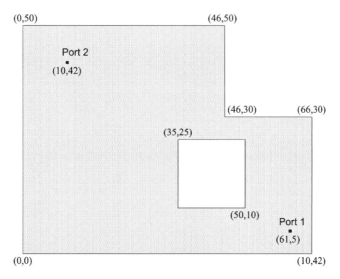

Figure 3.26   An irregularly shaped power-ground planes and cutout structure (unit: millimeter) [13].

dielectric material between the planes has a thickness of 1 mm, a relative permittivity of 2.65, and a loss tangent of 0.003. The source and observation SMA ports are located as shown in Figure 3.26. Some diagrams for the formation of the PMC cylinders in the FDCL, which will change in accordance with operating frequencies, are presented in Figures 3.27 and 3.28.

Figure 3.29 shows good agreement between the result of the SMM algorithm with FDCL and the measurement data [12] in a wide frequency band except first resonant frequency point has a slight shift. Again, the difference may be caused by the frequency-dependent dielectric material characteristics. However, the difference between the simulation result and the measurement data are within well acceptable margin of applications.

3.4.2.1 FDCL Method for Modeling of Finite-Domain P-G Planes

The conventional SMM using the model expansions of waves in PPWG is normally used to analyze the multiple scattering among the cylinders in the electronic package. However, the P-G planes of the package are assumed to be infinitely large, which cannot capture the resonant

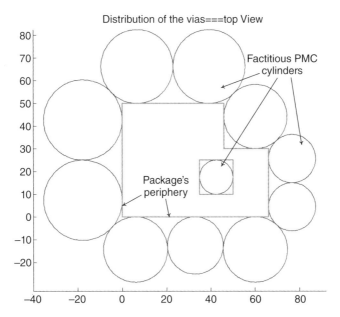

Figure 3.27 PMC cylinders formation in the FDCL at operation frequency of 1 GHz for the finite power-ground planes.

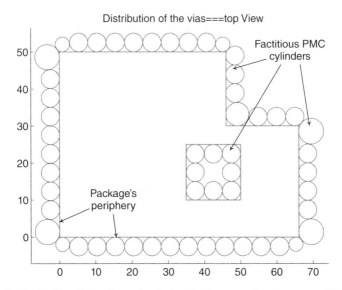

Figure 3.28 PMC cylinders formation in the FDCL at operation frequency of 5 GHz for the finite power-ground planes.

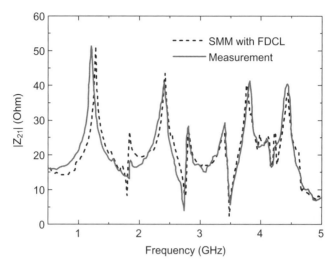

Figure 3.29   Comparison of the transfer impedance by the SMM simulation with the FDCL method against the measurement for the test board in Figure 3.26.

behavior of real-world packages. In this research work, we have contributed the important extension of the SMM using a novel factitious cylinder layer (FDCL) to simulate the finite-sized P-G planes in advanced electronic packages. In the FDCL approach, the radius of the PMC cylinders must be changed in accordance with operating frequencies. Numerical examples showed that the FDCL approach used in the SMM provides accurate prediction of the resonant frequencies of the electronic packages. By introducing the novel factitious layer of cylinders with dynamic radius at the periphery of an electronic package, the SMM is able to handle the real-world package structures. The simulated results of the SMM with FDCL show good agreement with the measured data. Finally, the proposed algorithm, SMM with FCDL, is also able to address the problem of multiple via coupling effects on the electrical performance for the entire electronic package in an efficient and correct way.

3.5 MODELING OF 3D ELECTRONIC PACKAGE STRUCTURE

The SMM with FDCL for analysis of multiple vias in the single layer package (a pair of P-G planes) has been presented in the previous sec-

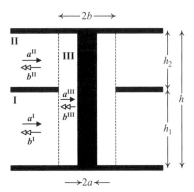

Figure 3.30   A though-hole via in two-layer structure and forming three PPWGs.

tions. Using several numerical examples, the developed algorithm is validated by comparing the simulated results with analytical solutions and measurement data. However, there are multiple layers (pairs of P-G planes) in practical structure of power distribution network for an advanced electronic package.

In this section, the formulae derivation for multilayered structure of P-G planes in an advanced electronic package is presented. The procedure is illustrated using the modal expansions of PPWG and the mode matching in the antipad region of the via. First, a case of two-layer structure of the P-G planes is considered for formulae derivation as shown in Figure 3.30. It has a case of three PPWGs—PPWG I, II, and III. Later, the formulation of the multilayered P-G planes is given for general case. Numerical simulations for the multilayered P-G planes with vias are presented and validated with full-wave numerical method.

3.5.1 Modal Expansions and Boundary Conditions

As discussed in Section 3.2, the tangential fields w.r.t. ρ inside the two-layer structure (Fig. 3.30) can be expressed by modal expansions as

$$\mathbf{E}_t^{(i)} = \sum_{n=-\infty}^{\infty} \sum_{m=0}^{\infty} \left\{ \left[a_{mn}^{e(i)} J_{mn}^{(i)} + b_{mn}^{e(i)} H_{mn}^{(i)} \right] \mathbf{e}_{t,mn}^{e(i)} + \left[a_{mn}^{h(i)} J_{mn}^{\prime(i)} + b_{mn}^{h(i)} H_{mn}^{\prime(i)} \right] \mathbf{e}_{t,mn}^{h(i)} \right\} e^{jn\phi},$$

(3.109)

$$\mathbf{H}_t^{(i)} = \sum_{n=-\infty}^{\infty} \sum_{m=0}^{\infty} \left\{ \left[a_{mn}^{e(i)} J_{mn}^{\prime(i)} + b_{mn}^{e(i)} H_{mn}^{\prime(i)} \right] \mathbf{h}_{t,mn}^{e(i)} \right. $$

$$\left. + \left[a_{mn}^{h(i)} J_{mn}^{(i)} + b_{mn}^{h(i)} H_{mn}^{(i)} \right] \mathbf{h}_{t,mn}^{h(i)} \right\} e^{jn\phi}, \tag{3.110}$$

where we have $a_{0n}^{h(i)} = b_{0n}^{h(i)} = 0$ for TE mode, and

$$\mathbf{e}_{t,mn}^{e(i)} = \cos\left(\beta_m^{(i)}(z-z_i)\right)\hat{z} - \frac{jn\beta_m^{(i)}}{\left(k_m^{(i)}\right)^2 \rho} \sin\left(\beta_m^{(i)}(z-z_i)\right)\hat{\varphi}$$

$$\mathbf{h}_{t,mn}^{e(i)} = -\frac{j\omega\varepsilon}{k_m^{(i)}} \cos\left(\beta_m^{(i)}(z-z_i)\right)\hat{\varphi}, \tag{3.111}$$

$$\mathbf{e}_{t,mn}^{h(i)} = \sin\left(\beta_m^{(i)}(z-z_i)\right)\hat{z} + \frac{jn\beta_m^{(i)}}{\left(k_m^{(i)}\right)^2 \rho} \cos\left(\beta_m^{(i)}(z-z_i)\right)\hat{\varphi}$$

$$\mathbf{h}_{t,mn}^{h(i)} = \frac{j\omega\mu}{k_m^{(i)}} \sin\left(\beta_m^{(i)}(z-z_i)\right)\hat{\varphi}, \tag{3.112}$$

$$k^2 = \omega^2 \mu\varepsilon = k_m^2 + \beta_m^2, \quad \text{and} \quad \beta_m = \frac{m\pi}{h_i}. \tag{3.113}$$

For the structure in Figure 3.30, the following boundary conditions are applied:

$$\mathbf{E}_t^{III}(\rho, \phi, z)\big|_{\rho=b} = \begin{cases} \mathbf{E}_t^{I}(\rho, \phi, z)\big|_{\rho=b}, & z \in [0, h_1] \\ \mathbf{E}_t^{II}(\rho, \phi, z)\big|_{\rho=b}, & z \in [h_1, h], \end{cases} \tag{3.114}$$

$$\mathbf{H}_t^{III}(\rho, \phi, z)\big|_{\rho=b} = \begin{cases} \mathbf{H}_t^{I}(\rho, \phi, z)\big|_{\rho=b}, & z \in [0, h_1] \\ \mathbf{H}_t^{II}(\rho, \phi, z)\big|_{\rho=b}, & z \in [h_1, h]. \end{cases} \tag{3.115}$$

Because of the decoupling of different modes n, we will only consider mode m in the following derivation. Substituting Equations (3.109) and (3.110) into Equations (3.114) and (3.115), respectively, we have

$$\sum_{m=0}^{\infty} \left(\begin{array}{l} \left[a_{mn}^{e,III} J_n(k_m^{III} b) + b_{mn}^{e,III} H_n^{(2)}(k_m^{III} b) \right] \mathbf{e}_{t,mn}^{e,III} \\ + \left[a_{mn}^{h,III} J_n'(k_m^{III} b) + b_{mn}^{h,III} H_n'^{(2)}(k_m^{III} b) \right] \mathbf{e}_{t,mn}^{h,III} \end{array} \right)$$

$$= \begin{cases} \sum_{m=0}^{\infty} \left(\begin{array}{l} \left[a_{mn}^{e,I} J_n(k_m^I b) + b_{mn}^{e,I} H_n^{(2)}(k_m^I b) \right] \mathbf{e}_{t,mn}^{e,I} \\ + \left[a_{mn}^{h,I} J_n'(k_m^I b) + b_{mn}^{h,I} H_n'^{(2)}(k_m^I b) \right] \mathbf{e}_{t,mn}^{h,I} \end{array} \right), & z \in [0, h_1] \\[2em] \sum_{m=0}^{\infty} \left(\begin{array}{l} \left[a_{mn}^{e,II} J_n(k_m^{II} b) + b_{mn}^{e,II} H_n^{(2)}(k_m^{II} b) \right] \mathbf{e}_{t,mn}^{e,II} \\ + \left[a_{mn}^{h,II} J_n'(k_m^{II} b) + b_{mn}^{h,II} H_n'^{(2)}(k_m^{II} b) \right] \mathbf{e}_{t,mn}^{h,II} \end{array} \right), & z \in [h_1, h] \end{cases} \tag{3.116}$$

For convenience, we drop all the subscripts n in the following derivation and use the following notations:

$$J_m^I = J_{mn}^I = J_n(k_m^I b), \quad \text{and} \quad H_m^I = H_{mn}^I = H_n^{(2)}(k_m^I b). \tag{3.117}$$

Then, Equation (3.116) can be written as

$$\sum_{m=0}^{\infty} \left(\begin{array}{l} \left[a_m^{e,III} J_m^{III} + b_m^{e,III} H_m^{III} \right] \mathbf{e}_{t,m}^{e,III} \\ + \left[a_m^{h,III} J_m'^{III} + b_m^{h,III} H_m'^{III} \right] \mathbf{e}_{t,m}^{h,III} \end{array} \right)$$

$$= \begin{cases} \sum_{m=0}^{\infty} \left(\begin{array}{l} \left[a_m^{e,I} J_m^I + b_m^{e,I} H_m^I \right] \mathbf{e}_{t,m}^{e,I} \\ + \left[a_m^{h,I} J_m'^I + b_m^{h,I} H_m'^I \right] \mathbf{e}_{t,m}^{h,I} \end{array} \right), & z \in [0, h_1] \\[2em] \sum_{m=0}^{\infty} \left(\begin{array}{l} \left[a_m^{e,II} J_m^{II} + b_m^{e,II} H_m^{II} \right] \mathbf{e}_{t,m}^{e,II} \\ + \left[a_m^{h,II} J_m'^{II} + b_m^{h,II} H_m'^{II} \right] \mathbf{e}_{t,m}^{h,II} \end{array} \right), & z \in [h_1, h]. \end{cases} \tag{3.118}$$

Similarly, the following equation is obtained from Equation (3.115) as

$$\sum_{m=0}^{\infty} \left(\begin{array}{l} \left[a_m^{e,III} J_m'^{III} + b_m^{e,III} H_m'^{III} \right] \mathbf{h}_{t,m}^{e,III} \\ + \left[a_m^{h,III} J_m^{III} + b_m^{h,III} H_m^{III} \right] \mathbf{h}_{t,m}^{h,III} \end{array} \right)$$

$$= \begin{cases} \sum_{m=0}^{\infty} \left(\begin{array}{l} \left[a_m^{e,I} J_m'^I + b_m^{e,I} H_m'^I \right] \mathbf{h}_{t,m}^{e,I} \\ + \left[a_m^{h,I} J_m^I + b_m^{h,I} H_m^I \right] \mathbf{h}_{t,m}^{h,I} \end{array} \right), & z \in [0, h_1] \\[2em] \sum_{m=0}^{\infty} \left(\begin{array}{l} \left[a_m^{e,II} J_m'^{II} + b_m^{e,II} H_m'^{II} \right] \mathbf{h}_{t,m}^{e,II} \\ + \left[a_m^{h,II} J_m^{II} + b_m^{h,II} H_m^{II} \right] \mathbf{h}_{t,m}^{h,II} \end{array} \right), & z \in [h_1, h]. \end{cases} \tag{3.119}$$

Replacing the tangential unit vectors in Equation (3.118) by those in Equations (3.111) and (3.112), we obtain for the L.H.S of Equation (3.118) as follows:

$$
\begin{aligned}
LHS\big|_{\mathbf{E}_t} &= \sum_{m=0}^{\infty} \left\{ \begin{aligned} &\left[a_m^{e,III} J_m^{III} + b_m^{e,III} H_m^{III} \right] \left[\cos\left(\beta_m^{III} z\right)\hat{z} - \frac{jn\beta_m^{III}}{\left(k_m^{III}\right)^2 b} \sin\left(\beta_m^{III} z\right)\hat{\varphi} \right] \\ &+ \left[a_m^{h,III} J_m'^{III} + b_m^{h,III} H_m'^{III} \right] \frac{j\omega\mu}{k_m^{III}} \sin\left(\beta_m^{III} z\right)\hat{\varphi} \end{aligned} \right\} \\
&= \sum_{m=0}^{\infty} \left[a_m^{e,III} J_m^{III} + b_m^{e,III} H_m^{III} \right] \cos\left(\beta_m^{III} z\right)\hat{z} \\
&+ \sum_{m=0}^{\infty} \left\{ \begin{aligned} &\left[a_m^{e,III} J_m^{III} + b_m^{e,III} H_m^{III} \right] \frac{-jn\beta_m^{III}}{\left(k_m^{III}\right)^2 b} \\ &+ \left[a_m^{h,III} J_m'^{III} + b_m^{h,III} H_m'^{III} \right] \frac{j\omega\mu}{k_m^{III}} \end{aligned} \right\} \sin\left(\beta_m^{III} z\right)\hat{\varphi}.
\end{aligned}
$$

(3.120)

Similarly, we can obtain for the L.H.S of Equation (3.119):

$$
\begin{aligned}
LHS\big|_{\mathbf{H}_t} &= \sum_{m=0}^{\infty} \left\{ \begin{aligned} &\left[a_m^{e,III} J_m'^{III} + b_m^{e,III} H_m'^{III} \right] \frac{-j\omega\varepsilon}{k_m^{III}} \cos\left(\beta_m^{III} z\right)\hat{\varphi} \\ &+ \left[a_m^{h,III} J_m^{III} + b_m^{h,III} H_m^{III} \right] \left[\sin\left(\beta_m^{III} z\right)\hat{z} + \frac{jn\beta_m^{III}}{\left(k_m^{III}\right)^2 b} \cos\left(\beta_m^{III} z\right)\hat{\varphi} \right] \end{aligned} \right\} \\
&= \sum_{m=0}^{\infty} \left[a_m^{h,III} J_m^{III} + b_m^{h,III} H_m^{III} \right] \sin\left(\beta_m^{III} z\right)\hat{z} \\
&+ \sum_{m=0}^{\infty} \left\{ \begin{aligned} &\left[a_m^{h,III} J_m^{III} + b_m^{h,III} H_m^{III} \right] \frac{jn\beta_m^{III}}{\left(k_m^{III}\right)^2 b} \\ &+ \left[a_m^{e,III} J_m'^{III} + b_m^{e,III} H_m'^{III} \right] \frac{-j\omega\varepsilon}{k_m^{III}} \end{aligned} \right\} \cos\left(\beta_m^{III} z\right)\hat{\varphi}.
\end{aligned}
$$

(3.121)

The $\hat{z}$ and $\hat{\varphi}$ components in Equations (3.118) and (3.119) are separated as shown in the following:

$$\sum_{m=0}^{\infty}\left[a_m^{e,III} J_m^{III} + b_m^{e,III} H_m^{III}\right]\cos\left(\beta_m^{III} z\right)$$

$$=\begin{cases}\displaystyle\sum_{m=0}^{\infty}\left[a_m^{e,I} J_m^{I} + b_m^{e,I} H_m^{I}\right]\cos\left(\beta_m^{I} z\right), & z\in[0,h_1] \\[2em] \displaystyle\sum_{m=0}^{\infty}\left[a_m^{e,II} J_m^{I} + b_m^{e,II} H_m^{II}\right]\cos\left(\beta_m^{II} z\right), & z\in[h_1,h],\end{cases} \qquad (3.122)$$

$$\sum_{m=0}^{\infty}\left\{\begin{array}{l}\left[a_m^{e,III} J_m^{III} + b_m^{e,III} H_m^{III}\right]\dfrac{-jn\beta_m^{III}}{\left(k_m^{III}\right)^2 b} \\[1.5em] +\left[a_m^{h,III} J_m'^{III} + b_m^{h,III} H_m'^{III}\right]\dfrac{j\omega\mu}{k_m^{III}}\end{array}\right\}\sin\left(\beta_m^{III} z\right)$$

$$=\begin{cases}\displaystyle\sum_{m=0}^{\infty}\left\{\left[a_m^{e,I} J_m^{I} + b_m^{e,I} H_m^{I}\right]\dfrac{-jn\beta_m^{I}}{\left(k_m^{I}\right)^2 b} + \left[a_m^{h,I} J_m'^{I} + b_m^{h,I} H_m'^{I}\right]\dfrac{j\omega\mu}{k_m^{I}}\right\} \\[1em] \sin\left(\beta_m^{I} z\right), \quad z\in[0,h_1] \\[1.5em] \displaystyle\sum_{m=0}^{\infty}\left\{\left[a_m^{e,II} J_m^{I} + b_m^{e,II} H_m^{II}\right]\dfrac{-jn\beta_m^{II}}{\left(k_m^{II}\right)^2 b} + \left[a_m^{h,II} J_m'^{II} + b_m^{h,II} H_m'^{II}\right]\dfrac{j\omega\mu}{k_m^{II}}\right\} \\[1em] \sin\left(\beta_m^{II} z\right), \quad z\in[0,h_1],\end{cases}$$

$$(3.123)$$

$$\sum_{m=0}^{\infty}\left[a_m^{h,III} J_m^{III} + b_m^{h,III} H_m^{III}\right]\sin\left(\beta_m^{III} z\right)=\begin{cases}\displaystyle\sum_{m=0}^{\infty}\left[a_m^{h,I} J_m^{I} + b_m^{h,I} H_m^{I}\right]\sin\left(\beta_m^{I} z\right), \\[1em] z\in[0,h_1] \\[1.5em] \displaystyle\sum_{m=0}^{\infty}\left[a_m^{h,II} J_m^{II} + b_m^{h,II} H_m^{II}\right]\sin\left(\beta_m^{II} z\right), \\[1em] z\in[0,h_1],\end{cases}$$

$$(3.124)$$

$$
= \begin{cases}
\sum_{m=0}^{\infty} \left\{ \begin{array}{l} \left[a_m^{h,III} J_m^{III} + b_m^{h,III} H_m^{III} \right] \dfrac{jn\beta_m^{III}}{\left(k_m^{III}\right)^2 b} \\[2ex] + \left[a_m^{e,III} J_m'^{III} + b_m^{e,III} H_m'^{III} \right] \dfrac{-j\omega\varepsilon}{k_m^{III}} \end{array} \right\} \cos\left(\beta_m^{III} z\right) \\[6ex]
\sum_{m=0}^{\infty} \left\{ \left[a_m^{h,I} J_m^{I} + b_m^{h,I} H_m^{I} \right] \dfrac{jn\beta_m^{I}}{\left(k_m^{III}\right)^2 b} + \left[a_m^{e,I} J_m'^{I} + b_m^{e,I} H_m'^{I} \right] \dfrac{-j\omega\varepsilon}{k_m^{I}} \right\} \\[2ex]
\cos\left(\beta_m^{I} z\right), \quad z \in [0, h_1] \\[4ex]
\sum_{m=0}^{\infty} \left\{ \left[a_m^{h,II} J_m^{II} + b_m^{h,II} H_m^{II} \right] \dfrac{jn\beta_m^{II}}{\left(k_m^{II}\right)^2 b} + \left[a_m^{e,II} J_m'^{II} + b_m^{e,II} H_m'^{II} \right] \dfrac{-j\omega\varepsilon}{k_m^{II}} \right\} \\[2ex]
\cos\left(\beta_m^{II} z\right), \quad z \in [h_1, h].
\end{cases}
$$

$$(3.125)$$

Equations (3.122–3.125) are correspondent to E_z, E_ϕ, H_z, and H_ϕ, respectively.

As referred to Section 3.2.2, for the PEC cylinder ($\rho = a$) in PPWG-III (Fig. 3.30), we have the following relation:

$$
b_m^{e,III} = -\frac{J_n\left(k_m^{III} a\right)}{H_n^{(2)}\left(k_m^{III} a\right)} a_m^{e,III}, \quad b_m^{h,III} = -\frac{J_n'\left(k_m^{III} a\right)}{H_n'^{(2)}\left(k_m^{III} a\right)} a_m^{h,III}. \quad (3.126)
$$

Then, we designate the following notations:

$$
a_m^{e,III} J_m^{III} + b_m^{e,III} H_m^{III} = \left(J_m^{III} - \frac{J_n\left(k_m^{III} a\right) H_m^{III}}{H_n^{(2)}\left(k_m^{III} a\right)} \right) a_m^{e,III} \triangleq J_{e,m}^{III} a_m^{e,III}, \quad (3.127)
$$

$$
a_m^{e,III} J_m'^{III} + b_m^{e,III} H_m'^{III} = \left(J_m'^{III} - \frac{J_n\left(k_m^{III} a\right) H_m'^{III}}{H_n^{(2)}\left(k_m^{III} a\right)} \right) a_m^{e,III} \triangleq J_{e,m}'^{III} a_m^{e,III}, \quad (3.128)
$$

$$
a_m^{h,III} J_m^{III} + b_m^{h,III} H_m^{III} = \left(J_m^{III} - \frac{J_n'\left(k_m^{III} a\right) H_m^{III}}{H_n'^{(2)}\left(k_m^{III} a\right)} \right) a_m^{h,III} \triangleq J_{h,m}^{III} a_m^{h,III}, \quad (3.129)
$$

$$
a_m^{h,III} J_m'^{III} + b_m^{h,III} H_m'^{III} = \left(J_m'^{III} - \frac{J_n'\left(k_m^{III} a\right) H_m'^{III}}{H_n'^{(2)}\left(k_m^{III} a\right)} \right) a_m^{h,III} \triangleq J_{h,m}'^{III} a_m^{h,III}, \quad (3.130)
$$

where J_m^{III} and H_m^{III} are defined as those in Equation (3.117).

We can now rewrite Equations (3.122–3.125) as follows:

$$\sum_{m=0}^{\infty} a_m^{e,III} J_{e,m}^{III} \cos\left(\beta_m^{III} z\right) = \begin{cases} \sum_{m=0}^{\infty} \left[a_m^{e,I} J_m^{I} + b_m^{e,I} H_m^{I}\right] \cos\left(\beta_m^{I} z\right), & z \in [0, h_1] \\ \sum_{m=0}^{\infty} \left[a_m^{e,II} J_m^{I} + b_m^{e,II} H_m^{II}\right] \cos\left(\beta_m^{II} z\right), & z \in [h_1, h], \end{cases}$$

(3.131)

$$\sum_{m=0}^{\infty} \left\{ a_m^{e,III} J_{e,m}^{III} \frac{-jn\beta_m^{III}}{\left(k_m^{III}\right)^2 b} + a_m^{h,III} J_{h,m}'^{III} \frac{j\omega\mu}{k_m^{III}} \right\} \sin\left(\beta_m^{III} z\right)$$

$$= \begin{cases} \sum_{m=0}^{\infty} \left\{ \left[a_m^{e,I} J_m^{I} + b_m^{e,I} H_m^{I}\right] \frac{-jn\beta_m^{I}}{\left(k_m^{I}\right)^2 b} + \left[a_m^{h,I} J_m'^{I} + b_m^{h,I} H_m'^{I}\right] \frac{j\omega\mu}{k_m^{I}} \right\} \\ \sin\left(\beta_m^{I} z\right), \quad z \in [0, h_1] \\ \sum_{m=0}^{\infty} \left\{ \left[a_m^{e,II} J_m^{II} + b_m^{e,II} H_m^{II}\right] \frac{-jn\beta_m^{II}}{\left(k_m^{II}\right)^2 b} + \left[a_m^{h,II} J_m'^{II} + b_m^{h,II} H_m'^{II}\right] \frac{j\omega\mu}{k_m^{II}} \right\} \\ \sin\left(\beta_m^{II} z\right), \quad z \in [0, h_1], \end{cases}$$

(3.132)

$$\sum_{m=0}^{\infty} a_m^{h,III} J_{h,m}^{III} \sin\left(\beta_m^{III} z\right) = \begin{cases} \sum_{m=0}^{\infty} \left[a_m^{h,I} J_m^{I} + b_m^{h,I} H_m^{I}\right] \sin\left(\beta_m^{I} z\right), & z \in [0, h_1] \\ \sum_{m=0}^{\infty} \left[a_m^{h,II} J_m^{II} + b_m^{h,II} H_m^{II}\right] \sin\left(\beta_m^{II} z\right), & z \in [0, h_1], \end{cases}$$

(3.133)

$$\sum_{m=0}^{\infty} \left\{ a_m^{h,III} J_{h,m}^{III} \frac{jn\beta_m^{III}}{\left(k_m^{III}\right)^2 b} + a_m^{e,III} J_{e,m}'^{III} \frac{-j\omega\varepsilon}{k_m^{III}} \right\} \cos\left(\beta_m^{III} z\right)$$

$$= \begin{cases} \sum_{m=0}^{\infty} \left\{ \left[a_m^{h,I} J_m^{I} + b_m^{h,I} H_m^{I}\right] \frac{jn\beta_m^{I}}{\left(k_m^{III}\right)^2 b} + \left[a_m^{e,I} J_m'^{I} + b_m^{e,I} H_m'^{I}\right] \frac{-j\omega\varepsilon}{k_m^{I}} \right\} \\ \cos\left(\beta_m^{I} z\right), \quad z \in [0, h_1] \\ \sum_{m=0}^{\infty} \left\{ \left[a_m^{h,II} J_m^{II} + b_m^{h,II} H_m^{II}\right] \frac{jn\beta_m^{II}}{\left(k_m^{II}\right)^2 b} + \left[a_m^{e,II} J_m'^{II} + b_m^{e,II} H_m'^{II}\right] \frac{-j\omega\varepsilon}{k_m^{II}} \right\} \\ \cos\left(\beta_m^{II} z\right), \quad z \in [h_1, h]. \end{cases}$$

(3.134)

3.5.2 Mode Matching in PPWGs

In this section, we focus on derivation of the following generalized T-matrix:

$$
\begin{bmatrix} b^{e,I} \\ b^{e,II} \\ b^{h,I} \\ b^{h,II} \end{bmatrix} = \begin{bmatrix} T_{I,I}^{ee} & T_{I,II}^{ee} & T_{I,I}^{eh} & T_{I,II}^{eh} \\ T_{II,I}^{ee} & T_{II,II}^{ee} & T_{II,I}^{eh} & T_{II,II}^{eh} \\ T_{I,I}^{he} & T_{I,II}^{he} & T_{I,I}^{hh} & T_{I,II}^{hh} \\ T_{II,I}^{he} & T_{II,II}^{he} & T_{II,I}^{hh} & T_{II,II}^{hh} \end{bmatrix} \begin{bmatrix} a^{e,I} \\ a^{e,II} \\ a^{h,I} \\ a^{h,II} \end{bmatrix}, \tag{3.135}
$$

where the size of the matrices depends on the number of modes used for each PPWGs.

The T-matrix in Equation (3.135) can be derived using the mode matching technique [15]. The orthogonality relations for the Fourier series used in the mode matching technique are given as

$$
\begin{cases} \int_0^a \cos\dfrac{n\pi x}{a}\cos\dfrac{m\pi x}{a}\,dx = \int_0^a \sin\dfrac{n\pi x}{a}\sin\dfrac{m\pi x}{a}\,dx = \dfrac{a}{2}\delta_{nm}, \\ \quad \text{except for } m = n = 0 \\ \int_0^a \cos\dfrac{n\pi x}{a}\sin\dfrac{m\pi x}{a}\,dx = 0. \end{cases} \tag{3.136}
$$

For numerical calculation, we also truncate the infinite summation to a finite one and the numbers of modes are M1, M2, and M3 for PPWG- I, II, and III, respectively.

For performing the mode matching, we can either test it over $[0, h]$, or $[0, h_1]$ and $[h_1, h]$. Here we choose the testing functions as being those of PPWG-III to enforcing E_z and H_z, and those of PPWG-I and II to enforcing E_ϕ and H_ϕ. Performing the testing by $\cos(\beta_p^{III} z)$ on Equation (3.131):

$$
\int_0^h (4.23) \times \cos\left(\beta_p^{III} z\right) dz \rightarrow \int_0^h (4.23) \times \cos\left(\frac{p\pi}{h} z\right) dz
$$

$$
\rightarrow a_p^{e,III} J_{e,p}^{III} \frac{h}{2} = \sum_{m=0}^{M1} \left[a_m^{e,I} J_m^I + b_m^{e,I} H_m^I \right] I_{mp}^{I,III} + \sum_{m=0}^{M2} \left[a_m^{e,II} J_m^{II} + b_m^{e,II} H_m^{II} \right] I_{mp}^{II,III},
$$

$$
\tag{3.137}
$$

where

$$I_{mp}^{I,III} = \int_0^{h_1} \cos\left(\beta_m^I z\right)\cos\left(\beta_m^{III} z\right)dz - \int_0^{h_1} \cos\left(\frac{m\pi}{h_1}z\right)\cos\left(\frac{p\pi}{h}z\right)dz,$$

(3.138)

$$I_{mp}^{II,III} = \int_0^{h_2} \cos\left(\beta_m^{II} z\right)\cos\left(\beta_m^{III} z\right)dz = \int_0^{h_2} \cos\left(\frac{m\pi}{h_2}z\right)\cos\left(\frac{p\pi}{h}z\right)dz.$$

(3.139)

Performing the testing on Equation (3.132) over $[0, h_1]$ and $[h_1, h]$ by $\sin(\beta_p^I z)$ and $\sin(\beta_p^{II} z)$, respectively:

$$\sum_{m=0}^{M3}\left\{a_m^{e,III} J_{e,m}^{III}\frac{-jn\beta_m^{III}}{(k_m^{III})^2 b} + a_m^{h,III} J_{h,m}'^{III}\frac{j\omega\mu}{k_m^{III}}\right\}I_{mq}'^{III,I}$$

$$= \left\{\left[a_q^{e,I} J_q^I + b_q^{e,I} H_q^I\right]\frac{-jn\beta_q^I}{(k_q^I)^2 b} + \left[a_q^{h,I} J_q'^I + b_q^{h,I} H_q'^I\right]\frac{j\omega\mu}{k_q^I}\right\}\frac{h_1}{2},$$

(3.140)

$$\sum_{m=0}^{M3}\left\{a_m^{e,III} J_{e,m}^{III}\frac{-jn\beta_m^{III}}{(k_m^{III})^2 b} + a_m^{h,III} J_{h,m}'^{III}\frac{j\omega\mu}{k_m^{III}}\right\}I_{mr}'^{III,II}$$

$$= \left\{\left[a_r^{e,II} J_r^{II} + b_r^{e,II} H_r^{II}\right]\frac{-jn\beta_r^{II}}{(k_r^{II})^2 b} + \left[a_r^{h,II} J_r'^{II} + b_r^{h,II} H_r'^{II}\right]\frac{j\omega\mu}{k_r^{II}}\right\}\frac{h_2}{2},$$

(3.141)

where

$$I_{mq}'^{III,I} = \int_0^{h_1} \sin\left(\beta_m^{III} z\right)\sin\left(\beta_q^I z\right)dz = \int_0^{h_1} \sin\left(\frac{m\pi}{h}z\right)\sin\left(\frac{q\pi}{h_1}z\right)dz, \quad (3.142)$$

$$I_{mr}'^{III,II} = \int_0^{h_2} \sin\left(\beta_m^{III} z\right)\sin\left(\beta_r^{II} z\right)dz = \int_0^{h_2} \sin\left(\frac{m\pi}{h}z\right)\sin\left(\frac{r\pi}{h_2}z\right)dz.$$

(3.143)

Performing the testing by $\sin(\beta_p^{III} z)$ on Equation (3.133):

$$\int_0^h (4.25)\times\sin\left(\beta_p^{III} z\right)dz \rightarrow \int_0^h (4.25)\times\sin\left(\frac{p\pi}{h}z\right)dz$$

$$\rightarrow a_p^{h,III} J_{h,p}^{III}\frac{h}{2} = \sum_{m=0}^{M1}\left[a_m^{h,I} J_m^I + b_m^{h,I} H_m^I\right]I_{mp}'^{I,III} + \sum_{m=0}^{M2}\left[a_m^{h,II} J_m^{II} + b_m^{h,II} H_m^{II}\right]I_{mp}'^{II,III},$$

(3.144)

where

$$I_{mp}^{\prime I,III} = \int_0^{h_1} \sin\left(\beta_m^I z\right)\sin\left(\beta_p^{III} z\right) dz = \int_0^{h_1} \sin\left(\frac{m\pi}{h_1}z\right)\sin\left(\frac{p\pi}{h}z\right) dz,$$

(3.145)

$$I_{mp}^{\prime II,III} = \int_0^{h_2} \sin\left(\beta_m^{II} z\right)\sin\left(\beta_p^{III} z\right) dz = \int_0^{h_2} \sin\left(\frac{m\pi}{h_2}z\right)\sin\left(\frac{p\pi}{h}z\right) dz.$$

(3.146)

Performing the testing on Equation (3.134) over $[0, h_1]$ and $[h_1, h]$ by $\cos(\beta_p^I z)$ and $\cos(\beta_p^{II} z)$, respectively,

$$\sum_{m=0}^{M3}\left\{a_m^{h,III} J_{h,m}^{III} \frac{jn\beta_m^{III}}{(k_m^{III})^2 b} + a_m^{e,III} J_{e,m}^{\prime III} \frac{-j\omega\varepsilon}{k_m^{III}}\right\} I_{mq}^{III,I}$$
$$= \left\{\left[a_q^{h,I} J_q^I + b_q^{h,I} H_q^I\right]\frac{jn\beta_q^I}{(k_q^I)^2 b} + \left[a_q^{e,I} J_q^{\prime I} + b_q^{e,I} H_q^{\prime I}\right]\frac{-j\omega\varepsilon}{k_m^I}\right\}\frac{h_1}{2}, \quad (3.147)$$

$$\sum_{m=0}^{M3}\left\{a_m^{h,III} J_{h,m}^{III} \frac{jn\beta_m^{III}}{(k_m^{III})^2 b} + a_m^{e,III} J_{e,m}^{\prime III} \frac{-j\omega\varepsilon}{k_m^{III}}\right\} I_{mr}^{III,II}$$
$$= \left\{\left[a_r^{h,II} J_r^{II} + b_r^{h,II} H_r^{II}\right]\frac{jn\beta_r^{II}}{(k_r^{II})^2 b} + \left[a_r^{e,II} J_r^{\prime II} + b_r^{e,II} H_r^{\prime II}\right]\frac{-j\omega\varepsilon}{k_r^{II}}\right\}\frac{h_2}{2},$$

(3.148)

where

$$I_{mq}^{III,I} = \int_0^{h_1} \cos\left(\beta_m^{III} z\right)\cos\left(\beta_q^I z\right) dz = \int_0^{h_1} \cos\left(\frac{m\pi}{h}z\right)\cos\left(\frac{q\pi}{h_1}z\right) dz,$$

(3.149)

$$I_{mr}^{III,II} = \int_0^{h_2} \cos\left(\beta_m^{III} z\right)\cos\left(\beta_r^{II} z\right) dz = \int_0^{h_2} \cos\left(\frac{m\pi}{h}z\right)\cos\left(\frac{r\pi}{h_2}z\right) dz.$$

(3.150)

We introduce the following notations to make the subsequent derivation concisely.

$$\frac{jn\beta_m^{III}}{(k_m^{III})^2 b} \triangleq \tau_m^{\beta,III}, \quad \frac{j\omega\mu}{k_m^{III}} \triangleq \tau_m^{\mu,III}, \quad \frac{j\omega\varepsilon}{k_m^{III}} \triangleq \tau_m^{\varepsilon,III}.$$

(3.151)

Thus, we have Equations (3.140), (3.141), (3.147), and (3.148):

$$\sum_{m=0}^{M3}\left\{a_m^{e,III}(-\tau_m^{\beta,III})J_{e,m}^{III}+a_m^{h,III}\tau_m^{\mu,III}J_{h,m}^{\prime III}\right\}I_{mq}^{\prime III,I}$$

$$=\left\{\left[a_q^{e,I}J_q^{I}+b_q^{e,I}H_q^{I}\right](-\tau_q^{\beta,I})+\left[a_q^{h,I}J_q^{\prime I}+b_q^{h,I}H_q^{\prime I}\right]\tau_q^{\mu,I}\right\}\frac{h_1}{2}, \qquad (3.152)$$

$$\sum_{m=0}^{M3}\left\{a_m^{e,III}(-\tau_m^{\beta,III})J_{e,m}^{III}+a_m^{h,III}\tau_m^{\mu,III}J_{h,m}^{\prime III}\right\}I_{mr}^{\prime III,II}$$

$$=\left\{\left[a_r^{e,II}J_r^{II}+b_r^{e,II}H_r^{II}\right](-\tau_r^{\beta,II})+\left[a_r^{h,II}J_r^{\prime II}+b_r^{h,II}H_r^{\prime II}\right]\tau_r^{\mu,II}\right\}\frac{h_2}{2}, \quad (3.153)$$

$$\sum_{m=0}^{M3}\left\{a_m^{h,III}\tau_m^{\beta,III}J_{h,m}^{III}+a_m^{e,III}(-\tau_m^{\varepsilon,III})J_{e,m}^{\prime III}\right\}I_{mq}^{III,I}$$

$$=\left\{\left[a_q^{h,I}J_q^{I}+b_q^{h,I}H_q^{I}\right]\tau_q^{\beta,I}+\left[a_q^{e,I}J_q^{\prime I}+b_q^{e,I}H_q^{\prime I}\right](-\tau_q^{\varepsilon,I})\right\}\frac{h_1}{2}, \qquad (3.154)$$

$$\sum_{m=0}^{M3}\left\{a_m^{h,III}\tau_m^{\beta,III}J_{h,m}^{III}+a_m^{e,III}(-\tau_m^{\varepsilon,III})J_{e,m}^{\prime III}\right\}I_{mr}^{III,II}$$

$$=\left\{\left[a_r^{h,II}J_r^{II}+b_r^{h,II}H_r^{II}\right]\tau_r^{\beta,II}+\left[a_r^{e,II}J_r^{\prime II}+b_r^{e,II}H_r^{\prime II}\right](-\tau_r^{\varepsilon,II})\right\}\frac{h_2}{2}. \quad (3.155)$$

The unknown coefficients are derived by manipulating Equations (3.137), (3.140), (3.141), (3.144), (3.147), and (3.148).

Equations (3.137) and (3.144) can be rewritten as follows:

$$a_p^{e,III}=\left(\sum_{m=0}^{M1}\left[a_m^{e,I}J_m^{I}+b_m^{e,I}H_m^{I}\right]I_{mp}^{I,III}+\sum_{m=0}^{M2}\left[a_m^{e,II}J_m^{II}+b_m^{e,II}H_m^{II}\right]I_{mp}^{II,III}\right)\frac{2}{hJ_{e,p}^{III}},$$

$$(3.156)$$

$$a_p^{h,III}=\left(\sum_{m=0}^{M1}\left[a_m^{h,I}J_m^{I}+b_m^{h,I}H_m^{I}\right]I_{mp}^{\prime I,III}+\sum_{m=0}^{M2}\left[a_m^{h,II}J_m^{II}+b_m^{h,II}H_m^{II}\right]I_{mp}^{\prime II,III}\right)\frac{2}{hJ_{h,p}^{III}}.$$

$$(3.157)$$

First changing the subscript m in Equation (3.152) to p, then

$$\sum_{p=0}^{M3}\left\{a_p^{e,III}(-\tau_p^{\beta,III})J_{e,p}^{III}+a_p^{h,III}\tau_p^{\mu,III}J_{h,p}^{\prime III}\right\}I_{pq}^{\prime III,I}$$

$$=\left\{\left[a_q^{e,I}J_q^{I}+b_q^{e,I}H_q^{I}\right](-\tau_q^{\beta,I})+\left[a_q^{h,I}J_q^{\prime I}+b_q^{h,I}H_q^{\prime I}\right]\tau_q^{\mu,I}\right\}\frac{h_1}{2}. \qquad (3.158)$$

Substituting Equations (3.156) and (3.157) into Equation (3.158), we obtain

$$
\sum_{p=0}^{M3} \left\{ \left(\frac{\sum_{m=0}^{M1} \left[a_m^{e,I} J_m^I + b_m^{e,I} H_m^I \right] I_{mp}^{I,III}}{+ \sum_{m=0}^{M2} \left[a_m^{e,II} J_m^{II} + b_m^{e,II} H_m^{II} \right] I_{mp}^{II,III}} \right) \frac{-2\tau_p^{\beta,III} I_{pq}^{\prime III,I}}{h} \right\}
$$
$$
+ \sum_{p=0}^{M3} \left\{ \left(\frac{\sum_{m=0}^{M1} \left[a_m^{h,I} J_m^I + b_m^{h,I} H_m^I \right] I_{mp}^{\prime I,III}}{+ \sum_{m=0}^{M2} \left[a_m^{h,II} J_m^{II} + b_m^{h,II} H_m^{II} \right] I_{mp}^{\prime II,III}} \right) \frac{2\tau_p^{\mu,III} J_{h,p}^{\prime III} I_{pq}^{\prime III,I}}{hJ_{h,p}^{III}} \right\}
$$
$$
= \left\{ \left[a_q^{e,I} J_q^I + b_q^{e,I} H_q^I \right] (-\tau_q^{\beta,I}) + \left[a_q^{h,I} J_q^{\prime I} + b_q^{h,I} H_q^{\prime I} \right] \tau_q^{\mu,I} \right\} \frac{h_1}{2}. \qquad (3.159)
$$

Similarly, we can eliminate $a_m^{e,III}$ and $a_m^{h,III}$ in Equations (3.153–3.155):

$$
\sum_{p=0}^{M3} \left\{ \left(\frac{\sum_{m=0}^{M1} \left[a_m^{e,I} J_m^I + b_m^{e,I} H_m^I \right] I_{mp}^{I,III}}{+ \sum_{m=0}^{M2} \left[a_m^{e,II} J_m^{II} + b_m^{e,II} H_m^{II} \right] I_{mp}^{II,III}} \right) \frac{-2\tau_p^{\beta,III} I_{pr}^{\prime III,II}}{h} \right\}
$$
$$
+ \sum_{p=0}^{M3} \left\{ \left(\frac{\sum_{m=0}^{M1} \left[a_m^{h,I} J_m^I + b_m^{h,I} H_m^I \right] I_{mp}^{\prime I,III}}{+ \sum_{m=0}^{M2} \left[a_m^{h,II} J_m^{II} + b_m^{h,II} H_m^{II} \right] I_{mp}^{\prime II,III}} \right) \frac{2\tau_p^{\mu,III} J_{h,p}^{\prime III} I_{pr}^{\prime III,II}}{hJ_{h,p}^{III}} \right\}
$$
$$
= \left\{ \left[a_r^{e,II} J_r^{II} + b_r^{e,II} H_r^{II} \right] (-\tau_r^{\beta,II}) + \left[a_r^{h,II} J_r^{\prime II} + b_r^{h,II} H_r^{\prime II} \right] \tau_r^{\mu,II} \right\} \frac{h_2}{2}, \qquad (3.160)
$$

$$
\sum_{p=0}^{M3} \left\{ \left(\frac{\sum_{m=0}^{M1} \left[a_m^{e,I} J_m^I + b_m^{e,I} H_m^I \right] I_{mp}^{I,III}}{+ \sum_{m=0}^{M2} \left[a_m^{e,II} J_m^{II} + b_m^{e,II} H_m^{II} \right] I_{mp}^{II,III}} \right) \frac{-2\tau_p^{\varepsilon,III} J_{e,p}^{\prime III} I_{pq}^{III,I}}{hJ_{e,p}^{III}} \right\}
$$
$$
+ \sum_{p=0}^{M3} \left\{ \left(\frac{\sum_{m=0}^{M1} \left[a_m^{h,I} J_m^I + b_m^{h,I} H_m^I \right] I_{mp}^{\prime I,III}}{+ \sum_{m=0}^{M2} \left[a_m^{h,II} J_m^{II} + b_m^{h,II} H_m^{II} \right] I_{mp}^{\prime II,III}} \right) \frac{2\tau_p^{\beta,III} I_{pq}^{III,I}}{h} \right\}
$$
$$
= \left\{ \left[a_q^{e,I} J_q^{\prime I} + b_q^{e,I} H_q^{\prime I} \right] (-\tau_q^{\varepsilon,I}) + \left[a_q^{h,I} J_q^I + b_q^{h,I} H_q^I \right] \tau_q^{\beta,I} \right\} \frac{h_1}{2}, \qquad (3.161)
$$

$$\sum_{p=0}^{M3}\left\{\left(\begin{array}{l}\displaystyle\sum_{m=0}^{M1}\left[a_m^{e,I}J_m^I+b_m^{e,I}H_m^I\right]I_{mp}^{I,III}\\[2mm]+\displaystyle\sum_{m=0}^{M2}\left[a_m^{e,II}J_m^{II}+b_m^{e,II}H_m^{II}\right]I_{mp}^{II,III}\end{array}\right)\frac{-2\tau_p^{\varepsilon,III}J_{e,p}'^{III}I_{pq}^{III,II}}{hJ_{e,p}^{III}}\right\}$$

$$+\sum_{p=0}^{M3}\left\{\left(\begin{array}{l}\displaystyle\sum_{m=0}^{M1}\left[a_m^{h,I}J_m^I+b_m^{h,I}H_m^I\right]I_{mp}'^{I,III}\\[2mm]+\displaystyle\sum_{m=0}^{M2}\left[a_m^{h,II}J_m^{II}+b_m^{h,II}H_m^{II}\right]I_{mp}'^{II,III}\end{array}\right)\frac{2\tau_p^{\beta,III}I_{pq}^{III,II}}{h}\right\}$$

$$=\left\{\left[a_r^{e,II}J_r'^{II}+b_r^{e,II}H_r'^{II}\right](-\tau_r^{\varepsilon,II})+\left[a_r^{h,II}J_r^{II}+b_r^{h,II}H_r^{II}\right]\tau_r^{\beta,II}\right\}\frac{h_2}{2}. \quad (3.162)$$

Reorganizing all the terms in Equation (3.159):

$$\sum_{p=0}^{M3}\sum_{m=0}^{M1}b_m^{e,I}\frac{-2\tau_p^{\beta,III}H_m^I I_{mp}^{I,III}I_{pq}'^{III,I}}{h}+\sum_{p=0}^{M3}\sum_{m=0}^{M1}a_m^{e,I}\frac{-2\tau_p^{\beta,III}J_m^I I_{mp}^{I,III}I_{pq}'^{III,I}}{h}$$

$$+\sum_{p=0}^{M3}\sum_{m=0}^{M2}b_m^{e,II}\frac{-2\tau_p^{\beta,III}H_m^{II}I_{mp}^{II,III}I_{pq}'^{III,I}}{h}+\sum_{p=0}^{M3}\sum_{m=0}^{M2}a_m^{e,II}\frac{-2\tau_p^{\beta,III}J_m^{II}I_{mp}^{II,III}I_{pq}'^{III,I}}{h}$$

$$+\sum_{p=0}^{M3}\sum_{m=0}^{M1}b_m^{h,I}\frac{2\tau_p^{\mu,III}H_m^I J_{h,p}'^{III}I_{mp}^{I,III}I_{pq}'^{III,I}}{J_{h,p}^{III}h}$$

$$+\sum_{p=0}^{M3}\sum_{m=0}^{M1}a_m^{h,I}\frac{2\tau_p^{\mu,III}J_m^I J_{h,p}'^{III}I_{mp}'^{I,III}I_{pq}'^{III,I}}{J_{h,p}^{III}h}$$

$$+\sum_{p=0}^{M3}\sum_{m=0}^{M2}b_m^{h,II}\frac{2\tau_p^{\mu,III}H_m^{II}J_{h,p}'^{III}I_{mp}'^{II,III}I_{pq}'^{III,I}}{J_{h,p}^{III}h}$$

$$+\sum_{p=0}^{M3}\sum_{m=0}^{M2}a_m^{h,II}\frac{2\tau_p^{\mu,III}J_m^{II}J_{h,p}'^{III}I_{mp}'^{II,III}I_{pq}'^{III,I}}{J_{h,p}^{III}h}$$

$$=a_q^{e,I}\frac{-\tau_q^{\beta,I}J_q^I h_1}{2}+b_q^{e,I}\frac{-\tau_q^{\beta,I}H_q^I h_1}{2}+a_q^{h,I}\frac{\tau_q^{\mu,I}J_q'^I h_1}{2}+b_q^{h,I}\frac{\tau_q^{\mu,I}H_q'^I h_1}{2}.$$

$$(3.163)$$

Then, all the terms with unknown coefficients for scattered fields are put to the L.H.S. of Equation (3.163) and all the remaining terms to the R.H.S.:

$$\sum_{p=0}^{M3}\sum_{m=0}^{M1}\left(b_m^{e,I}\frac{-2\tau_p^{\beta,III}H_m^I I_{mp}^{I,III} I_{pq}'^{III,I}}{h}+\delta_{mq}b_q^{e,I}\frac{\tau_q^{\beta,I}H_q^I h_1}{2}\right)$$

$$+\sum_{p=0}^{M3}\sum_{m=0}^{M2}b_m^{e,II}\frac{-2\tau_p^{\beta,III}H_m^{II} I_{mp}^{II,III} I_{pq}'^{III,I}}{h}$$

$$+\sum_{p=0}^{M3}\sum_{m=0}^{M1}\left(b_m^{h,I}\frac{2\tau_p^{\mu,III}H_m^I J_{h,p}'^{III} I_{mp}'^{I,III} I_{pq}'^{III,I}}{J_{h,p}^{III}h}+\delta_{mq}b_q^{h,I}\frac{-\tau_q^{\mu,I}H_q'^{I} h_1}{2}\right)$$

$$+\sum_{p=0}^{M3}\sum_{m=0}^{M2}b_m^{h,II}\frac{2\tau_p^{\mu,III}H_m^{II} J_{h,p}'^{III} I_{mp}^{II,III} I_{pq}'^{III,I}}{J_{h,p}^{III}h}$$

$$=\sum_{p=0}^{M3}\sum_{m=0}^{M1}\left(a_m^{e,I}\frac{2\tau_p^{\beta,III}J_m^I I_{mp}^{I,III} I_{pq}'^{III,I}}{h}+\delta_{mq}a_q^{e,I}\frac{-\tau_q^{\beta,I}J_q^I h_1}{2}\right)$$

$$+\sum_{p=0}^{M3}\sum_{m=0}^{M2}a_m^{e,II}\frac{2\tau_p^{\beta,III}J_m^{II} I_{mp}^{II,III} I_{pq}'^{III,I}}{h}$$

$$+\sum_{p=0}^{M3}\sum_{m=0}^{M1}\left(a_m^{h,I}\frac{-2\tau_p^{\mu,III}J_m^I J_{h,p}'^{III} I_{mp}'^{I,III} I_{pq}'^{III,I}}{J_{h,p}^{III}h}+\delta_{mq}a_q^{h,I}\frac{\tau_q^{\mu,I}J_q'^{I} h_1}{2}\right)$$

$$+\sum_{p=0}^{M3}\sum_{m=0}^{M2}a_m^{h,II}\frac{-2\tau_p^{\mu,III}J_m^{II} J_{h,p}'^{III} I_{mp}^{II,III} I_{pq}'^{III,I}}{J_{h,p}^{III}h}. \tag{3.164}$$

Similarly, we have for Equations (3.160–3.162):

$$\sum_{p=0}^{M3}\sum_{m=0}^{M1}b_m^{e,I}\frac{-2\tau_p^{\beta,III}H_m^I I_{mp}^{I,III} I_{pr}'^{III,II}}{h}$$

$$+\sum_{p=0}^{M3}\sum_{m=0}^{M2}\left(b_m^{e,II}\frac{-2\tau_p^{\beta,III}H_m^{II} I_{mp}^{II,III} I_{pr}'^{III,II}}{h}+\delta_{mr}b_r^{e,II}\frac{\tau_r^{\beta,II}H_r^{II} h_2}{2}\right)$$

$$+\sum_{p=0}^{M3}\sum_{m=0}^{M1}b_m^{h,I}\frac{2\tau_p^{\mu,III}H_m^I J_{h,p}'^{III} I_{mp}'^{I,III} I_{pr}'^{III,II}}{J_{h,p}^{III}h}$$

$$+\sum_{p=0}^{M3}\sum_{m=0}^{M2}\left(b_m^{h,II}\frac{2\tau_p^{\mu,III}H_m^{II} J_{h,p}'^{III} I_{mp}'^{II,III} I_{pr}'^{III,II}}{J_{h,p}^{III}h}+\delta_{mr}b_r^{h,II}\frac{-\tau_r^{\mu,II}H_r'^{II} h_2}{2}\right)$$

$$=\sum_{p=0}^{M3}\sum_{m=0}^{M1}a_m^{e,I}\frac{2\tau_p^{\beta,III}J_m^I I_{mp}^{I,III} I_{pr}'^{III,II}}{h}$$

$$+\sum_{p=0}^{M3}\sum_{m=0}^{M2}\left(a_m^{e,II}\frac{2\tau_p^{\beta,III}J_m^{II} I_{mp}^{II,III} I_{pr}'^{III,II}}{h}+\delta_{mr}a_r^{e,II}\frac{-\tau_r^{\beta,II}J_r^{II} h_2}{2}\right)$$

$$+\sum_{p=0}^{M3}\sum_{m=0}^{M1}a_m^{h,I}\frac{-2\tau_p^{\mu,III}J_m^I J_{h,p}'^{III} I_{mp}'^{I,III} I_{pr}'^{III,II}}{J_{h,p}^{III}h}$$

$$+\sum_{p=0}^{M3}\sum_{m=0}^{M2}\left(a_m^{h,II}\frac{-2\tau_p^{\mu,III}J_m^{II} J_{h,p}'^{III} I_{mp}'^{II,III} I_{pr}'^{III,II}}{J_{h,p}^{III}h}+\delta_{mr}a_r^{h,II}\frac{\tau_r^{\mu,II}J_r'^{II} h_2}{2}\right), \tag{3.165}$$

$$\sum_{p=0}^{M3}\sum_{m=0}^{M1}\left(b_m^{e,I}\frac{-2\tau_p^{\varepsilon,III}H_m^I J_{e,p}'^{III} I_{mp}^{I,III} I_{pq}^{III,I}}{J_{e,p}^{III}h}+\delta_{mq}b_q^{c,I}\frac{\tau_q^{\varepsilon,I}H_q'^I h_1}{2}\right)$$

$$+\sum_{p=0}^{M3}\sum_{m=0}^{M2}b_m^{e,II}\frac{-2\tau_p^{\varepsilon,III}H_m^{II}J_{e,p}'^{III}I_{mp}^{II,III}I_{pq}^{III,I}}{J_{e,p}^{III}h}$$

$$+\sum_{p=0}^{M3}\sum_{m=0}^{M1}\left(b_m^{h,I}\frac{2\tau_p^{\beta,III}H_m^I I_{mp}'^{I,III}I_{pq}^{III,I}}{h}+\delta_{mq}b_q^{h,I}\frac{-\tau_q^{\beta,I}H_q^I h_1}{2}\right)$$

$$+\sum_{p=0}^{M3}\sum_{m=0}^{M2}b_m^{h,II}\frac{2\tau_p^{\beta,III}H_m^{II}I_{mp}'^{II,III}I_{pq}^{III,I}}{h}$$

$$=\sum_{p=0}^{M3}\sum_{m=0}^{M1}\left(a_m^{e,I}\frac{2\tau_p^{\varepsilon,III}J_m^I J_{e,p}'^{III}I_{mp}^{I,III}I_{pq}^{III,I}}{J_{e,p}^{III}h}+\delta_{mq}a_q^{e,I}\frac{-\tau_q^{\varepsilon,I}J_q'^I h_1}{2}\right)$$

$$+\sum_{p=0}^{M3}\sum_{m=0}^{M2}a_m^{e,II}\frac{2\tau_p^{\varepsilon,III}J_m^{II}J_{e,p}'^{III}I_{mp}^{II,III}I_{pq}^{III,I}}{J_{e,p}^{III}h}$$

$$+\sum_{p=0}^{M3}\sum_{m=0}^{M1}\left(a_m^{h,I}\frac{-2\tau_p^{\beta,III}J_m^I I_{mp}'^{I,III}I_{pq}^{III,I}}{h}+\delta_{mq}a_q^{h,I}\frac{\tau_q^{\beta,I}J_q^I h_1}{2}\right)$$

$$+\sum_{p=0}^{M3}\sum_{m=0}^{M2}a_m^{h,II}\frac{-2\tau_p^{\beta,III}J_m^{II}I_{mp}'^{II,III}I_{pq}^{III,I}}{h},\tag{3.166}$$

$$\sum_{p=0}^{M3}\sum_{m=0}^{M1}b_m^{e,I}\frac{-2\tau_p^{\varepsilon,III}H_m^I J_{e,p}'^{III}I_{mp}^{I,III}I_{pr}^{III,II}}{J_{e,p}^{III}h}$$

$$+\sum_{p=0}^{M3}\sum_{m=0}^{M2}\left(b_m^{e,II}\frac{-2\tau_p^{\varepsilon,III}H_m^{II}J_{e,p}'^{III}I_{mp}^{II,III}I_{pr}^{III,II}}{J_{e,p}^{III}h}+\delta_{mr}b_r^{e,II}\frac{\tau_r^{\varepsilon,II}H_r'^{II}h_2}{2}\right)$$

$$+\sum_{p=0}^{M3}\sum_{m=0}^{M1}b_m^{h,I}\frac{2\tau_p^{\beta,III}H_m^I I_{mp}'^{I,III}I_{pr}^{III,II}}{h}$$

$$+\sum_{p=0}^{M3}\sum_{m=0}^{M2}\left(b_m^{h,II}\frac{2\tau_p^{\beta,III}H_m^{II}I_{mp}'^{II,III}I_{pr}^{III,II}}{h}+\delta_{mr}b_r^{h,II}\frac{-\tau_r^{\beta,II}H_r^{II}h_2}{2}\right)$$

$$=\sum_{p=0}^{M3}\sum_{m=0}^{M1}a_m^{e,I}\frac{2\tau_p^{\varepsilon,III}J_m^I J_{e,p}'^{III}I_{mp}^{I,III}I_{pr}^{III,II}}{J_{e,p}^{III}h}$$

$$+\sum_{p=0}^{M3}\sum_{m=0}^{M2}\left(a_m^{e,II}\frac{2\tau_p^{\varepsilon,III}J_m^{II}J_{e,p}'^{III}I_{mp}^{II,III}I_{pr}^{III,II}}{J_{e,p}^{III}h}+\delta_{mr}a_r^{e,II}\frac{-\tau_r^{\varepsilon,II}J_r'^{II}h_2}{2}\right)$$

$$+\sum_{p=0}^{M3}\sum_{m=0}^{M1}a_m^{h,I}\frac{-2\tau_p^{\beta,III}J_m^I I_{mp}'^{I,III}I_{pr}^{III,II}}{h}$$

$$+\sum_{p=0}^{M3}\sum_{m=0}^{M2}\left(a_m^{h,II}\frac{-2\tau_p^{\beta,III}J_m^{II}I_{mp}'^{II,III}I_{pr}^{III,II}}{h}+\delta_{mr}a_r^{h,II}\frac{\tau_r^{\beta,II}J_r^{II}h_2}{2}\right).\tag{3.167}$$

3.5.3 Generalized *T*-Matrix for Two-Layer Problem

As we have discussed the modal expansion and boundary conditions, and the mode matching in the previous two sections, we are ready to formulate the following generalized *T*-matrix (cf. Eq. 3.135):

$$
\begin{bmatrix} b^{e,I} \\ b^{e,II} \\ b^{h,I} \\ b^{h,II} \end{bmatrix} =
\begin{bmatrix}
T_{I,I}^{ee} & T_{I,II}^{ee} & T_{I,I}^{eh} & T_{I,II}^{eh} \\
T_{II,I}^{ee} & T_{II,II}^{ee} & T_{II,I}^{eh} & T_{II,II}^{eh} \\
T_{I,I}^{he} & T_{I,II}^{he} & T_{I,I}^{hh} & T_{I,II}^{hh} \\
T_{II,I}^{he} & T_{II,II}^{he} & T_{II,I}^{hh} & T_{II,II}^{hh}
\end{bmatrix}
\begin{bmatrix} a^{e,I} \\ a^{e,II} \\ a^{h,I} \\ a^{h,II} \end{bmatrix}.
\tag{3.168}
$$

We will derive the elements of the generalized *T*-matrix in Equation (3.168) one column by another. For such case, we first let $a^{e,I} \neq 0$ and $a^{e,II} = a^{h,I} = a^{h,II} = 0$ in Equations (3.164–3.167). Then we can project Equations (3.164–3.167) into a linear system of equation:

$$
[A]\{b\} = \left[P^{el}\right]^{C1}\{a^{e,I}\},
\tag{3.169}
$$

or

$$
\begin{bmatrix}
A_{M1\times M1}^{11} & A_{M1\times M2}^{12} & A_{M1\times M1}^{13} & A_{M1\times M2}^{14} \\
A_{M2\times M1}^{21} & A_{M2\times M2}^{22} & A_{M2\times M1}^{23} & A_{M2\times M2}^{24} \\
A_{M1\times M1}^{31} & A_{M1\times M2}^{32} & A_{M1\times M1}^{33} & A_{M1\times M2}^{34} \\
A_{M2\times M1}^{41} & A_{M2\times M2}^{42} & A_{M2\times M1}^{43} & A_{M2\times M2}^{44}
\end{bmatrix}_{N\times N}
\begin{Bmatrix} b^{e,I} \\ b^{e,II} \\ b^{h,I} \\ b^{h,II} \end{Bmatrix}_{N}
$$

$$
=
\begin{bmatrix}
P_{M1\times M1}^{el,1} \\
P_{M2\times M1}^{el,2} \\
P_{M1\times M1}^{el,3} \\
P_{M2\times M1}^{el,4}
\end{bmatrix}_{N\times M1}
\{a^{e,I}\}_{M1},
\tag{3.170}
$$

where $N = 2(M1 + M2)$. The superscript *C1* in Equation (3.169) indicates that the entries in the vector *b* are corresponding to the entries in the first column of the *T*-matrix in Equation (3.168), and

$$
\begin{aligned}
\{b^{e(h),I}\} &= \{b^{e(h),I}\}^{T}, \quad m = 1, \cdots M1 \\
\{b^{e(h),II}\} &= \{b^{e(h),II}\}^{T}, \quad m = 1, \cdots M2.
\end{aligned}
\tag{3.171}
$$

The entries in Row 1 of the matrices $[A]$ and $[P]$ in Equation (3.170) are given as

$$A_{qm}^{11} = \sum_{p=0}^{M3} \left(\frac{-2\tau_p^{\beta,III} H_m^I I_{mp}^{I,III} I_{pq}'^{III,I}}{h} + \delta_{mq} \frac{\tau_q^{\beta,I} H_q^I h_1}{2} \right), \quad q(m) = 1, \cdots M1,$$

(3.172)

$$A_{qm}^{12} = \sum_{p=0}^{M3} \frac{-2\tau_p^{\beta,III} H_m^{II} I_{mp}^{II,III} I_{pq}'^{III,I}}{h}, \quad q(m) = 1, \cdots M1(M2),$$

(3.173)

$$A_{qm}^{13} = \sum_{p=0}^{M3} \left(\frac{2\tau_p^{\mu,III} H_m^I J_{h,p}'^{III} I_{mp}'^{I,III} I_{pq}'^{III,I}}{J_{h,p}^{III} h} + \delta_{mq} \frac{-\tau_q^{\mu,I} H_q'^I h_1}{2} \right), \quad q(m) = 1, \cdots M1,$$

(3.174)

$$A_{qm}^{14} = \sum_{p=0}^{M3} \frac{2\tau_p^{\mu,III} H_m^{II} J_{h,p}'^{III} I_{mp}'^{II,III} I_{pq}'^{III,I}}{J_{h,p}^{III} h}, \quad q(m) = 1, \cdots M1(M2),$$

(3.175)

$$P_{qm}^{eI,1} = \sum_{p=0}^{M3} \left(\frac{2\tau_p^{\beta,III} J_m^I I_{mp}^{I,III} I_{pq}'^{III,I}}{h} + \delta_{mq} \frac{-\tau_q^{\beta,I} J_q^I h_1}{2} \right), \quad q(m) = 1, \cdots M1.$$

(3.176)

Those entries in Row 2 of the matrices $[A]$ and $[P]$ are given as follows:

$$A_{rm}^{21} = \sum_{p=0}^{M3} \frac{-2\tau_p^{\beta,III} H_m^I I_{mp}^{I,III} I_{pr}'^{III,II}}{h}, \quad r(m) = 1, \cdots M2(M1),$$

(3.177)

$$A_{rm}^{22} = \sum_{p=0}^{M3} \left(\frac{-2\tau_p^{\beta,III} H_m^{II} I_{mp}^{II,III} I_{pr}'^{III,II}}{h} + \delta_{mr} \frac{\tau_r^{\beta,II} H_r^{II} h_2}{2} \right),$$

(3.178)

$$r(m) = 1, \cdots M2(M2),$$

$$A_{rm}^{23} = \sum_{p=0}^{M3} \frac{2\tau_p^{\mu,III} H_m^I J_{h,p}'^{III} I_{mp}'^{I,III} I_{pr}'^{III,II}}{J_{h,p}^{III} h}, \quad r(m) = 1, \cdots M2(M1),$$

(3.179)

$$A_{rm}^{24} = \sum_{p=0}^{M3} \left(\frac{2\tau_p^{\mu,III} H_m^{II} J_{h,p}'^{III} I_{mp}'^{II,III} I_{pr}'^{III,II}}{J_{h,p}^{III} h} + \delta_{mr} \frac{-\tau_r^{\mu,II} H_r'^{II} h_2}{2} \right),$$

(3.180)

$$r(m) = 1, \cdots M2(M2),$$

$$P_{rm}^{el,2} = \sum_{p=0}^{M3} \frac{2\tau_p^{\beta,III} J_m^I I_{mp}^{I,III} I_{pr}^{\prime III,II}}{h}, \quad r(m) = 1, \cdots M2(M1). \qquad (3.181)$$

Those entries in Row 3 of the matrices $[A]$ and $[P]$ are given as follows:

$$A_{qm}^{31} = \sum_{p=0}^{M3} \left(\frac{-2\tau_p^{\varepsilon,III} H_m^I J_{e,p}^{\prime III} I_{mp}^{I,III} I_{pq}^{III,I}}{J_{e,p}^{III} h} + \delta_{mq} \frac{\tau_q^{\varepsilon,I} H_q^{\prime I} h_1}{2} \right), \quad q(m) = 1, \cdots M1,$$

$$(3.182)$$

$$A_{qm}^{32} = \sum_{p=0}^{M3} \frac{-2\tau_p^{\varepsilon,III} H_m^{II} J_{e,p}^{\prime III} I_{mp}^{II,III} I_{pq}^{III,I}}{J_{e,p}^{III} h}, \quad q(m) = 1, \cdots M1(M2), \qquad (3.183)$$

$$A_{qm}^{33} = \sum_{p=0}^{M3} \left(\frac{2\tau_p^{\beta,III} H_m^I I_{mp}^{\prime I,III} I_{pq}^{III,I}}{h} + \delta_{mq} \frac{-\tau_q^{\beta,I} H_q^I h_1}{2} \right), \quad q(m) = 1, \cdots M1,$$

$$(3.184)$$

$$A_{qm}^{34} = \sum_{p=0}^{M3} \frac{2\tau_p^{\beta,III} H_m^{II} I_{mp}^{\prime II,III} I_{pq}^{III,I}}{h}, \quad q(m) = 1, \cdots M1(M2), \qquad (3.185)$$

$$P_{qm}^{el,3} = \sum_{p=0}^{M3} \left(\frac{2\tau_p^{\varepsilon,III} J_m^I J_{e,p}^{\prime III} I_{mp}^{I,III} I_{pq}^{III,I}}{J_{e,p}^{III} h} + \delta_{mq} \frac{-\tau_q^{\varepsilon,I} J_q^{\prime I} h_1}{2} \right), \quad q(m) = 1, \cdots M1.$$

$$(3.186)$$

Those entries in Row 4 of the matrices $[A]$ and $[P]$ are given as follows:

$$A_{rm}^{41} = \sum_{p=0}^{M3} \frac{-2\tau_p^{\varepsilon,III} H_m^I J_{e,p}^{\prime III} I_{mp}^{I,III} I_{pr}^{III,II}}{J_{e,p}^{III} h}, \quad r(m) = 1, \cdots M2(M1), \qquad (3.187)$$

$$A_{rm}^{42} = \sum_{p=0}^{M3} \left(\frac{-2\tau_p^{\varepsilon,III} H_m^{II} J_{e,p}^{\prime III} I_{mp}^{II,III} I_{pr}^{III,II}}{J_{e,p}^{III} h} + \delta_{mr} \frac{\tau_r^{\varepsilon,II} H_r^{\prime II} h_2}{2} \right),$$

$$r(m) = 1, \cdots M2(M2), \qquad (3.188)$$

$$A_{rm}^{43} = \sum_{p=0}^{M3} \frac{2\tau_p^{\beta,III} H_m^I I_{mp}^{\prime I,III} I_{pr}^{III,II}}{h}, \quad r(m) = 1, \cdots M2(M1), \qquad (3.189)$$

$$A_{rm}^{44} = \sum_{p=0}^{M3} \left(\frac{2\tau_p^{\beta,III} H_m^{II} I_{mp}^{\prime II,III} I_{pr}^{III,II}}{h} + \delta_{mr} \frac{-\tau_r^{\beta,II} H_r^{II} h_2}{2} \right),$$

(3.190)

$$r(m) = 1, \cdots M2(M2),$$

$$P_{rm}^{el,4} = \sum_{p=0}^{M3} \frac{2\tau_p^{\varepsilon,III} J_m^{I} J_{e,p}^{\prime III} I_{mp}^{I,III} I_{pr}^{III,II}}{J_{e,p}^{III} h}, \quad r(m) = 1, \cdots M2(M1).$$

(3.191)

We can now obtain from Equation (3.169):

$$\{b\} = [A]^{-1} [P^{el}]\{a^{e,I}\},$$

(3.192)

where $[A]^{-1}[P^{el}]$ correspond to the first column of the **T**-matrix in Equation (3.168).

The derivation for the other three columns of the **T**-matrix in Equation (3.168) follows exactly the same procedure as that for Column 1 of the **T**-matrix. We can easily notice that the matrix $[A]$ for deriving all the other three columns of the **T**-matrix in Equation (3.168) is identical to the one in Equation (3.192) for deriving the first column of the **T**-matrix in Equation (3.168). The only difference is the matrix $[P]$, so we only present here the entries of the matrix $[P]$.

The Column 2 of the **T**-matrix in Equation (3.168) can be derived by setting $a^{e,II} \neq 0$ and $a^{e,I} = a^{h,I} = a^{h,II} = 0$ in Equations (3.164–3.167). The expression is as follows:

$$\{b\} = [A]^{-1} [P^{ell}]\{a^{e,II}\},$$

(3.193)

where $[A]^{-1}[P^{ell}]$ correspond to the second column of the **T**-matrix in Equation (3.168), and

$$[P^{ell}] = \begin{bmatrix} P_{M1\times M2}^{ell,1} \\ P_{M2\times M2}^{ell,2} \\ P_{M1\times M2}^{ell,3} \\ P_{M2\times M2}^{ell,4} \end{bmatrix}_{N\times M2} \{a^{e,II}\}_{M2},$$

(3.194)

where

$$P_{qm}^{ell,1} = \sum_{p=0}^{M3} \frac{2\tau_p^{\beta,III} J_m^{II} I_{mp}^{II,III} I_{pq}^{\prime III,I}}{h}, \quad q(m) = 1, \cdots M1(M2),$$

(3.195)

$$P_{rm}^{eII,2} = \sum_{p=0}^{M3} \left(\frac{2\tau_p^{\beta,III} J_m^{II} I_{mp}^{II,III} I_{pr}'^{III,II}}{h} + \delta_{mr} \frac{-\tau_q^{\beta,II} J_q^{II} h_2}{2} \right), \quad r(m) = 1, \cdots M2,$$

(3.196)

$$P_{qm}^{eII,3} = \sum_{p=0}^{M3} \frac{2\tau_p^{\varepsilon,III} J_m^{II} J_{e,p}'^{III} I_{mp}^{II,III} I_{pq}^{III,I}}{J_{e,p}^{III} h}, \quad q(m) = 1, \cdots M1(M2),$$

(3.197)

$$P_{rm}^{eII,4} = \sum_{p=0}^{M3} \left(\frac{2\tau_p^{\varepsilon,III} J_m^{II} J_{e,p}'^{III} I_{mp}^{II,III} I_{pr}^{III,II}}{J_{e,p}^{III} h} + \delta_{mr} \frac{-\tau_r^{\varepsilon,II} J_r'^{II} h_2}{2} \right),$$

(3.198)

$$r(m) = 1, \cdots M2.$$

Similarly, the Column 3 of the **T**-matrix in Equation (3.168) is derived by setting $a^{h,I} \neq 0$ and $a^{e,I} = a^{e,II} = a^{h,II} = 0$ in Equations (3.164–3.167). The formulae relevant to the third column of the **T**-matrix in Equation (3.168) are:

$$\{b\} = [A]^{-1} [P^{hI}] \{a^{h,I}\},$$

(3.199)

where $[A]^{-1}[P^{hI}]$ correspond to the third column of the **T**-matrix in Equation (3.168), and

$$[P^{hI}] = \begin{bmatrix} P_{M1 \times M1}^{hI,1} \\ P_{M2 \times M1}^{hI,2} \\ P_{M1 \times M1}^{hI,3} \\ P_{M2 \times M1}^{hI,4} \end{bmatrix}_{N \times M1} \{a^{h,I}\}_{M1},$$

(3.200)

where

$$P_{qm}^{hI,1} = \sum_{p=0}^{M3} \left(\frac{-2\tau_p^{\mu,III} J_m^I J_{h,p}'^{III} I_{mp}'^{I,III} I_{pq}'^{III,I}}{J_{h,p}^{III} h} + \delta_{mq} \frac{\tau_q^{\mu,I} J_q'^I h_1}{2} \right), \quad q(m) = 1, \cdots M1,$$

(3.201)

$$P_{rm}^{hI,2} = \sum_{p=0}^{M3} \frac{-2\tau_p^{\mu,III} J_m^I J_{h,p}'^{III} I_{mp}'^{I,III} I_{pr}'^{III,II}}{J_{h,p}^{III} h}, \quad r(m) = 1, \cdots M2(M1),$$

(3.202)

$$P_{qm}^{hI,3} = \sum_{p=0}^{M3} \left(\frac{-2\tau_p^{\beta,III} J_m^I I_{mp}^{\prime I,III} I_{pq}^{III,I}}{h} + \delta_{mq} \frac{\tau_q^{\beta,I} J_q^I h_1}{2} \right), \quad q(m) = 1, \cdots M1,$$

$$(3.203)$$

$$P_{rm}^{hI,4} = \sum_{p=0}^{M3} \frac{-2\tau_p^{\beta,III} J_m^I I_{mp}^{\prime I,III} I_{pr}^{III,II}}{h}, \quad r(m) = 1, \cdots M2(M1).$$

$$(3.204)$$

The Column 4 of the *T*-matrix in Equation (3.168) is derived by setting $a^{h,II} \neq 0$ and $a^{e,I} = a^{e,II} = a^{h,I} = 0$ in Equations (3.164–3.167). The formulae relevant to the fourth column of the *T*-matrix in Equation (3.168) are:

$$\{b\} = [A]^{-1} [P^{hII}] \{a^{h,II}\},$$

$$(3.205)$$

where $[A]^{-1}[P^{hII}]$ correspond to the fourth column of the *T*-matrix in Equation (3.168), and

$$[P^{hII}] = \begin{bmatrix} P_{M1 \times M2}^{hII,1} \\ P_{M2 \times M2}^{hII,2} \\ P_{M1 \times M2}^{hII,3} \\ P_{M2 \times M2}^{hII,4} \end{bmatrix}_{N \times M2} \{a^{h,II}\}_{M2},$$

$$(3.206)$$

where

$$P_{qm}^{hII,1} = \sum_{p=0}^{M3} \frac{-2\tau_p^{\mu,III} J_m^{II} J_{h,p}^{\prime III} I_{mp}^{\prime II,III} I_{pq}^{\prime III,I}}{J_{h,p}^{III} h}, \quad q(m) = 1, \cdots M1(M2),$$

$$(3.207)$$

$$P_{rm}^{hII,2} = \sum_{p=0}^{M3} \left(\frac{-2\tau_p^{\mu,III} J_m^{II} J_{h,p}^{\prime III} I_{mp}^{\prime II,III} I_{pr}^{\prime III,II}}{J_{h,p}^{III} h} + \delta_{mr} \frac{\tau_r^{\mu,II} J_r^{II} h_2}{2} \right),$$

$$(3.208)$$

$$r(m) = 1, \cdots M2,$$

$$P_{qm}^{hII,3} = \sum_{p=0}^{M3} \frac{-2\tau_p^{\beta,III} J_m^{II} I_{mp}^{\prime II,III} I_{pq}^{III,I}}{h}, \quad q(m) = 1, \cdots M1(M2),$$

$$(3.209)$$

$$P_{rm}^{hII,4} = \sum_{p=0}^{M3} \left(\frac{-2\tau_p^{\beta,III} J_m^{II} I_{mp}^{\prime II,III} I_{pr}^{III,II}}{h} + \delta_{mr} \frac{\tau_r^{\beta,II} J_r^{II} h_2}{2} \right), \quad r(m) = 1, \cdots M2.$$

$$(3.210)$$

3.5.4 Formulae Summary for Two-Layer Problem

Generalized T-matrix:

$$
\begin{bmatrix} b^{e,I} \\ b^{e,II} \\ b^{h,I} \\ b^{h,II} \end{bmatrix} =
\begin{bmatrix}
T_{I,I}^{ee} & T_{I,II}^{ee} & T_{I,I}^{eh} & T_{I,II}^{eh} \\
T_{II,I}^{ee} & T_{II,II}^{ee} & T_{II,I}^{eh} & T_{II,II}^{eh} \\
T_{I,I}^{he} & T_{I,II}^{he} & T_{I,I}^{hh} & T_{I,II}^{hh} \\
T_{II,I}^{he} & T_{II,II}^{he} & T_{II,I}^{hh} & T_{II,II}^{hh}
\end{bmatrix}
\begin{bmatrix} a^{e,I} \\ a^{e,II} \\ a^{h,I} \\ a^{h,II} \end{bmatrix}
$$

$$
= \begin{bmatrix} T^{C1} & T^{C2} & T^{C3} & T^{C4} \end{bmatrix}
\begin{bmatrix} a^{e,I} \\ a^{e,II} \\ a^{h,I} \\ a^{h,II} \end{bmatrix}, \tag{3.211}
$$

Column 1 of T-Matrix:

$$
\begin{bmatrix} T^{C1} \end{bmatrix} = \begin{bmatrix} A \end{bmatrix}^{-1} \begin{bmatrix} P^{el} \end{bmatrix}, \tag{3.212}
$$

where

$$
[A] =
\begin{bmatrix}
A_{M1\times M1}^{11} & A_{M1\times M2}^{12} & A_{M1\times M1}^{13} & A_{M1\times M2}^{14} \\
A_{M2\times M1}^{21} & A_{M2\times M2}^{22} & A_{M2\times M1}^{23} & A_{M2\times M2}^{24} \\
A_{M1\times M1}^{31} & A_{M1\times M2}^{32} & A_{M1\times M1}^{33} & A_{M1\times M2}^{34} \\
A_{M2\times M1}^{41} & A_{M2\times M2}^{42} & A_{M2\times M1}^{43} & A_{M2\times M2}^{44}
\end{bmatrix}_{N\times N}. \tag{3.213}
$$

Row 1 of Matrix $[A]$:

$$
A_{qm}^{11} = \sum_{p=0}^{M3} \left(\frac{-2\tau_p^{\beta,III} H_m^I I_{mp}^{I,III} I_{pq}^{\prime III,I}}{h} + \delta_{mq} \frac{\tau_q^{\beta,I} H_q^I h_1}{2} \right), \quad q(m) = 1, \cdots M1, \tag{3.214}
$$

$$
A_{qm}^{12} = \sum_{p=0}^{M3} \frac{-2\tau_p^{\beta,III} H_m^{II} I_{mp}^{II,III} I_{pq}^{\prime III,I}}{h}, \quad q(m) = 1, \cdots M1(M2), \tag{3.215}
$$

$$
A_{qm}^{13} = \sum_{p=0}^{M3} \left(\frac{2\tau_p^{\mu,III} H_m^I J_{h,p}^{\prime III} I_{mp}^{\prime I,III} I_{pq}^{\prime III,I}}{J_{h,p}^{III} h} + \delta_{mq} \frac{-\tau_q^{\mu,I} H_q^{\prime I} h_1}{2} \right), \quad q(m) = 1, \cdots M1, \tag{3.216}
$$

$$
A_{qm}^{14} = \sum_{p=0}^{M3} \frac{2\tau_p^{\mu,III} H_m^{II} J_{h,p}^{\prime III} I_{mp}^{\prime II,III} I_{pq}^{\prime III,I}}{J_{h,p}^{III} h}, \quad q(m) = 1, \cdots M1(M2). \tag{3.217}
$$

Row 2 of Matrix $[A]$:

$$A_{rm}^{21} = \sum_{p=0}^{M3} \frac{-2\tau_p^{\beta,III} H_m^I I_{mp}^{I,III} I_{pr}'^{III,II}}{h}, \quad r(m) = 1, \cdots M2(M1), \tag{3.218}$$

$$A_{rm}^{22} = \sum_{p=0}^{M3} \left(\frac{-2\tau_p^{\beta,III} H_m^{II} I_{mp}^{II,III} I_{pr}'^{III,II}}{h} + \delta_{mr} \frac{\tau_r^{\beta,II} H_r^{II} h_2}{2} \right), \tag{3.219}$$

$$r(m) = 1, \cdots M2(M2),$$

$$A_{rm}^{23} = \sum_{p=0}^{M3} \frac{2\tau_p^{\mu,III} H_m^I J_{h,p}'^{III} I_{mp}'^{I,III} I_{pr}'^{III,II}}{J_{h,p}^{III} h}, \quad r(m) = 1, \cdots M2(M1), \tag{3.220}$$

$$A_{rm}^{24} = \sum_{p=0}^{M3} \left(\frac{2\tau_p^{\mu,III} H_m^{II} J_{h,p}'^{III} I_{mp}'^{II,III} I_{pr}'^{III,II}}{J_{h,p}^{III} h} + \delta_{mr} \frac{-\tau_r^{\mu,II} H_r'^{II} h_2}{2} \right), \tag{3.221}$$

$$r(m) = 1, \cdots M2(M2).$$

Row 3 of Matrix $[A]$:

$$A_{qm}^{31} = \sum_{p=0}^{M3} \left(\frac{-2\tau_p^{\varepsilon,III} H_m^I J_{e,p}'^{III} I_{mp}^{I,III} I_{pq}^{III,I}}{J_{e,p}^{III} h} + \delta_{mq} \frac{\tau_q^{\varepsilon,I} H_q'^I h_1}{2} \right), \quad q(m) = 1, \cdots M1, \tag{3.222}$$

$$A_{qm}^{32} = \sum_{p=0}^{M3} \frac{-2\tau_p^{\varepsilon,III} H_m^{II} J_{e,p}'^{III} I_{mp}^{II,III} I_{pq}^{III,I}}{J_{e,p}^{III} h}, \quad q(m) = 1, \cdots M1(M2), \tag{3.223}$$

$$A_{qm}^{33} = \sum_{p=0}^{M3} \left(\frac{2\tau_p^{\beta,III} H_m^I I_{mp}'^{I,III} I_{pq}^{III,I}}{h} + \delta_{mq} \frac{-\tau_q^{\beta,I} H_q^I h_1}{2} \right), \quad q(m) = 1, \cdots M1, \tag{3.224}$$

$$A_{qm}^{34} = \sum_{p=0}^{M3} \frac{2\tau_p^{\beta,III} H_m^{II} I_{mp}'^{II,III} I_{pq}^{III,I}}{h}, \quad q(m) = 1, \cdots M1(M2). \tag{3.225}$$

Row 4 of Matrix $[A]$:

$$A_{rm}^{41} = \sum_{p=0}^{M3} \frac{-2\tau_p^{\varepsilon,III} H_m^I J_{e,p}'^{III} I_{mp}^{I,III} I_{pr}^{III,II}}{J_{e,p}^{III} h}, \quad r(m) = 1, \cdots M2(M1), \tag{3.226}$$

$$A_{rm}^{42} = \sum_{p=0}^{M3} \left(\frac{-2\tau_p^{\varepsilon,III} H_m^{II} J_{e,p}^{\prime III} I_{mp}^{II,III} I_{pr}^{III,II}}{J_{e,p}^{III} h} + \delta_{mr} \frac{\tau_r^{\varepsilon,II} H_r^{\prime II} h_2}{2} \right), \qquad (3.227)$$
$$r(m) = 1, \cdots M2(M2),$$

$$A_{rm}^{43} = \sum_{p=0}^{M3} \frac{2\tau_p^{\beta,III} H_m^I I_{mp}^{\prime I,III} I_{pr}^{III,II}}{h}, \quad r(m) = 1, \cdots M2(M1), \qquad (3.228)$$

$$A_{rm}^{44} = \sum_{p=0}^{M3} \left(\frac{2\tau_p^{\beta,III} H_m^{II} I_{mp}^{\prime II,III} I_{pr}^{III,II}}{h} + \delta_{mr} \frac{-\tau_r^{\beta,II} H_r^{II} h_2}{2} \right), \qquad (3.229)$$
$$r(m) = 1, \cdots M2(M2).$$

Matrix [P]:

$$\left[P^{el} \right]_{N \times M1} = \left[P_{M1 \times M1}^{el,1} \quad P_{M2 \times M1}^{el,2} \quad P_{M1 \times M1}^{el,3} \quad P_{M2 \times M1}^{el,4} \right]^T, \qquad (3.230)$$

$$P_{qm}^{el,1} = \sum_{p=0}^{M3} \left(\frac{2\tau_p^{\beta,III} J_m^I I_{mp}^{I,III} I_{pq}^{\prime III,I}}{h} + \delta_{mq} \frac{-\tau_q^{\beta,I} J_q^I h_1}{2} \right), \quad q(m) = 1, \cdots M1, \qquad (3.231)$$

$$P_{rm}^{el,2} = \sum_{p=0}^{M3} \frac{2\tau_p^{\beta,III} J_m^I I_{mp}^{I,III} I_{pr}^{\prime III,II}}{h}, \quad r(m) = 1, \cdots M2(M1), \qquad (3.232)$$

$$P_{qm}^{el,3} = \sum_{p=0}^{M3} \left(\frac{2\tau_p^{\varepsilon,III} J_m^I J_{e,p}^{\prime III} I_{mp}^{I,III} I_{pq}^{III,I}}{J_{e,p}^{III} h} + \delta_{mq} \frac{-\tau_q^{\varepsilon,I} J_q^{\prime I} h_1}{2} \right), \quad q(m) = 1, \cdots M1, \qquad (3.233)$$

$$P_{rm}^{el,4} = \sum_{p=0}^{M3} \frac{2\tau_p^{\varepsilon,III} J_m^I J_{e,p}^{\prime III} I_{mp}^{I,III} I_{pr}^{III,II}}{J_{e,p}^{III} h}, \quad r(m) = 1, \cdots M2(M1). \qquad (3.234)$$

Column 2 of T-Matrix:

$$\left[T^{C2} \right] = \left[A \right]^{-1} \left[P^{ell} \right], \qquad (3.235)$$

where

$$\left[P^{ell} \right]_{N \times M2} = \left[P_{M1 \times M2}^{ell,1} \quad P_{M2 \times M2}^{ell,2} \quad P_{M1 \times M2}^{ell,3} \quad P_{M2 \times M2}^{ell,4} \right]^T, \qquad (3.236)$$

$$P_{qm}^{ell,1} = \sum_{p=0}^{M3} \frac{2\tau_p^{\beta,III} J_m^{II} I_{mp}^{II,III} I_{pq}^{\prime III,I}}{h}, \quad q(m) = 1, \cdots M1(M2), \qquad (3.237)$$

$$P_{rm}^{eII,2} = \sum_{p=0}^{M3} \left(\frac{2\tau_p^{\beta,III} J_m^{II} I_{mp}^{II,III} I_{pr}'^{III,II}}{h} + \delta_{mr} \frac{-\tau_r^{\beta,II} J_r^{II} h_2}{2} \right), \quad r(m) = 1, \cdots M2,$$

(3.238)

$$P_{qm}^{eII,3} = \sum_{p=0}^{M3} \frac{2\tau_p^{\varepsilon,III} J_m^{II} J_{e,p}'^{III} I_{mp}^{II,III} I_{pq}^{III,I}}{J_{e,p}^{III} h}, \quad q(m) = 1, \cdots M1(M2),$$

(3.239)

$$P_{rm}^{eII,4} = \sum_{p=0}^{M3} \left(\frac{2\tau_p^{\varepsilon,III} J_m^{II} J_{e,p}'^{III} I_{mp}^{II,III} I_{pr}^{III,II}}{J_{e,p}^{III} h} + \delta_{mr} \frac{-\tau_r^{\varepsilon,II} J_r'^{II} h_2}{2} \right),$$

(3.240)

$$r(m) = 1, \cdots M2.$$

Column 3 of *T*-Matrix:

$$\left[T^{C3} \right] = \left[A \right]^{-1} \left[P^{hI} \right],$$

(3.241)

$$\left[P^{hI} \right]_{N \times M1} = \left[P_{M1 \times M1}^{hI,1} \quad P_{M2 \times M1}^{hI,2} \quad P_{M1 \times M1}^{hI,3} \quad P_{M2 \times M1}^{hI,4} \right]^T,$$

(3.242)

$$P_{qm}^{hI,1} = \sum_{p=0}^{M3} \left(\frac{-2\tau_p^{\mu,III} J_m^{I} J_{h,p}'^{III} I_{mp}'^{I,III} I_{pq}'^{III,I}}{J_{h,p}^{III} h} + \delta_{mq} \frac{\tau_q^{\mu,I} J_q'^{I} h_1}{2} \right), \quad q(m) = 1, \cdots M1,$$

(3.243)

$$P_{rm}^{hI,2} = \sum_{p=0}^{M3} \frac{-2\tau_p^{\mu,III} J_m^{I} J_{h,p}'^{III} I_{mp}'^{I,III} I_{pr}'^{III,II}}{J_{h,p}^{III} h}, \quad r(m) = 1, \cdots M2(M1),$$

(3.244)

$$P_{qm}^{hI,3} = \sum_{p=0}^{M3} \left(\frac{-2\tau_p^{\beta,III} J_m^{I} I_{mp}'^{I,III} I_{pq}^{III,I}}{h} + \delta_{mq} \frac{\tau_q^{\beta,I} J_q^{I} h_1}{2} \right), \quad q(m) = 1, \cdots M1,$$

(3.245)

$$P_{rm}^{hI,4} = \sum_{p=0}^{M3} \frac{-2\tau_p^{\beta,III} J_m^{I} I_{mp}'^{I,III} I_{pr}^{III,II}}{h}, \quad r(m) = 1, \cdots M2(M1).$$

(3.246)

Column 4 of *T*-Matrix:

$$\left[T^{C4} \right] = \left[A \right]^{-1} \left[P^{hII} \right],$$

(3.247)

$$\left[P^{hII} \right]_{N \times M2} = \left[P_{M1 \times M2}^{hII,1} \quad P_{M2 \times M2}^{hII,2} \quad P_{M1 \times M2}^{hII,3} \quad P_{M2 \times M2}^{hII,4} \right]^T,$$

(3.248)

$$P_{qm}^{hII,1} = \sum_{p=0}^{M3} \frac{-2\tau_p^{\mu,III} J_m^{II} J_{h,p}^{\prime III} I_{mp}^{\prime II,III} I_{pq}^{\prime III,I}}{J_{h,p}^{III} h}, \quad q(m) = 1, \cdots M1(M2), \quad (3.249)$$

$$P_{rm}^{hII,2} = \sum_{p=0}^{M3} \left(\frac{-2\tau_p^{\mu,III} J_m^{II} J_{h,p}^{\prime III} I_{mp}^{\prime II,III} I_{pr}^{\prime III,II}}{J_{h,p}^{III} h} + \delta_{mr} \frac{\tau_r^{\mu,II} J_r^{\prime II} h_2}{2} \right), \quad (3.250)$$

$$r(m) = 1, \cdots M2,$$

$$P_{qm}^{hII,3} = \sum_{p=0}^{M3} \frac{-2\tau_p^{\beta,III} J_m^{II} I_{mp}^{\prime II,III} I_{pq}^{III,I}}{h}, \quad q(m) = 1, \cdots M1(M2), \quad (3.251)$$

$$P_{rm}^{hII,4} = \sum_{p=0}^{M3} \left(\frac{-2\tau_p^{\beta,III} J_m^{II} I_{mp}^{\prime II,III} I_{pr}^{III,II}}{h} + \delta_{mr} \frac{\tau_r^{\beta,II} J_r^{II} h_2}{2} \right), \quad r(m) = 1, \cdots M2.$$

$$(3.252)$$

Summary of all the variables in the above equations:

$$\beta_m^{I(II)(III)} = \frac{m\pi}{h_{1(2)(3)}}, \quad \left(k_m^{I(II)(III)} \right)^2 = k^2 - \left(\beta_m^{I(II)(III)} \right)^2 = \omega^2 \mu \varepsilon - \left(\frac{m\pi}{h_{1(2)(3)}} \right)^2,$$

$$(3.253)$$

$$\tau_m^{\beta,I(II)(III)} = \frac{jn\beta_m^{I(II)(III)}}{\left(k_m^{I(II)(III)} \right)^2 b}, \quad \tau_m^{\mu,I(II)(III)} = \frac{j\omega\mu}{k_m^{I(II)(III)}}, \quad \tau_m^{\varepsilon,I(II)(III)} = \frac{j\omega\varepsilon}{k_m^{I(II)(III)}},$$

$$(3.254)$$

$$J_m^{I(II)} = J_{mn}^{I(II)} = J_n \left(k_m^{I(II)} b \right), \quad H_m^{I(II)} = H_{mn}^{I(II)} = H_n^{(2)} \left(k_m^{I(II)} b \right)$$

$$J_m^{\prime I(II)} = J_{mn}^{\prime I(II)} = \frac{\partial J_n \left(k_m^{I(II)} b \right)}{\partial \left(k_m^{I(II)} b \right)}, \quad H_m^{\prime I(II)} = H_{mn}^{\prime I(II)} = \frac{\partial H_n^{(2)} \left(k_m^{I(II)} b \right)}{\partial \left(k_m^{I(II)} b \right)},$$

$$(3.255)$$

$$J_{e,m}^{III} = \left(J_m^{III} - \frac{J_n \left(k_m^{III} a \right) H_m^{III}}{H_n^{(2)} \left(k_m^{III} a \right)} \right), \quad J_{e,m}^{\prime III} = \left(J_m^{\prime III} - \frac{J_n \left(k_m^{III} a \right) H_m^{\prime III}}{H_n^{(2)} \left(k_m^{III} a \right)} \right), \quad (3.256)$$

$$J_{h,m}^{III} = \left(J_m^{III} - \frac{J_n^{\prime} \left(k_m^{III} a \right) H_m^{III}}{H_n^{\prime(2)} \left(k_m^{III} a \right)} \right), \quad J_{h,m}^{\prime III} = \left(J_m^{\prime III} - \frac{J_n^{\prime} \left(k_m^{III} a \right) H_m^{\prime III}}{H_n^{\prime(2)} \left(k_m^{III} a \right)} \right). \quad (3.257)$$

Formulae to be used for computing the integrals such as $I_{mp}^{I,III}$, $I_{mp}^{II,III}$:

$$I_1 = \int \cos(az)\cos(bz)dz = \begin{cases} \dfrac{\sin[(a-b)z]}{2(a-b)} + \dfrac{\sin[(a+b)z]}{2(a+b)}, & \text{for } |a| \neq |b| \\[3mm] \dfrac{z}{2} + \dfrac{\sin(2az)}{4a}, & \text{for } |a| = |b|, \end{cases}$$

(3.258)

$$I_2 = \int \sin(az)\sin(bz)dz = \begin{cases} \dfrac{\sin[(a-b)z]}{2(a-b)} - \dfrac{\sin[(a+b)z]}{2(a+b)}, & \text{for } |a| \neq |b| \\[3mm] \dfrac{z}{2} - \dfrac{\sin(2az)}{4a}, & \text{for } |a| = |b| \,\&\, ab \geq 0 \\[3mm] -\dfrac{z}{2} + \dfrac{\sin(2az)}{4a}, & \text{for } |a| = |b| \,\&\, ab \leq 0. \end{cases}$$

(3.259)

3.5.5 Formulae Summary for 3D Structure Problem

In this section, we have summarized the formulae of generalized T-matrix for multilayered 3D structure in Figure 3.31. The following

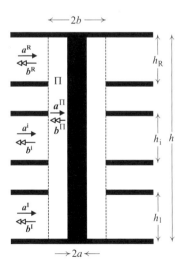

Figure 3.31   A through-hole via in 3D multilayer structure and forming PPWGs [14].

formulae are consolidated for source-free via and source via comprised in multiple layer.

3.5.5.1 For Source-free Via

Generalized T-matrix for a multilayered 3D structure:

$$
\begin{bmatrix} b^{e,I} \\ \vdots \\ b^{e,R} \\ b^{h,I} \\ \vdots \\ b^{h,R} \end{bmatrix} = \begin{bmatrix} T_{I,I}^{ee} & \cdots & T_{I,R}^{ee} & T_{I,I}^{eh} & \cdots & T_{I,R}^{eh} \\ \vdots & \ddots & \vdots & \vdots & \ddots & \vdots \\ T_{R,I}^{ee} & \cdots & T_{R,R}^{ee} & T_{R,I}^{eh} & \cdots & T_{R,R}^{eh} \\ T_{I,I}^{he} & \cdots & T_{I,R}^{he} & T_{I,I}^{hh} & \cdots & T_{I,R}^{hh} \\ \vdots & \ddots & \vdots & \vdots & \ddots & \vdots \\ T_{R,I}^{he} & \cdots & T_{R,R}^{he} & T_{R,I}^{hh} & \cdots & T_{R,R}^{hh} \end{bmatrix} \begin{bmatrix} a^{e,I} \\ \vdots \\ a^{e,R} \\ a^{h,I} \\ \vdots \\ a^{h,R} \end{bmatrix}
$$

$$
= \begin{bmatrix} T^{e,C_1} & \cdots & T^{e,C_R} & T^{h,C_1} & \cdots & T^{h,C_R} \end{bmatrix} \begin{bmatrix} a^{e,I} \\ \vdots \\ a^{e,R} \\ a^{h,I} \\ \vdots \\ a^{h,R} \end{bmatrix}. \quad (3.260)
$$

Columns of T-matrix:

$$
\left[T^{e(h),C_\kappa} \right] = \left[A \right]^{-1} \left[P^{e(h),\kappa} \right], \quad (3.261)
$$

where Matrix $[A]$:

$$
[A] = \begin{bmatrix} A^{ee} & A^{eh} \\ A^{he} & A^{hh} \end{bmatrix} = \begin{bmatrix} A_{M_1 \times M_1}^{ee,I:I} & \cdots & A_{M_1 \times M_R}^{ee,I:R} & A_{M_1 \times M_1}^{eh,I:I} & \cdots & A_{M_1 \times M_R}^{eh,I:R} \\ \vdots & \ddots & \vdots & \vdots & \ddots & \vdots \\ A_{M_R \times M_1}^{ee,R:I} & \cdots & A_{M_R \times M_R}^{ee,R:R} & A_{M_R \times M_1}^{eh,R:I} & \cdots & A_{M_R \times M_R}^{eh,R:R} \\ A_{M_1 \times M_1}^{he,I:I} & \cdots & A_{M_1 \times M_R}^{he,I:R} & A_{M_1 \times M_1}^{hh,I:I} & \cdots & A_{M_1 \times M_R}^{hh,I:R} \\ \vdots & \ddots & \vdots & \vdots & \ddots & \vdots \\ A_{M_R \times M_1}^{he,R:I} & \cdots & A_{M_R \times M_R}^{he,R:R} & A_{M_R \times M_1}^{hh,R:I} & \cdots & A_{M_R \times M_R}^{hh,R:R} \end{bmatrix},
$$

$$
(3.262)
$$

and the size of Matrix $[A]$ is $2\sum_{k=1}^{R}M_k$ by $2\sum_{k=1}^{R}M_k$. The elements are as follows:

$$A_{qm}^{ee,\Upsilon:\Psi} = \sum_{p=0}^{M3}\left(\frac{-2\tau_p^{\beta,\Pi}H_m^{\Psi}I_{mp}^{\Psi,\Pi}I_{pq}'^{\Pi,\Upsilon}}{h} + \delta_{\Upsilon\Psi}\delta_{mq}\frac{\tau_q^{\beta,\Upsilon}H_q^{\Upsilon}h_{\Upsilon}}{2}\right), \qquad (3.263)$$

$$q(m) = 1, \cdots M_{\Upsilon}(M_{\Psi}),$$

$$A_{qm}^{eh,\Upsilon:\Psi} = \sum_{p=0}^{M_{\Pi}}\left(\frac{2\tau_p^{\mu,\Pi}H_m^{\Psi}J_{h,p}'^{\Pi}I_{mp}'^{\Psi,\Pi}I_{pq}'^{\Pi,\Upsilon}}{J_{h,p}^{\Pi}h} + \delta_{\Upsilon\Psi}\delta_{mq}\frac{-\tau_q^{\mu,\Upsilon}H_q'^{\Upsilon}h_{\Upsilon}}{2}\right), \qquad (3.264)$$

$$q(m) = 1, \cdots M_{\Upsilon}(M_{\Psi}),$$

$$A_{qm}^{he,\Upsilon:\Psi} = \sum_{p=0}^{M_{\Pi}}\left(\frac{-2\tau_p^{\varepsilon,\Pi}H_m^{\Psi}J_{e,p}'^{\Pi}I_{mp}^{\Psi,\Pi}I_{pq}^{\Pi,\Upsilon}}{J_{e,p}^{\Pi}h} + \delta_{\Upsilon\Psi}\delta_{mq}\frac{\tau_q^{\varepsilon,\Upsilon}H_q'^{\Upsilon}h_{\Upsilon}}{2}\right), \qquad (3.265)$$

$$q(m) = 1, \cdots M_{\Upsilon}(M_{\Psi}),$$

$$A_{qm}^{hh,\Upsilon:\Psi} = \sum_{p=0}^{M3}\left(\frac{2\tau_p^{\beta,\Pi}H_m^{\Psi}I_{mp}'^{\Psi,\Pi}I_{pq}^{\Pi,\Upsilon}}{h} + \delta_{\Upsilon\Psi}\delta_{mq}\frac{-\tau_q^{\beta,\Upsilon}H_q^{\Upsilon}h_{\Upsilon}}{2}\right), \qquad (3.266)$$

$$q(m) = 1, \cdots M_{\Upsilon}(M_{\Psi}),$$

Matrix $[P]$:

$$\left[P^{e,\kappa}\right] = \begin{bmatrix}P^{ee,\kappa}\\P^{he,\kappa}\end{bmatrix} = \left[P^{ee,\kappa:1}\quad\cdots\quad P^{ee,\kappa:R}\quad P^{he,\kappa:1}\quad\cdots\quad P^{he,\kappa:R}\right]^{T}, \quad (3.267)$$

$$\left[P^{h,\kappa}\right] = \begin{bmatrix}P^{eh,\kappa}\\P^{hh,\kappa}\end{bmatrix} = \left[P^{eh,\kappa:1}\quad\cdots\quad P^{eh,\kappa:R}\quad P^{hh,\kappa:1}\quad\cdots\quad P^{hh,\kappa:R}\right]^{T}, \quad (3.268)$$

where

$$P_{qm}^{ee,\kappa:\Omega} = \sum_{p=0}^{M_{\Pi}}\left(\frac{2\tau_p^{\beta,\Pi}J_m^{\kappa}I_{mp}^{\kappa,\Pi}I_{pq}'^{\Pi,\Omega}}{h} + \delta_{\kappa\Omega}\delta_{mq}\frac{-\tau_q^{\beta,\kappa}J_q^{\kappa}h_{\kappa}}{2}\right), \qquad (3.269)$$

$$q(m) = 1, \cdots M_{\kappa}(M_{\Omega}),$$

$$P_{qm}^{he,\kappa:\Omega} = \sum_{p=0}^{M_{\Pi}}\left(\frac{2\tau_p^{\varepsilon,\Pi}J_m^{\kappa}J_{e,p}'^{\Pi}I_{mp}^{\kappa,\Pi}I_{pq}^{\Pi,\Omega}}{J_{e,p}^{\Pi}h} + \delta_{\kappa\Omega}\delta_{mq}\frac{-\tau_q^{\varepsilon,\kappa}J_q'^{\kappa}h_{\kappa}}{2}\right), \qquad (3.270)$$

$$q(m) = 1, \cdots M_{\kappa}(M_{\Omega}),$$

$$P_{qm}^{eh,\kappa:\Omega} = \sum_{p=0}^{M_\Pi} \left(\frac{-2\tau_p^{\mu,\Pi} J_m^{\kappa} J_{h,p}^{\prime\Pi} I_{mp}^{\prime\kappa,\Pi} I_{pq}^{\prime\Pi,\Omega}}{J_{h,p}^{\Pi} h} + \delta_{\kappa\Omega}\delta_{mq} \frac{\tau_q^{\mu,\kappa} J_q^{\prime\kappa} h_{\kappa}}{2} \right), \quad (3.271)$$

$$q(m) = 1, \cdots M_\kappa (M_\Omega),$$

$$P_{qm}^{hh,\kappa:\Omega} = \sum_{p=0}^{M_\Pi} \left(\frac{-2\tau_p^{\beta,\Pi} J_m^{\kappa} I_{mp}^{\prime\kappa,\Pi} I_{pq}^{\Pi,\Omega}}{h} + \delta_{\kappa\Omega}\delta_{mq} \frac{\tau_q^{\beta,\kappa} J_q^{\kappa} h_{\kappa}}{2} \right), \quad (3.272)$$

$$q(m) = 1, \cdots M_\kappa (M_\Omega).$$

3.5.5.2 For Source Via

If the via is a source via, then we have $a^i = 0$ ($i = 1, \cdots, R$).

Two-Layer Problem. For the two-layer problem, we can rewrite Equations (3.152–3.155) as two set of linear equations:

$$\begin{bmatrix} \dfrac{H_q^I(-\tau_q^{\beta,I})h_1}{2} & \dfrac{H_q^{\prime I} \tau_q^{\mu,I} h_1}{2} \\ \dfrac{H_q^{\prime I}(-\tau_q^{\varepsilon,I})h_1}{2} & \dfrac{H_q^I \tau_q^{\beta,I} h_1}{2} \end{bmatrix} \begin{bmatrix} b_q^{e,I} \\ b_q^{h,I} \end{bmatrix}$$

$$= \begin{bmatrix} \displaystyle\sum_{p=0}^{M3} [a_p^{e,III}(-\tau_p^{\beta,III})J_{e,p}^{III} + a_p^{h,III}\tau_p^{\mu,III} J_{h,p}^{\prime III}] I_{pq}^{\prime III,I} \\ \displaystyle\sum_{p=0}^{M3} [a_p^{e,III}(-\tau_p^{\varepsilon,III})J_{e,p}^{\prime III} + a_p^{h,III}\tau_p^{\beta,III} J_{h,p}^{III}] I_{pq}^{III,I} \end{bmatrix}, \quad (3.273)$$

$$\begin{bmatrix} \dfrac{H_r^{II}(-\tau_r^{\beta,II})h_2}{2} & \dfrac{H_r^{\prime II} \tau_r^{\mu,II} h_2}{2} \\ \dfrac{H_r^{\prime II}(-\tau_r^{\varepsilon,II})h_2}{2} & \dfrac{H_r^{II} \tau_r^{\beta,II} h_2}{2} \end{bmatrix} \begin{bmatrix} b_r^{e,II} \\ b_r^{h,II} \end{bmatrix}$$

$$= \begin{bmatrix} \displaystyle\sum_{p=0}^{M3} [a_p^{e,III}(-\tau_p^{\beta,III})J_{e,p}^{III} + a_p^{h,III}\tau_p^{\mu,III} J_{h,p}^{\prime III}] I_{pr}^{\prime III,II} \\ \displaystyle\sum_{p=0}^{M3} [a_p^{e,III}(-\tau_p^{\varepsilon,III})J_{e,p}^{\prime III} + a_p^{h,III}\tau_p^{\beta,III} J_{h,p}^{III}] I_{pr}^{III,II} \end{bmatrix}. \quad (3.274)$$

We can solve for b in the above two matrix equations.

Multilayer-Layer Problem. We can find the reflection coefficients of the κth PPWG by solving the following linear system of equations.

$$
\begin{bmatrix}
\dfrac{H_r^{\kappa}(-\tau_r^{\beta,\kappa})h_{\kappa}}{2} & \dfrac{H_r'^{\kappa}\tau_r^{\mu,\kappa}h_{\kappa}}{2} \\[3mm]
\dfrac{H_r'^{\kappa}(-\tau_r^{\varepsilon,\kappa})h_{\kappa}}{2} & \dfrac{H_r^{\kappa}\tau_r^{\beta,\kappa}h_{\kappa}}{2}
\end{bmatrix}
\begin{bmatrix}
b_r^{e,\kappa} \\[2mm]
b_r^{h,\kappa}
\end{bmatrix}
$$

$$
=
\begin{bmatrix}
\displaystyle\sum_{p=0}^{M_{\Pi}}\left[a_p^{e,\Pi}(-\tau_p^{\beta,\Pi})J_{e,p}^{\Pi}+a_p^{h,\Pi}\tau_p^{\mu,\Pi}J_{h,p}'^{\Pi}\right]I_{pr}'^{\Pi,\kappa} \\[4mm]
\displaystyle\sum_{p=0}^{M_{\Pi}}\left[a_p^{e,\Pi}(-\tau_p^{\varepsilon,\Pi})J_{e,p}'^{\Pi}+a_p^{h,\Pi}\tau_p^{\beta,\Pi}J_{h,p}^{\Pi}\right]I_{pr}^{\Pi,\kappa}
\end{bmatrix},
\quad r=1,\cdots M_{\kappa}. \quad (3.275)
$$

Recall that we have the following formulae to solve the linear system of equations:

$$
\begin{bmatrix}
a_1 & b_1 \\
a_2 & b_2
\end{bmatrix}
\begin{bmatrix}
x \\
y
\end{bmatrix}
=
\begin{bmatrix}
c_1 \\
c_2
\end{bmatrix},
\qquad (3.276)
$$

where

$$
x=\begin{bmatrix} c_1 & b_1 \\ c_2 & b_2 \end{bmatrix}\Big/\begin{bmatrix} a_1 & b_1 \\ a_2 & b_2 \end{bmatrix},\quad
y=\begin{bmatrix} a_1 & c_1 \\ a_2 & c_2 \end{bmatrix}\Big/\begin{bmatrix} a_1 & b_1 \\ a_2 & b_2 \end{bmatrix}. \quad (3.277)
$$

Reorganizing Equation (3.275) into a generalized T-matrix:

$$
\begin{bmatrix}
\dfrac{H_r^{\kappa}(-\tau_r^{\beta,\kappa})h_{\kappa}}{2} & \dfrac{H_r'^{\kappa}\tau_r^{\mu,\kappa}h_{\kappa}}{2} \\[3mm]
\dfrac{H_r'^{\kappa}(-\tau_r^{\varepsilon,\kappa})h_{\kappa}}{2} & \dfrac{H_r^{\kappa}\tau_r^{\beta,\kappa}h_{\kappa}}{2}
\end{bmatrix}
\begin{bmatrix}
b_r^{e,\kappa} \\[2mm]
b_r^{h,\kappa}
\end{bmatrix}
=
$$

$$
\begin{bmatrix}
(-\tau_0^{\beta,\Pi})J_{e,0}^{\Pi}I_{0r}'^{\Pi,\kappa} & \cdots & (-\tau_{M_{\Pi}}^{\beta,\Pi})J_{e,M_{\Pi}}^{\Pi}I_{M_{\Pi}r}'^{\Pi,\kappa} & \tau_1^{\mu,\Pi}J_{h,1}'^{\Pi}I_{1r}'^{\Pi,\kappa} & \cdots & \tau_{M_{\Pi}}^{\mu,\Pi}J_{h,M_{\Pi}}'^{\Pi}I_{M_{\Pi}r}'^{\Pi,\kappa} \\[2mm]
(-\tau_0^{\varepsilon,\Pi})J_{e,0}'^{\Pi}I_{0r}^{\Pi,\kappa} & \cdots & (-\tau_{M_{\Pi}}^{\varepsilon,\Pi})J_{e,M_{\Pi}}'^{\Pi}I_{M_{\Pi}r}^{\Pi,\kappa} & \tau_1^{\beta,\Pi}J_{h,1}^{\Pi}I_{1r}^{\Pi,\kappa} & \cdots & \tau_{M_{\Pi}}^{\beta,\Pi}J_{h,M_{\Pi}}^{\Pi}I_{M_{\Pi}r}^{\Pi,\kappa}
\end{bmatrix}
$$

$$
\cdot
\begin{bmatrix}
a_0^{e,\Pi} \\
\vdots \\
a_{M_{\Pi}}^{e,\Pi} \\
a_1^{h,\Pi} \\
\vdots \\
a_{M_{\Pi}}^{h,\Pi}
\end{bmatrix},
\quad r=1,\cdots M_{\kappa}, \qquad (3.278)
$$

$$
\begin{bmatrix}
\dfrac{H_0^{\kappa}(-\tau_0^{\beta,\kappa})h_\kappa}{2} & & & 0 \\
& [H_1] & & \\
& & \ddots & \\
0 & & & [H_{M_\kappa}]
\end{bmatrix}
\begin{bmatrix}
b_0^{e,\kappa} \\
\begin{bmatrix} b_1^{e,\kappa} \\ b_1^{h,\kappa} \end{bmatrix} \\
\vdots \\
\begin{bmatrix} b_{M_\kappa}^{e,\kappa} \\ b_{M_\kappa}^{h,\kappa} \end{bmatrix}
\end{bmatrix}
=
\begin{bmatrix} C^e \\ C^h \end{bmatrix}
\begin{bmatrix}
a_0^{e,\Pi} \\
\vdots \\
a_{M_\Pi}^{e,\Pi} \\
a_0^{h,\Pi} \\
\vdots \\
a_{M_\Pi}^{h,\Pi}
\end{bmatrix},
\qquad (3.279)
$$

where

$$
[H_r] =
\begin{bmatrix}
\dfrac{H_r^{\kappa}(-\tau_r^{\beta,\kappa})h_\kappa}{2} & \dfrac{H_r'^{\kappa}\tau_r^{\mu,\kappa}h_\kappa}{2} \\
\dfrac{H_r'^{\kappa}(-\tau_r^{\varepsilon,\kappa})h_\kappa}{2} & \dfrac{H_r^{\kappa}\tau_r^{\beta,\kappa}h_\kappa}{2}
\end{bmatrix},
\qquad (3.280)
$$

$$
\begin{bmatrix} C^e \\ C^h \end{bmatrix}
=
$$

$$
\begin{bmatrix}
(-\tau_0^{\beta,\Pi})J_{e,0}^{\Pi}I_{0r}'^{\Pi,\kappa} & \cdots & (-\tau_{M_\Pi}^{\beta,\Pi})J_{e,M_\Pi}^{\Pi}I_{M_\Pi r}'^{\Pi,\kappa} & \tau_1^{\mu,\Pi}J_{h,1}'^{\Pi}I_{1r}'^{\Pi,\kappa} & \cdots & \tau_{M_\Pi}^{\mu,\Pi}J_{h,M_\Pi}'^{\Pi}I_{M_\Pi r}'^{\Pi,\kappa} \\
(-\tau_0^{\varepsilon,\Pi})J_{e,0}'^{\Pi}I_{0r}^{\Pi,\kappa} & \cdots & (-\tau_{M_\Pi}^{\varepsilon,\Pi})J_{e,M_\Pi}'^{\Pi}I_{M_\Pi r}^{\Pi,\kappa} & \tau_1^{\beta,\Pi}J_{h,1}^{\Pi}I_{1r}^{\Pi,\kappa} & \cdots & \tau_{M_\Pi}^{\beta,\Pi}J_{h,M_\Pi}^{\Pi}I_{M_\Pi r}^{\Pi,\kappa}
\end{bmatrix}.
$$
$$
(3.281)
$$

As discussed in Section 3.2.3, we recall that the magnetic frill current source for the packaging problem will not excite H-mode around the source via. Hence, Equation (3.278) is reduced to the following:

$$
\begin{bmatrix}
\dfrac{H_r^{\kappa}(-\tau_r^{\beta,\kappa})h_\kappa}{2} & \dfrac{H_r'^{\kappa}\tau_r^{\mu,\kappa}h_\kappa}{2} \\
\dfrac{H_r'^{\kappa}(-\tau_r^{\varepsilon,\kappa})h_\kappa}{2} & \dfrac{H_r^{\kappa}\tau_r^{\beta,\kappa}h_\kappa}{2}
\end{bmatrix}
\begin{bmatrix} b_r^{e,\kappa} \\ b_r^{h,\kappa} \end{bmatrix}
$$

$$
=
\begin{bmatrix}
(-\tau_0^{\beta,\Pi})J_{e,0}^{\Pi}I_{0r}'^{\Pi,\kappa} & \cdots & (-\tau_{M_\Pi}^{\beta,\Pi})J_{e,M_\Pi}^{\Pi}I_{M_\Pi r}'^{\Pi,\kappa} \\
(-\tau_0^{\varepsilon,\Pi})J_{e,0}'^{\Pi}I_{0r}^{\Pi,\kappa} & \cdots & (-\tau_{M_\Pi}^{\varepsilon,\Pi})J_{e,M_\Pi}'^{\Pi}I_{M_\Pi r}^{\Pi,\kappa}
\end{bmatrix}
\begin{bmatrix}
a_0^{e,\Pi} \\
\vdots \\
a_{M_\Pi}^{e,\Pi}
\end{bmatrix},
\qquad r = 1, \cdots M_\kappa,
$$

$$
(3.282)
$$

$$[H_r] = \begin{bmatrix} \dfrac{H_r^{\kappa}(-\tau_r^{\beta,\kappa})h_{\kappa}}{2} & \dfrac{H_r^{\prime\kappa}\tau_r^{\mu,\kappa}h_{\kappa}}{2} \\[3mm] \dfrac{H_r^{\prime\kappa}(-\tau_r^{\varepsilon,\kappa})h_{\kappa}}{2} & \dfrac{H_r^{\kappa}\tau_r^{\beta,\kappa}h_{\kappa}}{2} \end{bmatrix} = \dfrac{h_{\kappa}}{2}\begin{bmatrix} H_r^{\kappa}(-\tau_r^{\beta,\kappa}) & H_r^{\prime\kappa}\tau_r^{\mu,\kappa} \\[2mm] H_r^{\prime\kappa}(-\tau_r^{\varepsilon,\kappa}) & H_r^{\kappa}\tau_r^{\beta,\kappa} \end{bmatrix}$$

$$= \frac{h_{\kappa}}{2}\begin{bmatrix} -H_n^{(2)}\!\left(k_r^{\kappa}b\right)\dfrac{jn\beta_r^{\kappa}}{(k_r^{\kappa})^2 b} & H_n^{\prime(2)}\!\left(k_r^{\kappa}b\right)\dfrac{jn\mu}{k_r^{\kappa}} \\[4mm] -H_n^{\prime(2)}\!\left(k_r^{\kappa}b\right)\dfrac{jn\varepsilon}{k_r^{\kappa}} & H_n^{(2)}\!\left(k_r^{\kappa}b\right)\dfrac{jn\beta_r^{\kappa}}{(k_r^{\kappa})^2 b} \end{bmatrix}, \qquad (3.283)$$

$$C^e = \left[(-\tau_0^{\beta,\Pi})J_{e,0}^{\Pi}I_{0r}^{\prime\Pi,\kappa} \quad \cdots \quad (-\tau_p^{\beta,\Pi})J_{e,p}^{\Pi}I_{pr}^{\prime\Pi,\kappa} \quad \cdots \quad (-\tau_{M_\Pi}^{\beta,\Pi})J_{e,M_\Pi}^{\Pi}I_{M_\Pi r}^{\prime\Pi,\kappa}\right], \tag{3.284}$$

$$C^h = \left[(-\tau_0^{\varepsilon,\Pi})J_{e,0}^{\prime\Pi}I_{0r}^{\Pi,\kappa} \quad \cdots \quad (-\tau_p^{\varepsilon,\Pi})J_{e,p}^{\prime\Pi}I_{pr}^{\Pi,\kappa} \quad \cdots \quad (-\tau_{M_\Pi}^{\varepsilon,\Pi})J_{e,M_\Pi}^{\prime\Pi}I_{M_\Pi r}^{\Pi,\kappa}\right], \tag{3.285}$$

$$C_p^e = (-\tau_p^{\beta,\Pi})J_{e,p}^{\Pi}I_{pr}^{\prime\Pi,\kappa} = -\frac{jn\beta_p^{\Pi}}{(k_p^{\Pi})^2 b}\left(J_n\!\left(k_p^{\Pi}b\right) - \frac{J_n\!\left(k_p^{\Pi}a\right)H_n^{(2)}\!\left(k_p^{\Pi}b\right)}{H_n^{(2)}\!\left(k_p^{\Pi}a\right)}\right)I_{pr}^{\prime\Pi,\kappa}$$

$$= -\frac{jn\beta_p^{\Pi}}{(k_p^{\Pi})^2 b}\left(\frac{J_n\!\left(k_p^{\Pi}b\right)H_n^{(2)}\!\left(k_p^{\Pi}a\right) - J_n\!\left(k_p^{\Pi}a\right)H_n^{(2)}\!\left(k_p^{\Pi}b\right)}{H_n^{(2)}\!\left(k_p^{\Pi}a\right)}\right)I_{pr}^{\prime\Pi,\kappa}, \tag{3.286}$$

$$C_p^h = (-\tau_p^{\varepsilon,\Pi})J_{e,p}^{\prime\Pi}I_{pr}^{\Pi,\kappa} = -\frac{jn\varepsilon}{k_p^{\Pi}}\left(J_n'\!\left(k_p^{\Pi}b\right) - \frac{J_n\!\left(k_p^{\Pi}a\right)H_n^{\prime(2)}\!\left(k_p^{\Pi}b\right)}{H_n^{(2)}\!\left(k_p^{\Pi}a\right)}\right)I_{pr}^{\Pi,\kappa}$$

$$= -\frac{jn\varepsilon}{k_p^{\Pi}}\left(\frac{J_n'\!\left(k_p^{\Pi}b\right)H_n^{(2)}\!\left(k_p^{\Pi}a\right) - J_n\!\left(k_p^{\Pi}a\right)H_n^{\prime(2)}\!\left(k_p^{\Pi}b\right)}{H_n^{(2)}\!\left(k_p^{\Pi}a\right)}\right)I_{pr}^{\Pi,\kappa}, \tag{3.287}$$

$$I_{pr}^{\prime\Pi,\kappa} = \int_0^{h_{\kappa}} \sin\!\left(\frac{p\pi}{h_{\Pi}}z\right)\sin\!\left(\frac{r\pi}{h_{\kappa}}z\right)dz, \tag{3.288}$$

$$I_{pr}^{\Pi,\kappa} = \int_0^{h_{\kappa}} \cos\!\left(\frac{p\pi}{h_{\Pi}}z\right)\cos\!\left(\frac{r\pi}{h_{\kappa}}z\right)dz. \tag{3.289}$$

The integration can be computed by using Equations (3.258) and (3.259).

3.5.6 Numerical Simulations for Multilayered Power-Ground Planes with Multiple Vias

An example is given to demonstrate the modal expansion with SMM combined with the FDCL boundary modeling method and the generalized *T*-matrix approach for the analysis of multiple via coupling in multilayered parallel-plate structures. The geometry of the multilayered parallel-plate structure is shown in Figure 3.32. It has three conductor P-G planes. The relative permittivity of the substrate is 4.2 with a loss tangent of 0.02. The total of 100 vias are distributed as 64 vias in center block and 36 vias in four corner blocks as shown in Figure 3.32. The simulation results of the input impedance seen from the top end of the

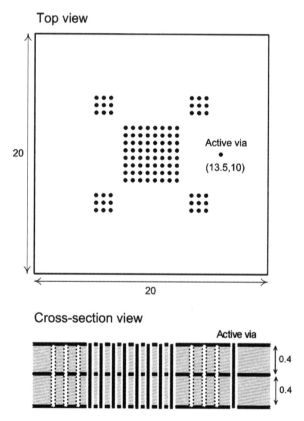

Figure 3.32 Example 1—a multilayered parallel-plate structure with three conductor power-ground planes and 101 vias (unit: millimeter).

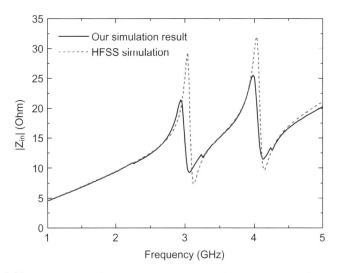

Figure 3.33   Comparison of the input impedance seen from the top end of the active via in Example 1: our algorithm versus HFSS simulation.

Table 3.1
Comparison of Memory Usage and Computing Time for Example 1

	Ansoft HFSS*	SMM with FDCL*
No. of vias	101	101
No. of unknowns	41,467 tetrahedrons	404 modes
Memory usage	420 MB	70 MB
CPU time	5 hours 21 minutes	20 minutes 14 seconds

* Simulated on the machine of Intel Centrino 1.3 GHz, 512 MB.

active via by our method agree quite well with those by the Ansoft HFSS software as shown in Figure 3.33. In the Table 3.1, the comparison of the memory usage and computing time is presented for the extended SMM algorithm with the FDCL and the Ansoft HFSS simulation. The simulation time of our SMM algorithm is much faster than one of the HFSS and the memory usage is also much lesser. Hence, the developed algorithm is very much efficient compared to the full-wave simulation tools and still provides the correct solution.

Another example considered is to discuss a bottleneck of the conventional full-wave simulation methods. Figure 3.34 shows the

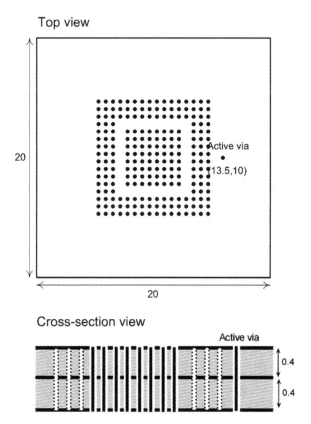

Top view

Cross-section view

Figure 3.34   Example 2—a multilayered parallel-plate structure with three conductor power-ground planes and 221 vias (unit: millimeter).

geometry of the three conductor P-G planes which has more vias, compared to Example 1. The signal via is also at the same location as in the previous model, and the total of 221 vias are distributed as 64 vias in center block and 156 vias in outer ring block as shown in the figure. For this example, the HFSS simulation cannot be performed due to the memory insufficient while the SMM algorithm with FDCL can simulate with no difficulty. The simulated result for the input impedance is shown in Figure 3.35. The comparison of the memory usage and computing time between the algorithm of the SMM algorithm with FDCL and the Ansoft HFSS simulation is presented in Table 3.2.

The example of two active vias in multilayered parallel-plate structure is also considered as shown in Figure 3.36. It has six conductor

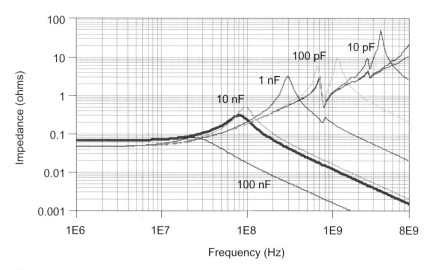

Figure 1.3 Example of antiresonances in total PDN impedances for various on-chip capacitance values [3].

Electrical Modeling and Design for 3D System Integration: 3D Integrated Circuits and Packaging, Signal Integrity, Power Integrity and EMC, First Edition. Er-Ping Li.
© 2012 Institute of Electrical and Electronics Engineers. Published 2012 by John Wiley & Sons, Inc.

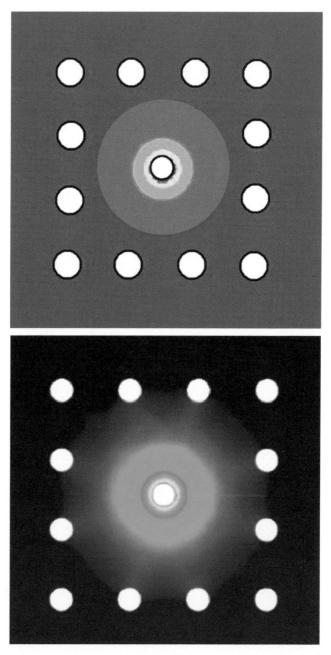

Figure 3.6 Comparison of the E_z field distribution at 1 GHz: our simulation results (left) versus Ansoft HFSS simulation results (right). The vias are drawn as white dots [11].

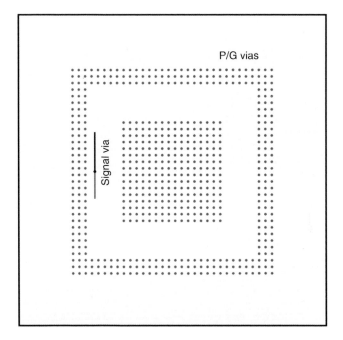

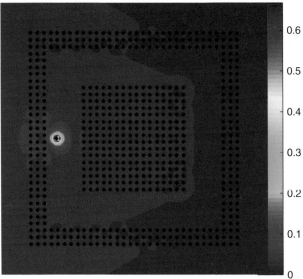

Figure 3.9 Field distribution of multiple scattering among the shorting vias.

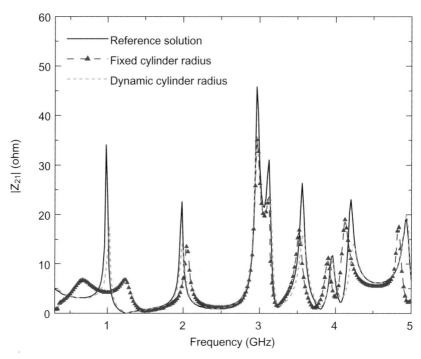

Figure 3.16 Comparison of the extended SMM results with fixed and dynamic radius of the PMC cylinders in the FDCL and the reference solution [14].

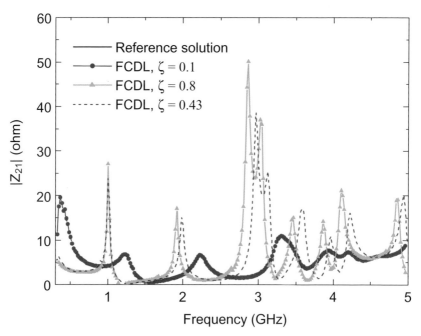

Figure 3.17 Effects of the different values of ζ on the accuracy of the simulation results by the FDCL boundary modeling method [14].

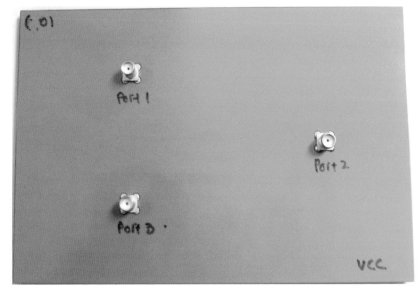

Figure 3.18 Test printed circuit board (PCB) for measurement.

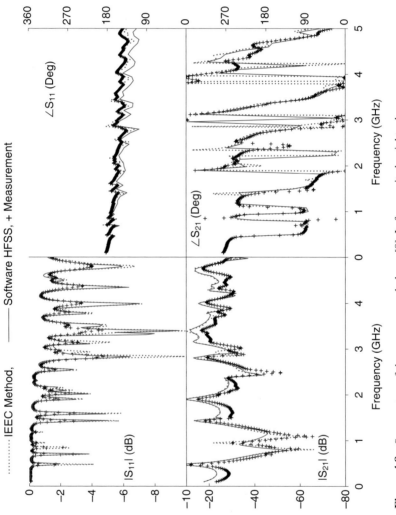

Figure 4.8 *S*-parameters of the power-ground planes [8]. Left: magnitude; right: phase.

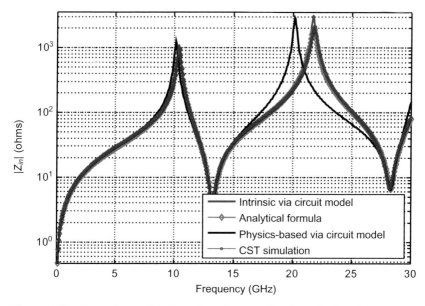

Figure 5.18 Comparisons of the input impedance results of a probe in a finite circular plate pair ($r = a = 0.127$, $b = 0.3810$, $h = 1.016$, unit: mm; $\varepsilon_\infty = 4.2$, $\varepsilon_s = 4.4$, $\tau = 1.6 \times 10^{-11}$ s; $R = 5.08$ mm with PEC boundary).

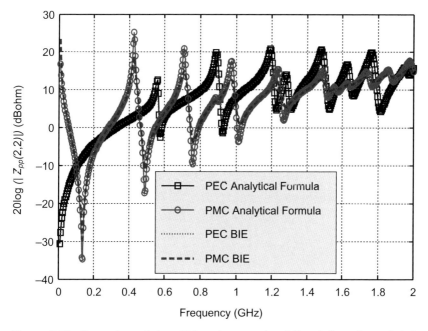

Figure 5.29 Comparison of the self impedance results of Port 2 from the analytical method and the boundary integral-equation method for the geometry shown in Figure 5.26.

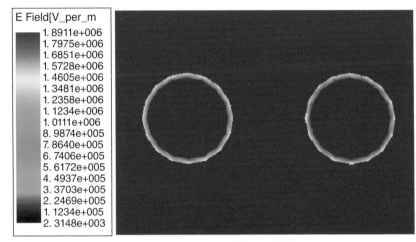

(a) 0.1 GHz

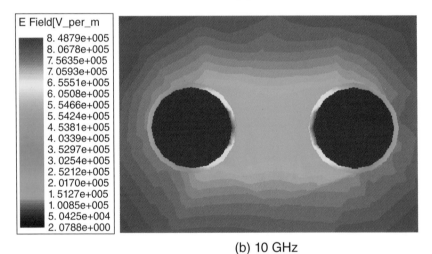

(b) 10 GHz

Figure 6.11 Skin and proximity effects of TSVs, demonstrated by the electromagnetic field plots with HFSS software [29].

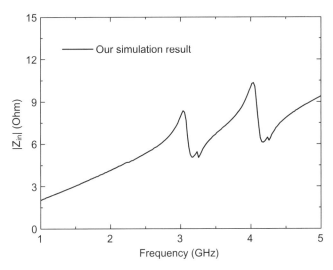

Figure 3.35   Input impedance seen from the top end of the active via in Example 2.

Table 3.2
Comparison of Memory Usage and Computing Time for Example 2

	Ansoft HFSS*	SMM with FDCL*
No. of vias	221	221
No. of unknowns	74,111 tetrahedrons	884 modes
Memory usage	Insufficient	180 MB
CPU time	–	1 hour 16 seconds

* Simulated on the machine of Intel Centrino 1.3 GHz, 512 MB.

Table 3.3
Comparison of Memory Usage and Computing Time for Example 3

	Ansoft HFSS*	SMM with FDCL*
No. of vias	18	18
No. of planes	6	6
No. of unknowns	22,487 tetrahedrons	180 modes
Memory usage	294 MB	52 MB
CPU time	3 hours 54 minutes	12 minutes 17 seconds

* Simulated on the machine of Intel Centrino 1.3 GHz, 512 MB.

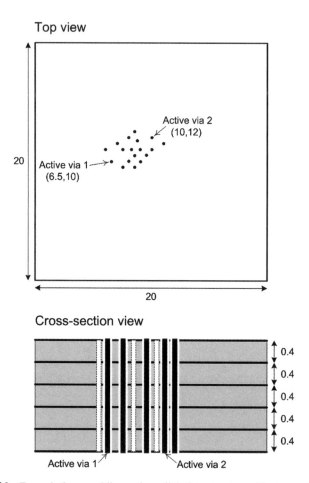

Figure 3.36 Example 3—a multilayered parallel-plate structure with six conductor power-ground planes (unit: millimeter).

P-G planes. The relative permittivity of the substrate is 4.2 with a loss tangent of 0.02. The two active vias' locations are given in the figure, and the rest of 16 P-G vias are located at (7.5,11), (8.5,10.5), (9,10), (9,11), (8,12), (8,10), (8.25,11), (7,11.5), (9.5,10.5), (8.75,11.75), (7.5,9.5), (10,11), (8.5,12.5), (6,11), (8.5,9.5), and (11,11.5), all in millimeter. In Figure 3.37, the S-parameters simulation results by our algorithm implemented for analysis of multilayered P-G planes with multiple vias are plotted and compared with those from the HFSS simulation. And Table 3.3 shows the comparison of memory usage and computing time.

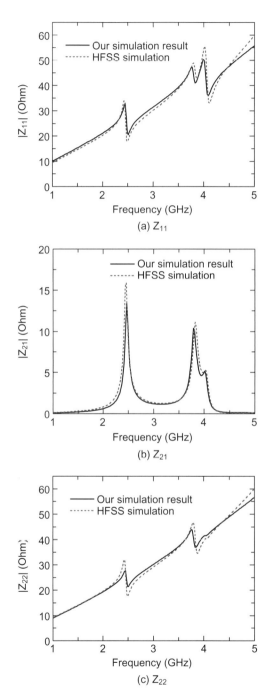

Figure 3.37 Comparison of the S-parameters simulated results for multilayered structure of Example 3: our algorithm versus HFSS simulation.

3.6 CONCLUSION

The SMM of the parallel-plate waveguide modes are developed to analyze the electrical performance of P-G planes due to the multiple scattering among the vias in a 3D electronic package integration. The scattering T-matrices for coupling among the P-G vias and the outgoing wave coefficients for each via are computed to model the equivalent admittance (Y) matrix of the P-G planes including the multiple scattering effects of the P-G vias with signal vias. The integral equation method is used to extract the equivalent circuit model for the coupling between the external signal traces and the vertical vias. The SPICE-like simulation is made using the extracted circuit model and the Y-matrix to analyze the signal response of the system.

The novel methods in this chapter are also developed to transform the modal expansion with the SMM into a viable and efficient method for the analysis of multiple via coupling in finite-sized multilayered parallel-plate structures. The FDCL method is a simple yet powerful method for boundary modeling. The radii of the cylinders in the FDCL vary according to the simulation frequency, which ensure the accuracy of the simulation results. All the cylinders in the FDCL can have the same radii for each frequency, which makes the FDCL a uniform layer; they can also have slightly different radii at different regions along the boundary, which facilitate modeling sharp corners at the boundary and enhance the accuracy of the simulation results. Numerical examples show that the proposed FDCL method used in the SMM provides accurate prediction of the resonant frequencies of electronic packages and then the extended SMM algorithm is able to analyze the real-world package structures. The simulated results of the extended algorithm with the FDCL show good agreement with the measured data.

Unlike the absorbing boundary or radiation boundary used in other computational electromagnetic methods, the FDCL utilizes physical structures, that is, cylinders with finite radii to enforce the boundary. Although all the examples in this research employ the FDCL for the PMC boundary, the FDCL is not limited to that, and it is straightforward to implement it for PEC boundary and other types of boundaries.

A generalized T-matrix model for the SMM is derived by the mode matching technique to analyze the vias penetrating more than two conductor planes. The generalized T-matrix model obviates the use of multiple equivalent magnetic sources to model the plated-through vias. And it facilitates modeling the coupling of multilayered vias. The modal expansion with the SMM incorporating the FDCL boundary modeling method and the generalized T-matrix approach is a powerful numerical method. Its simulation time and memory usage is greatly reduced as compared to full-wave methods, and it still yields accurate simulation results.

REFERENCES

[1] D. FELBACQ, G. TAYEB, and D. MAYSTRE, Scattering by a random set of parallel cylinders, *J Opt Soc Am A Opt Image Sci Vis*, vol. 11, no. 9, pp. 2526–2538, 1994.

[2] L. TSANG, J. A. KONG, K. H. DING, and C. AO, *Scattering of Electromagnetic Waves: Numerical simulations*, Wiley Interscience, New York, 2001.

[3] L. TSANG, H. CHEN, C.-C. HUANG, and V. JANDHYALA, Modeling of multiple scattering among vias in planar waveguides using foldy-lax equations, *Microw. Opt. Technol. Lett.*, vol. 31, pp. 201–208, 2001.

[4] H. CHEN, Q. LI, L. TSANG, C. C. HUANG, and V. JANDHYALA, Analysis of a large number of vias and differential signal in multilayered structures, *IEEE Trans. Microw. Theory Tech.*, vol. 51, no. 3, pp. 818–829, 2003.

[5] C. A. BALANIS, *Advanced Engineering Electromagnetics*, Wiley, New York, 1989.

[6] C. WEI, R. F. HARRINGTON, J. R. MAUTZ, and T. K. SARKAR, Multiconductor transmission lines in multilayered dielectric media, *IEEE Trans. Microw. Theory Tech.*, vol. 32, pp. 439–449, 1984.

[7] M. B. BAZDAR, A. R. DJORDJEVIC, R. F. HARRINGTON, and T. K. SARKAR, Evaluation of quasi-static matrix parameters for multiconductor transmission lines using Galerkin's method, *IEEE Trans. Microw. Theory Tech*, vol. MTT-42, no. 7, pp. 1293–1295, 1994.

[8] F. JUN, J. L. DREWNIAK, J. L. KNIGHTEN, N. W. SMITH, A. ORLANDI, T. P. VAN DOREN, T. H. HUBING, and R. E. DUBROFF, Quantifying SMT decoupling capacitor placement in DC power-bus design for multilayer PCBs, *IEEE Trans. Electromagn. Compat.*, vol. 43, pp. 588–599, 2001.

[9] T. OKOSHI, *Planar Circuits for Microwaves and Lightwaves*, chapter 2, Springer-Verlag, New York, 1985.

[10] Z. O. ZAW, E.-X. LIU, X. WEI, Y. C. MARK TAN, E.-P. LI, Y. ZHANG, and L.-W. LI, Hybrid of scattering matrix method and integral equation used for co-simulation of power integrity and EMI in electronic package with large number of P/G vias, in *Proc. 57th Electronic Components and Technology Conference*, May 2007, pp. 815–820.

[11] Z. O. ZAW, E.-X. LIU, E.-P. LI, X. WEI, Y. ZHANG, M. TAN, L.-W. J. LI, and R. VAHLDIECK, A semi-analytical approach for system-level electrical modeling of electronic packages with large number of vias, *IEEE Trans. Adv. Packag.*, vol. 31, no. 2, pp. 267–274, 2008.

[12] P. LIU and Z.-F. LI, An efficient method for calculating bounces in the irregular power/ground plane structure with holes in high-speed PCBs, *IEEE Trans. Electromagn. Compat.*, vol. 47, no. 4, pp. 889–898, 2005.

[13] E.-X. LIU, X. WEI, Z. O. ZAW, and E.-P. LI, An efficient method for power integrity and EMI analysis of irregular-shaped power/ground planes in packages, in *Proc. 16th Topical Meeting on Electrical Performance of Electronic Packaging*, October 2007, pp. 263–266.

[14] E.-X. LIU, E.-P. LI, Z. O. ZAW, X. WEI, and R. VAHLDIECK, Novel methods for analysis of multiple via coupling in multilayered parallel-plate structures, *IEEE Trans. Microw. Theory Tech.*, vol. 57, no. 7, pp. 1724–1733, 2009.

[15] D. M. POZAR, *Microwave Engineering*, 2nd ed., Wiley, New York, 1998.

CHAPTER 4

Hybrid Integral Equation Modeling Methods for 3D Integration

4.1 INTRODUCTION

In order to provide a lower impedance path and reduce the interference between the circuits, power-ground planes (PGPs) are widely employed in the power distribution network (PDN) of the electronic packages and printed circuit boards. However, the PGPs also introduce an additional electromagnetic interference problems due to higher operating frequency and power density. The signal traces are often laid out in different layers of PGPs. Their return currents flow on the PGPs just below them. When the traces pass through different layers, their return currents also exchange from one plane to another plane. Accordingly, a vertical displacement current is induced between different planes for the continuity of the return currents. This displacement current will excite electromagnetic field noise, which then propagates inside the PGPs and couples to other signal traces passing through the same layer. At the same time, this noise also leaks into the surrounding area of the electronic package through the periphery and gaps of the PGPs. These interferences will be further amplified if the noise's spectrum covers any inherent resonant frequency of the PGPs.

Electrical Modeling and Design for 3D System Integration: 3D Integrated Circuits and Packaging, Signal Integrity, Power Integrity and EMC, First Edition. Er-Ping Li.
© 2012 Institute of Electrical and Electronics Engineers. Published 2012 by John Wiley & Sons, Inc.

To tackle the electromagnetic compatibility (EMC) issue, efficient electrical modeling technologies are required, and moreover, it is also a very interesting topic in the computational electromagnetics. The cavity mode theory [1] gives a quick analytic simulation of the PGPs. However, it is limited to the PGPs with regular shapes, such as the rectangle and triangle. Other semianalytic methods, such as the Foldy–Lax multiple scattering method [2–4], are proposed, where the electromagnetic field between the power and ground planes are expanded by using cylindrical waves, and the reflections between those waves and the vias are considered by using the scattering matrix method. However, this method also has the difficulty to model the PGPs with arbitrary shapes. More accurate modeling of the PGPs requires directly solving the electromagnetic field by using three-dimensional (3D) numerical methods, such as the partial element equivalent circuit method [5] and the finite differential method [6]. Although the overall electrical size of the PGPs is small enough to apply the 3D methods, the high respect ratio of the PGPs results in a very dense meshing. This makes these 3D methods very expensive in terms of computing time and memory requirements.

To alleviate the computational cost of these 3D methods, algorithms that exploit geometrical features of the PGPs are the better choice. In this chapter, two-dimensional (2D) and 3D integral equation methods are employed for the analysis of the complex PDN. The 2D integral equation method [7–10] provides a comprehensive way for one to quickly extract the equivalent circuits of the PDN, and then substitute them into a Simulation Program with Integrated Circuit Emphasis (SPICE)-like simulator to perform the signal and power integrity (PI) analysis. The 3D integral equation method [11] provides a more accurate solution for both the emission and susceptibility issues of the PDN. Both of the 2D and 3D integral equation methods are optimized by making a full use of the structural features of the PDN.

4.2 2D INTEGRAL EQUATION EQUIVALENT CIRCUIT (IEEC) METHOD

4.2.1 Overview of the Algorithm

It has been demonstrated that the electromagnetic field between a pair of conducting planes can be decomposed and treated separately [12].

In the following, a simple yet accurate modal decoupling method is introduced to simulate the PGPs with arbitrary shapes. The total electromagnetic field of the PDN is decoupled into the parallel plate mode and the transmission line mode. The transmission line mode includes the stripline mode and the microstrip line mode. The parallel plate mode is solved by using an efficient integral equation method. Meanwhile, the discontinuity due to the through-hole via is also considered by an analytic formula. The equivalent circuits for the parallel plate mode, transmission mode, and the discontinuities are extracted separately. The whole equivalent circuit of the PDN is obtained by the connections of each individual equivalent circuit. The final circuit model can be substituted into a circuit simulator to perform the system-level EMC analysis of the PDN.

This proposed method is named as the IEEC method. The advantages of the IEEC method over available 3D methods are:

(a) It decouples the complex 3D problem into several simple one-dimensional (1D) and 2D problems. Therefore, it greatly reduces the computing time and still keeps a good accuracy.

(b) It extracts the equivalent circuit from the complex electromagnetic field distribution. This equivalent circuit provides a more comprehensive solution for the real industrial applications than the purely electromagnetic field solvers.

4.2.2 Modal Decoupling inside the Power Distribution Network (PDN)

Figure 4.1 shows the typical structure of the PDN. The antipad is a clearance hole between the via and the metal planes. The ground and power planes are highly conducting metal planes, which provide a low-impedance path for the power supply. Usually, the substrate sandwiched between metal planes is uniform, isotropic, and with a thickness much smaller than the interesting wavelength. Based on this, we decouple the total electromagnetic field into two kinds of independent modes: the parallel plate mode and transmission line mode.

(a) The parallel plate mode represents the standing-wave field constrained between the ground and power planes. It is due to the reflections from the edges of the cavity-like PGPs pair. Due to

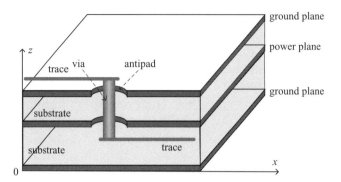

Figure 4.1   The typical structure of the power distribution network [8].

the thin substrate, it is assumed that the electromagnetic field does not change along the z-direction. The vector electric field E is in z-direction, and vector magnetic field H lies in the horizontal plane.

(b) The signal traces sandwiched between PGPs are taken as the striplines while the signal traces above or below the PGPs are taken as the microstrip lines. They support the transmission line modes. They are transverse electromagnetic (TEM) modes (transverse to the traces' directions) and propagate along the traces.

These two kinds of modes can be solved in xoy plane and the cross section of traces separately. After that, the parallel plate mode converses with the microstrip line mode at the through-hole via region, while the parallel plate mode is distributively accumulated to the stripline mode along the stripline. How to calculate the through-hole via's equivalent circuit is a very interesting topic of a long history [13–15]. In this chapter, a simple but accurate analytic formula is derived based on the work of Reference 16, in order to calculate the parasitic capacitance of the through-hole via.

According to above modal decoupling, the whole PDN is decomposed into three subdomains as in Figure 4.2: the PGPs pairs, the microstrip lines and striplines, and the through-hole vias. Their electromagnetic field distributions are also plotted in Figure 4.2.

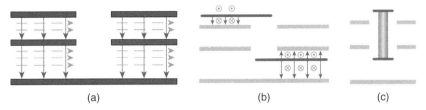

Figure 4.2 Three subdomains used for the modal decoupling: (a) power-ground plane pairs, (b) signal traces, and (c) through-hole via [8]. →: Electric field, —➤: Magnetic field.

The transmission line modes can be easily solved by using available transmission line solvers. In the following, we will focus on the integral equation solution of PGPs, the recombination of parallel plate mode and transmission line mode, and the whole equivalent circuit of the PDN including the through-hole via's effect.

4.2.3 2D Integral Equation Solution of Parallel Plate Mode in Power-Ground Planes (PGPs)

Figure 4.3a shows a pair of PGPs with the antipad. Since the electromagnetic field is independent with z, the PGPs can be modeled as a 2D problem. A 2D region D is defined in Figure 4.3b, where the periphery of the PGPs and all antipad's perimeters form the boundary C with $\hat{n}'$ as its outward unit normal vector. According to the directions of the electromagnetic fields, we define the voltage between the top and bottom planes as $V = -d*E_z$ with d being the thickness of the substrate, and the horizontal current density as $J = -\hat{z} \times H$ on the top plane and $J = \hat{z} \times H$ on the bottom plane. J on top and bottom planes have the same values and opposite directions.

By using the integral equation theory [1], the voltage at certain point of the contour C can be expressed as the integral of voltages and current densities along the whole contour C as

$$V(r) = \frac{k}{2j} \oint_C \left[\frac{R}{R} \cdot \hat{n} H_1^{(2)}(kR) V(r') + j\eta d H_0^{(2)}(kR) J_n(r') \right] dl', \quad (4.1)$$

where $H_0^{(2)}$ and $H_1^{(2)}$ are the zero-order and first-order Hankel functions of the second kind, respectively. $R = r' - r$ and R represents the length

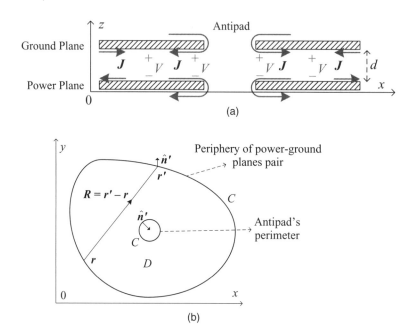

Figure 4.3   A pair of power-ground planes with the antipad. (a) Cross section and (b) top view [8].

of $\mathbf{R}$. $\mathbf{r}$ and $\mathbf{r'}$ are the observation and source points, respectively. The prime on dl emphasizes that the integration is over $\mathbf{r'}$. Both $\mathbf{r}$ and $\mathbf{r'}$ are located on C. k and η are the wavenumber and wave-impedance of the substrate. $k = \omega\sqrt{\mu\varepsilon}\,(1-j(\tan\delta + t/d)/2)$ and $\eta = \omega\mu/k$, where $\tan\delta$, ε, and μ are the loss tangent, permittivity, and permeability of the substrate, respectively, and t is the skin depth of the metal planes. $j = \sqrt{-1}$. $J_n = \hat{n}' \cdot (\hat{z} \times \mathbf{H})$ means the current density flowing into/from the region D on top/bottom plane. It should be noted that this method can solve PGPs with arbitrary shapes since C in Equation (4.1) can be arbitrarily shaped.

(V, J_n) in Equation (4.1) are classified into (V$_p$, J$_p$) (along the periphery) and (V$_a$, J$_a$) (along the perimeters of the antipads). The physical meaning of J_a is explained in Figure 4.4. I_{trace} denotes the current along the signal trace. J_a starts from the top plane, passes through the distributed resistance (R), inductance (L), capacitance (C), and conductance (G) between the top and bottom planes, and arrives at the bottom plane. Therefore, J_a represents the rerouted return current density of I_{trace}. By

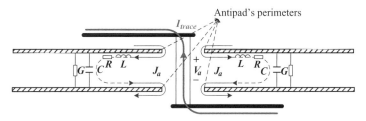

Figure 4.4 The rerouted return current of the traces (side view) [8].

this physical meaning, the electromagnetic field problem can be converted to a comprehensive equivalent circuit. In the following, the relationship between the defined V and J_n are obtained by solving the integral Equation (4.1).

The periphery of the PGPs is divided into many straight segments as those in Reference 1. Since the antipad's perimeter is much small in terms of the interesting wavelength, each of them is taken as one circle and (V_a, J_a) are assumed to be constant along each circle. After that, (V, J_n) of Equation (4.1) are expanded by using unit pulse functions defined on these straight segments/circles as

$$V_p(\boldsymbol{r}) = \sum_{i=1}^{N_p} V_{p,i} P_{p,i}(\boldsymbol{r}) \quad \text{and} \quad V_a(\boldsymbol{r}) = \sum_{i=1}^{N_a} V_{a,i} P_{a,i}(\boldsymbol{r}), \tag{4.2}$$

$$J_p(\boldsymbol{r}) = \sum_{i=1}^{N_p} I_{p,i} P_{p,i}(\boldsymbol{r})/w_{p,i} \quad \text{and} \quad J_a(\boldsymbol{r}) = \sum_{i=1}^{N_a} I_{a,i} P_{a,i}(\boldsymbol{r})/w_{a,i}, \tag{4.3}$$

with the unit pulse function defined as

$$P_{p/a,i}(\boldsymbol{r}) = \begin{cases} 1, & \boldsymbol{r} \in w_{p/a,i} \\ 0, & \boldsymbol{r} \notin w_{p/a,i}. \end{cases}$$

$w_{p/a,i}$ represents the ith straight segment/circle. In the following, $w_{p/a,i}$ is also used to represent the length of ith straight segment/circle. $I_{p/a,i}$ is the current flowing on ith straight segment/circle. $N_{p/a}$ is the number of peripheral segments/antipads. The total number of segments is $N = N_a + N_p$.

After this expansion, we define ports on each peripheral segment and antipad's perimeter as in Figure 4.5. Let $[V_p \ V_a]$ and $[I_p \ I_a]$ represent the expansion coefficients in Equations (4.2) and (4.3), the whole PGPs

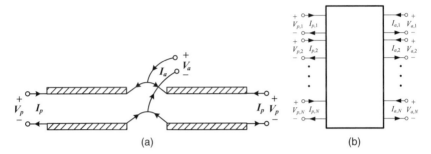

Figure 4.5   Definitions of (a) port voltages and currents (cross section) and (b) the equivalent *N*-port network.

pair is equivalent to a *N*-port circuit network, with $[V_p \; V_a]$ and $[I_p \; I_a]$ as its port voltages and currents. It should be noted that the reference planes of these ports are defined on the bottom surface of the top plane and top surface of the bottom plane.

Substituting Equations (4.2) and (4.3) into Equation (4.1), and matching both sides of Equation (4.1) with r at the center of each straight segment and circle, the following *N* by *N* linear equations are obtained

$$[U] \cdot \begin{bmatrix} V_p \\ V_a \end{bmatrix} = [H] \cdot \begin{bmatrix} I_p \\ I_a \end{bmatrix}, \qquad (4.4)$$

where $[U] = \begin{bmatrix} U^{pp} & U^{pa} \\ U^{ap} & U^{aa} \end{bmatrix}$ and $[H] = \begin{bmatrix} H^{pp} & H^{pa} \\ H^{ap} & H^{aa} \end{bmatrix}.$

The elements in $[U]$ and $[H]$ are calculated as

$$U_{ij} = \delta_{ij} - \frac{k}{2j} \int_{w_{pla,j}} \frac{R}{R} \cdot \hat{n}'_j H_1^{(2)}(kR) dl'_j, \qquad (4.5)$$

$$H_{ij} = \frac{k\eta d}{2w_{pla,j}} \int_{w_{pla,j}} H_0^{(2)}(kR) dl'_j, \qquad (4.6)$$

with $\hat{n}'_j$ being the normal vector of *j*th straight segment/circle, and δ_{ij} is the Kronecker delta. For $i = j$, the Hankel functions in Equations (4.5) and (4.6) are singular. In this case, the above integrals are analytically calculated in the following:

(a) ith testing point falls on ith straight segment.

Since $\boldsymbol{R} \perp \hat{\boldsymbol{n}}_i'$, $U_{ii}^{pp} = 1$.

Considering that $kw_{p,i}/2 \ll 1$, by using asymptotic expansions of Hankel functions, we get

$$H_{ii}^{pp} \approx (k\eta d/2)[1-(2j/\pi)\cdot(\ln(kw_{p,i}/4)-0.4228)]. \qquad (4.7)$$

(b) ith testing point falls on ith circle.

Without loss of the generality, the local coordinates are chosen as in Figure 4.6. We get $\boldsymbol{R} = \boldsymbol{r}_i' - \boldsymbol{r} = a_i[(\cos\phi-1)\hat{\boldsymbol{x}}+\sin\phi\hat{\boldsymbol{y}}]$ and $\hat{\boldsymbol{n}}_i' = -(\cos\phi\hat{\boldsymbol{x}}+\sin\phi\hat{\boldsymbol{y}})$, so that

$$U_{ii}^{aa} = 1-(k/2j)\int_{w_{a,i}} \boldsymbol{R}\cdot\hat{\boldsymbol{n}}_i'H_1^{(2)}(kR)/R dl_i' \approx 2, \qquad (4.8)$$

$$H_{ii}^{aa} \approx (k\eta d/2)[1-(2j/\pi)(\ln(ka_i/2)+0.5772)]. \qquad (4.9)$$

A perfect magnetic wall is assumed along the periphery due to the thin substrate. This means the magnetic field tangential to the periphery is zero. Therefore, $J_p = \hat{\boldsymbol{n}}'\cdot(\hat{z}\times\boldsymbol{H}) = 0$ along the periphery, and all ports on the left side of Figure 4.5b are open. Substituting $[I_p] = 0$ into Equation (4.4), we get

$$[\boldsymbol{Z}^a]\cdot[\boldsymbol{I}_a] = [\boldsymbol{V}_a], \qquad (4.10)$$

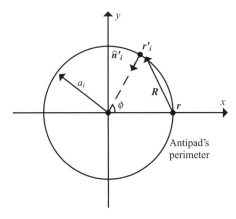

Figure 4.6   Local coordinates used for the calculation of self-elements [8].

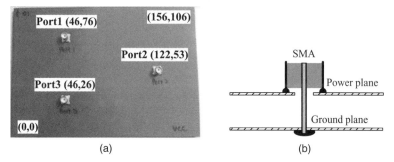

(a)   (b)

Figure 4.7 A pair of power-ground planes with three SMAs mounted on it. (unit: mm) (a) Top view and (b) cross section [8].

$$[Z^a] = \left([U^{aa}] - [U^{ap}] \cdot [U^{pp}]^{-1} \cdot [U^{pa}] \right)^{-1}$$

$$\left([H^{aa}] - [U^{ap}] \cdot [U^{pp}]^{-1} \cdot [H^{pa}] \right). \quad\quad (4.11)$$

The elements in $[Z^a]$ represent the self and mutual ground impedances of the PGPs. In this case, each PGPs pair is equivalent to an N_a-port network with $[Z^a]$ as its impedance matrix.

In the following, the ground impedance matrix $[Z^a]$ obtained from Equation (4.11) is validated. The PGPs under study are shown in Figure 4.7. The substrate between the metal planes has a thickness of 1.2 mm, a dielectric constant of 4.1, and a loss tangent of 0.015. Three subminiature version A (SMA) connectors are mounted on the up plane. The inner conductors of the SMAs pass through the antipads on the up plane and are soldered to the down plane. The outer conductors of the SMAs are soldered to the up plane. A network analyzer is then connected to these SMAs to measure the S-parameters. Figure 4.8 shows S_{11} and S_{21} obtained by the IEEC method, measurement, and the 3D numerical software HFSS [17]. Good agreement can be observed from the figures. For the IEEC method, the $[Z^a]$ is converted to S-parameters by using the reference impedance of 50 Ω.

4.2.4 Combinations of Transmission and Parallel Plate Modes

In this section, the above obtained equivalent network of PGPs is connected with the equivalent network of the signal traces (includes the

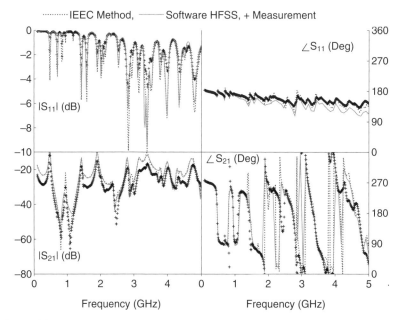

Figure 4.8 S-parameters of the power-ground planes [8]. Left: magnitude; right: phase. (See color insert.)

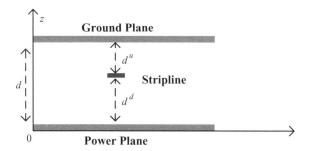

Figure 4.9   Cross section of a stripline [8].

striplines and microstrip lines), where the discontinuities of the via is considered by a PI circuit.

First of all, for the sake of simplification, a single stripline sandwiched between a pair of PGPs as shown in Figure 4.9 is used here to demonstrate the recombination of the parallel plate mode and the

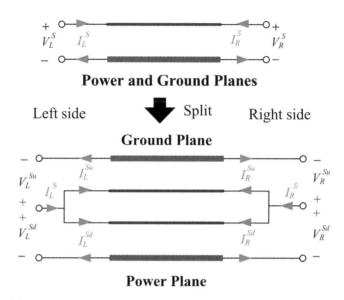

Figure 4.10 One stripline is split into up and down striplines [8].

stripline mode. The proposed recombination method is then extended to multi stripline cases later.

For the stripline commonly used in microwave engineering, its two parallel reference planes are shorted and hence equipotential. For the stripline sandwiched between PGPs, the parallel plate modes are excited which accumulate voltage drop between the power and ground planes. To consider this potential difference between the power and ground planes, we split the stripline and accordingly divide the total current into the currents flowing on the up and down surfaces of the stripline separately. Figure 4.10 shows such splitting. The subscripts L and R denote the left and right ports of the transmission lines, respectively. The superscripts S denotes the unsplit stripline, S_u and S_d denote the split up and down striplines, respectively.

For the unsplit stripline, its admittance matrix is defined as

$$\begin{bmatrix} I_L^S \\ I_R^S \end{bmatrix} = [Y^S] \begin{bmatrix} V_L^S \\ V_R^S \end{bmatrix}. \tag{4.12}$$

For the split stripline, the up and down admittance matrices are defined as

$$\begin{bmatrix} I_L^{S_u} \\ I_R^{S_u} \end{bmatrix} = \begin{bmatrix} Y^{S_u} \end{bmatrix} \begin{bmatrix} V_L^{S_u} \\ V_R^{S_u} \end{bmatrix}, \tag{4.13}$$

$$\begin{bmatrix} I_L^{S_d} \\ I_R^{S_d} \end{bmatrix} = \begin{bmatrix} Y^{S_d} \end{bmatrix} \begin{bmatrix} V_L^{S_d} \\ V_R^{S_d} \end{bmatrix}, \tag{4.14}$$

where $V_{R/L}^{S/S_u/S_d}$ and $I_{R/L}^{S/S_u/S_d}$ are port voltages and currents defined in Figure 4.10. $I_{R/L}^{S_u}$ and $I_{R/L}^{S_d}$ flow on the up and down surfaces of the stripline, respectively.

Now we need to derive the relationship between $[Y^{S_u}]$, $[Y^{S_d}]$, and $[Y^S]$. This is obtained by the split stripline returns to the unsplit stripline when the power and ground planes are shorted. By shorting the power and ground planes of the split stripline, we get

$$\begin{bmatrix} V_L^{S_u} \\ V_R^{S_u} \end{bmatrix} = \begin{bmatrix} V_L^{S_d} \\ V_R^{S_d} \end{bmatrix} = \begin{bmatrix} V_L^{S} \\ V_R^{S} \end{bmatrix}. \tag{4.15}$$

When the stripline approaches the ground plane ($d^u \to 0$ in Fig. 4.9), $I_{R/L}^{S_u} \to I_{R/L}^{S}$ and $I_{R/L}^{S_d} \to 0$. Conversely, when the stripline approaches the power plane ($d^u \to d$ in Fig. 4.9), $I_{R/L}^{S_d} \to I_{R/L}^{S}$ and $I_{R/L}^{S_u} \to 0$. Since d is much smaller in terms of the wavelength, it is reasonable to assume that $I_{R/L}^{S_u}$ and $I_{R/L}^{S_d}$ are linearly dependent with d. Based on these observations, we get that

$$\begin{bmatrix} I_L^{S_u} \\ I_R^{S_u} \end{bmatrix} = \frac{d^d}{d} \begin{bmatrix} I_L^{S} \\ I_R^{S} \end{bmatrix} \quad \text{and} \quad \begin{bmatrix} I_L^{S_d} \\ I_R^{S_d} \end{bmatrix} = \frac{d^u}{d} \begin{bmatrix} I_L^{S} \\ I_R^{S} \end{bmatrix}, \tag{4.16}$$

Substituting Equations (4.15) and (4.16) into Equations (4.12–4.14), we get

$$\begin{bmatrix} Y^{S_u} \end{bmatrix} = \frac{d^d}{d} \begin{bmatrix} Y^s \end{bmatrix}, \tag{4.17}$$

$$\begin{bmatrix} Y^{S_d} \end{bmatrix} = \frac{d^u}{d} \begin{bmatrix} Y^s \end{bmatrix}. \tag{4.18}$$

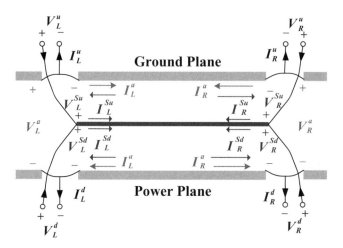

Figure 4.11 Port voltages and currents of three equivalent networks (longitudinal cross section) [8].

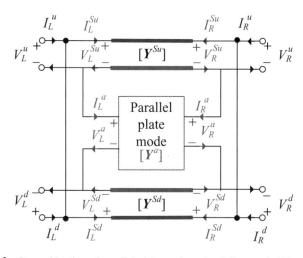

Figure 4.12 Recombination of parallel plate mode and stripline mode [8].

Observe that $[Y^S] = [Y^{Su}] + [Y^{Sd}]$.

The combination of stripline mode and parallel plate mode can be equivalent to the connection of three networks: the up split stripline, down split stripline, and the equivalent network of PGPs. Figure 4.11 shows the port voltages and currents of these three networks. Figure 4.12

shows their connections. The combined network is a four-port network with the admittance matrix defined as

$$
\begin{bmatrix} I_L^u \\ I_R^u \\ I_L^d \\ I_R^d \end{bmatrix} = [Y] \begin{bmatrix} V_L^u \\ V_R^u \\ V_L^d \\ V_R^d \end{bmatrix},
\tag{4.19}
$$

with port voltages $V_{R/L}^{u/d}$ and currents $I_{R/L}^{u/d}$ defined in Figures 4.11 and 4.12. These four ports account for the possible connections between this stripline and other signal traces (microstrip lines or striplines).

For the PGPs, the 2 by 2 admittance matrix is defined as

$$
\begin{bmatrix} I_L^a \\ I_R^a \end{bmatrix} = [Y^a] \begin{bmatrix} V_L^a \\ V_R^a \end{bmatrix},
\tag{4.20}
$$

with $[Y^a] = [Z^a]^{-1}$.

From Figures 4.11 and 4.12 we can find the relationship between port voltages and currents of different networks,

$$
\begin{bmatrix} \begin{bmatrix} V_L^{S_u} \\ V_R^{S_u} \end{bmatrix} \\ \begin{bmatrix} V_L^{S_d} \\ V_R^{S_d} \end{bmatrix} \\ \begin{bmatrix} V_L^a \\ V_R^a \end{bmatrix} \end{bmatrix} = [T] \cdot \begin{bmatrix} \begin{bmatrix} V_L^u \\ V_R^u \end{bmatrix} \\ \begin{bmatrix} V_L^d \\ V_R^d \end{bmatrix} \end{bmatrix} \quad \text{and} \quad \begin{bmatrix} \begin{bmatrix} I_L^u \\ I_R^u \end{bmatrix} \\ \begin{bmatrix} I_L^d \\ I_R^d \end{bmatrix} \end{bmatrix} = [T]^t \cdot \begin{bmatrix} \begin{bmatrix} I_L^{S_u} \\ I_R^{S_u} \end{bmatrix} \\ \begin{bmatrix} I_L^{S_d} \\ I_R^{S_d} \end{bmatrix} \\ \begin{bmatrix} I_L^a \\ I_R^a \end{bmatrix} \end{bmatrix},
\tag{4.21}
$$

with

$$
[T] = \begin{bmatrix} U_2 & 0 \\ 0 & U_2 \\ -U_2 & U_2 \end{bmatrix},
$$

and $[U_2]$ is a 2 by 2 unit matrix. Superscript t in Equation (4.21) means the transpose.

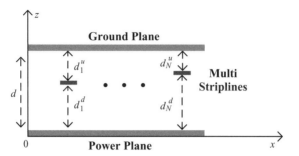

Figure 4.13   Cross section of N-striplines.

Substituting Equations (4.13), (4.14), (4.20), and (4.21) into Equation (4.19), we get the admittance matrix of the recombined network as

$$[Y] = [T]^t \cdot \begin{bmatrix} Y^{S_u} & 0 & 0 \\ 0 & Y^{S_d} & 0 \\ 0 & 0 & Y^a \end{bmatrix} \cdot [T] = \begin{bmatrix} Y^{S_u} + Y^a & -Y^a \\ -Y^a & Y^{S_d} + Y^a \end{bmatrix}. \quad (4.22)$$

The proposed recombination method can be easily extended to multi striplines case, where scalars $V_{R/L}^{S/S_u/S_d/u/d/a}$ and $I_{R/L}^{S/S_u/S_d/u/d/a}$ will be extended to vectors in order to consider the multiport of the striplines. Figure 4.13 shows the cross section of N-striplines. Similar to the single stripline, these multi striplines can be split as shown in Figure 4.14.

The admittance matrices of the unsplit striplines, up split striplines, and down split striplines are

$$\begin{bmatrix} I_L^S \\ I_R^S \end{bmatrix} = [Y^S] \begin{bmatrix} V_L^S \\ V_R^S \end{bmatrix}, \quad (4.23)$$

$$\begin{bmatrix} I_L^{S_u} \\ I_R^{S_u} \end{bmatrix} = [Y^{S_u}] \begin{bmatrix} V_L^{S_u} \\ V_R^{S_u} \end{bmatrix}, \quad (4.24)$$

and

$$\begin{bmatrix} I_L^{S_d} \\ I_R^{S_d} \end{bmatrix} = [Y^{S_d}] \begin{bmatrix} V_L^{S_d} \\ V_R^{S_d} \end{bmatrix}, \quad (4.25)$$

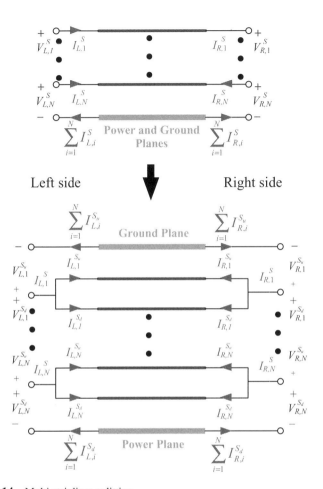

Figure 4.14   Multi striplines splitting.

where

$$\left[\boldsymbol{V}_L^{S/S_u/S_d}\right]=\begin{bmatrix}V_{L,1}^{S/S_u/S_d}\\V_{L,2}^{S/S_u/S_d}\\\vdots\\V_{L,N}^{S/S_u/S_d}\end{bmatrix},\quad\left[\boldsymbol{V}_R^{S/S_u/S_d}\right]=\begin{bmatrix}V_{R,1}^{S/S_u/S_d}\\V_{R,2}^{S/S_u/S_d}\\\vdots\\V_{R,N}^{S/S_u/S_d}\end{bmatrix},\quad\left[\boldsymbol{I}_L^{S/S_u/S_d}\right]=\begin{bmatrix}I_{L,1}^{S/S_u/S_d}\\I_{L,2}^{S/S_u/S_d}\\\vdots\\I_{L,N}^{S/S_u/S_d}\end{bmatrix},$$

and

$$\left[I_R^{S/S_u/S_d} \right] = \begin{bmatrix} I_{R,1}^{S/S_u/S_d} \\ I_{R,2}^{S/S_u/S_d} \\ \vdots \\ I_{R,N}^{S/S_u/S_d} \end{bmatrix}.$$

The relationship between $[Y^{S_u}]$, $[Y^{S_d}]$, and $[Y^S]$ are

$$[Y^{s_u}] = \begin{bmatrix} K_u & 0 \\ 0 & K_u \end{bmatrix} [Y^s], \tag{4.26}$$

$$[Y^{s_d}] = \begin{bmatrix} K_d & 0 \\ 0 & K_d \end{bmatrix} [Y^s], \tag{4.27}$$

where

$$[K_u] = \frac{1}{d} \begin{bmatrix} d_1^d & & 0 \\ & \ddots & \\ 0 & & d_N^d \end{bmatrix} \quad \text{and} \quad [K_d] = \frac{1}{d} \begin{bmatrix} d_1^u & & 0 \\ & \ddots & \\ 0 & & d_N^u \end{bmatrix},$$

with d, d_i^u, and d_i^d as shown in Figure 4.13.

The PGPs together with the multi striplines structure can be equivalent to a 4N-port network. It includes three 2N-port subnetworks: the up split N-striplines $[Y^{S_u}]$, down split N-stripline $[Y^{S_d}]$, and the equivalent network of PGPs$[Y^a]$ with 2N possible antipads. Figure 4.15 shows their connections. The admittance matrix of the combined network is defined as

$$\begin{bmatrix} I_L^u \\ I_R^u \\ I_L^d \\ I_R^d \end{bmatrix} = [Y] \begin{bmatrix} V_L^u \\ V_R^u \\ V_L^d \\ V_R^d \end{bmatrix}. \tag{4.28}$$

For the PGPs, the admittance matrix is defined as

$$\begin{bmatrix} I_L^a \\ I_R^a \end{bmatrix} = [Y^a] \begin{bmatrix} V_L^a \\ V_R^a \end{bmatrix}, \tag{4.29}$$

with $[Y^a] = [Z^a]^{-1}$.

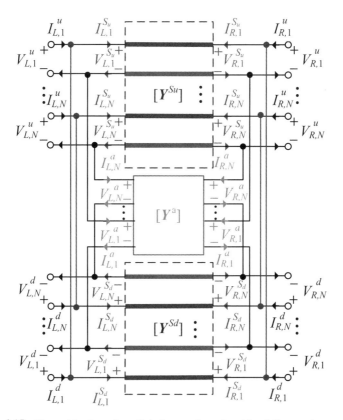

Figure 4.15   Recombination of parallel plate mode and multi stripline modes.

In Equations (4.28) and (4.29), the port voltages $V_{R/L}^{u/d/a}$ and currents $I_{R/L}^{u/d/a}$ are defined in Figure 4.15:

$$
\left[V_L^{u/d/a} \right] = \begin{bmatrix} V_{L,1}^{u/d/a} \\ V_{L,2}^{u/d/a} \\ \vdots \\ V_{L,N}^{u/d/a} \end{bmatrix}, \quad
\left[V_R^{u/d/a} \right] = \begin{bmatrix} V_{R,1}^{u/d/a} \\ V_{R,2}^{u/d/a} \\ \vdots \\ V_{R,N}^{u/d/a} \end{bmatrix},
$$

$$
\left[I_L^{u/d/a} \right] = \begin{bmatrix} I_{L,1}^{u/d/a} \\ I_{L,2}^{u/d/a} \\ \vdots \\ I_{L,N}^{u/d/a} \end{bmatrix}, \quad \text{and} \quad
\left[I_R^{u/d/a} \right] = \begin{bmatrix} I_{R,1}^{u/d/a} \\ I_{R,2}^{u/d/a} \\ \vdots \\ I_{R,N}^{u/d/a} \end{bmatrix}.
$$

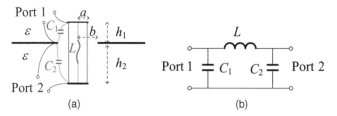

Figure 4.16 (a) Through-hole via (cross section) and (b) its equivalent PI circuit.

Similarly, we can get

$$[Y] = \begin{bmatrix} Y^{S_u} + Y^a & -Y^a \\ -Y^a & Y^{S_d} + Y^a \end{bmatrix}. \tag{4.30}$$

Above we had shown the combination of parallel plate mode with one of the transmission mode: the stripline mode. The combination of parallel plate mode with the microstrip mode is easier and much straightforward. The microstrip modes are connected to the parallel plate mode through the through-hole via, which is equivalent to a PI circuit.

The through-hole via is an important discontinued structure in the PDN. Figure 4.16 shows a through-hole via and its equivalent PI circuit.

The closed form of L in Figure 4.16 can be obtained from Reference 15 as

$$L = \frac{\mu}{2\pi}\left[h_1 \ln\left(0.5413\frac{h_1}{a}\right) + h_2 \ln\left(0.5413\frac{h_2}{a}\right)\right]. \tag{4.31}$$

Measurements had indicated that the effect of through-hole via is mainly capacitive, so the parasitic capacitors C_1 and C_2 should be accurately calculated. By assuming $R \to \infty$ in Reference 16 and using the asymptotic expansions of Bessel functions, we get

$$C_{1/2} = 4\pi\varepsilon / [h_{1/2} \ln(b/a)] \sum_{n=1,3,5,\cdots}^{\infty} [1 - K_0(k_n b)/K_0(k_n a)]/k_n^2, \tag{4.32}$$

where K_0 are the modified Bessel function of the second kind with zero order. $k_n = \sqrt{(n\pi/2h_{1/2})^2 - 1/\lambda_g^2}$ with λ_g being the wavelength in the

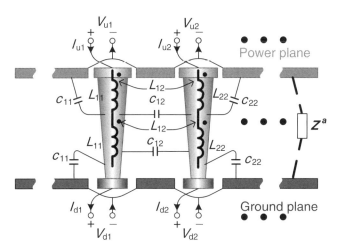

Figure 4.17   The cross section of the multi through-hole vias [9].

substrate. The convergence of Equation (4.32) is very fast, so that a few terms inside Σ are enough for the accurate calculation of C_1 and C_2.

For a more general simulation of the multi-vias, we propose the *nonequipotential* multi transmission line modeling method. Figure 4.17 shows such N-through-hole vias. The multi-vias are divided into small L-C sections, where L means the series inductors, C means the shunt capacitors, and the substrate between the power and ground planes serves as the "reference conductor." Unlike a traditional transmission line model, here, z^a obtained from Equation (4.10) is inserted between the reference conductors of adjacent L-C sections. Therefore, the reference conductors of each L-C section are nonequipotential.

The self and mutual inductors and capacitors in Figure 4.17 are the parameters to be extracted. These parameters can be calculated under the quasi-static assumption, or by analytical or full-wave de-embedding methods [9].

4.2.5 Cascade Connections of Equivalent Networks

In the proposed IEEC method, the whole PDN is decoupled into different parts: PGPs, vias, striplines, and microstrip lines. Each part supports different electromagnetic field distribution. Based on this, the equivalent circuit had been extracted for each part. The final equivalent circuit of the whole PDN (includes the PGPs and signal traces) can be

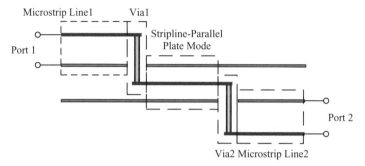

Figure 4.18 A power distribution network (longitudinal cross section) [8].

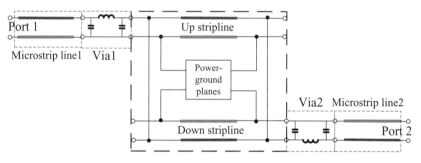

Figure 4.19 Equivalent circuit of the power distribution network of Figure 4.18 [8].

obtained by cascading the individual equivalent circuits. Figure 4.18 shows a typical PDN. To get its equivalent circuit,

1. Divide the whole structure into five parts: microstrip line1, via1, stripline-parallel plate mode, via2, and microstrip line2;
2. Each part is equivalent to a network;
3. Calculate the *ABCD* matrices of each network and then cascade them to get the final equivalent circuit as in Figure 4.19.

Here, we suggest using the transmission (*ABCD*) matrix to represent these individual equivalent networks, because the cascade connection can be easily found by multiplying the *ABCD* matrices of the individual network. Therefore, it is easy to develop the corresponding computer code. For the detailed descriptions of the microwave network matrices, reader can refer to Reference 18.

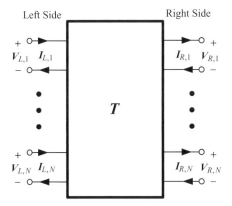

Figure 4.20   2N-port network.

The definition of $ABCD$ matrix of a 2N-port network as shown in Figure 4.20 is

$$\begin{bmatrix} V_L \\ I_L \end{bmatrix} = \begin{bmatrix} A & B \\ C & D \end{bmatrix} \begin{bmatrix} V_R \\ I_R \end{bmatrix},$$ (4.33)

where

$$[V_{L/R}] = \begin{bmatrix} V_{L/R,1} \\ V_{L/R,2} \\ \vdots \\ V_{L/R,N} \end{bmatrix} \quad \text{and} \quad [I_{L/R}] = \begin{bmatrix} I_{L/R,1} \\ I_{L/R,2} \\ \vdots \\ I_{L/R,N} \end{bmatrix}.$$

The subscripts L and R denote the left and right sides of the network, respectively.

The conversions between the transmission, impedance, admittance, and scattering matrices of a 2N-port network are summarized in Table 4.1.

The $ABCD$ matrix of the via's PI circuit as shown in Figure 4.16b is

$$\begin{bmatrix} A & B \\ C & D \end{bmatrix}_{Via} = \begin{bmatrix} 1 - \omega^2 C_2 L & j\omega L \\ j\omega(C_1 + C_2)\left(1 - \omega^2 \dfrac{C_1 C_2}{C_1 + C_2}L\right) & 1 - \omega^2 C_1 L \end{bmatrix}.$$ (4.34)

Table 4.1
Conversions between 2N-port Network Parameters

	S	Z	Y	ABCD
S	Definition: $\begin{bmatrix} V_L^- \\ V_R^- \end{bmatrix} = [S]\begin{bmatrix} V_L^+ \\ V_R^+ \end{bmatrix}$ with $[S] = \begin{bmatrix} S_{LL} & S_{LR} \\ S_{RL} & S_{RR} \end{bmatrix}$	$([Z]/Z_0 + [U_{2N}])^{-1}$ $([Z]/Z_0 - [U_{2N}])$	$([U_{2N}] + [Y]/Y_0)^{-1}$ $([U_{2N}] - [Y]/Y_0)$	$\begin{bmatrix} -U_N & A+B/Z_0 \\ U_N & CZ_0+D \end{bmatrix}^{-1}\begin{bmatrix} U_N & B/Z_0-A \\ U_N & D-CZ_0 \end{bmatrix}$
Z	$([U_{2N}] + [S])([U_{2N}] - [S])^{-1}Z_0$	Definition: $\begin{bmatrix} V_L \\ V_R \end{bmatrix} = [Z]\begin{bmatrix} I_L \\ -I_R \end{bmatrix}$ with $[Z] = \begin{bmatrix} Z_{LL} & Z_{LR} \\ Z_{RL} & Z_{RR} \end{bmatrix}$	$[Y]^{-1}$	$\begin{bmatrix} AC^{-1} & AC^{-1}D-B \\ C^{-1} & C^{-1}D \end{bmatrix}$

Y $([U_{2N}] - [S])([U_{2N}] + [S])^{-1})Y_0$

$[Z]^{-1}$

$$\begin{bmatrix} DB^{-1} & C - DB^{-1}A \\ -B^{-1} & B^{-1}A \end{bmatrix}$$

Definition:

$$\begin{bmatrix} I_L \\ -I_R \end{bmatrix} = [Y]\begin{bmatrix} V_L \\ V_R \end{bmatrix}$$

with

$$[Y] = \begin{bmatrix} Y_{LL} & Y_{LR} \\ Y_{RL} & Y_{RR} \end{bmatrix}$$

$$\begin{bmatrix} Z_{LL}Z_{RL}^{-1} & Z_{LL}Z_{RL}^{-1}Z_{RR} - Z_{LR} \\ Z_{RL}^{-1} & Z_{RL}^{-1}Z_{RR} \end{bmatrix}$$

$$\begin{bmatrix} -Y_{RL}^{-1}Y_{RR} & -Y_{RL}^{-1} \\ Y_{LR} - Y_{LL}Y_{RL}^{-1}Y_{RR} & -Y_{LL}Y_{RL}^{-1} \end{bmatrix}$$

ABCD $$\begin{bmatrix} S_{LL} - U_N & (S_{LL} + U_N)Z_0 \\ S_{RL} & S_{RL}Z_0 \end{bmatrix}^{-1} \begin{bmatrix} -S_{LR} & S_{LR}Z_0 \\ U_N - S_{RR} & (U_N + S_{RR})Z_0 \end{bmatrix}$$

Definition:

$$\begin{bmatrix} V_L \\ I_L \end{bmatrix} = \begin{bmatrix} A & B \\ C & D \end{bmatrix}\begin{bmatrix} V_R \\ I_R \end{bmatrix}$$

1. $[V_L]$ and $[V_R]$ are defined in Figure 4.20. $[V_L] = [V_L^+] + [V_L^-]$. $[V_R] = [V_R^+] + [V_R^-]$. $[I_L] = \frac{[V_L^+] - [V_L^-]}{Z_0}$, and $[I_R] = -\frac{[V_R^+] - [V_R^-]}{Z_0}$. $[V_L^+]$ and $[V_R^+]$ denote the incoming wave into left and right ports, respectively. $[V_L^-]$ and $[V_R^-]$ denote the outgoing wave from left and right ports, respectively. $Z_0(Y_0)$ is the reference impedance (admittance) for scattering parameters.

2. $[U_N]$ is an N by N unit matrix, and $[U_{2N}]$ is an $2N$ by $2N$ unit matrix.

3. Assume there exists one and only one C^{-1}, B^{-1}, Y_{RL}^{-1}, and Z_{RL}^{-1}.

209

For the frequencies much lower than the resonant frequency of the via, that is, $\omega^2(C_1 + C_2)L \ll 1$, above equation can be approximated as

$$\begin{bmatrix} A & B \\ C & D \end{bmatrix}_{Via} \approx \begin{bmatrix} 1 & j\omega L \\ j\omega(C_1 + C_2) & 1 \end{bmatrix}. \tag{4.35}$$

From Table 4.1, the *ABCD* matrix of [*Y*] in Equation (4.30) including *N*-striplines can be obtained as

$$\begin{bmatrix} A & B \\ C & D \end{bmatrix}_{PPS} = \begin{bmatrix} U_{2N} + Z^a Y^{Sd} & Z^a \\ Y^{Su} Z^a Y^{Sd} + Y^S & U_{2N} + Y^{Su} Z^a \end{bmatrix}, \tag{4.36}$$

where the subscript *PPS* means the parallel plate-stripline mode, [U_{2N}] is a 2*N* by 2*N* unit matrix. [Z^a], [Y^S], [Y^{Su}], and [Y^{Sd}] are defined in Equations (4.11), (4.23), (4.26), and (4.27), respectively. [Y^S] = [Y^{Su}] + [Y^{Sd}] are considered in Equation (4.36). It should be noted that for the *ABCD* matrix obtained in Equation (4.36), the ports are grouped into up and down ports, but not into the left and right ports as those in the definition of Equation (4.33).

Now the last question is how to get the *ABCD* matrix of the multi microstrip lines as in Figure 4.19 and how to get the [Y^S] of multi striplines. Both of the multi microstrip lines and striplines belong to the multiconductor transmission lines. Below we will derive their *ABCD* and admittance matrices from the telegrapher equations.

Figure 4.21 shows the multi transmission lines, which include *N*-signal conductors and one reference conductor. The telegrapher equations in frequency domain is

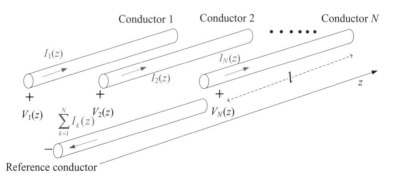

Figure 4.21 *N*-conductor transmission lines and their reference conductor.

$$\frac{\partial [V(z)]}{\partial z} = -[Z][I(z)], \tag{4.37}$$

$$\frac{\partial [I(z)]}{\partial z} = -[Y][V(z)], \tag{4.38}$$

where

$$[V(z)] = \begin{bmatrix} V_1(z) \\ V_2(z) \\ \vdots \\ V_N(z) \end{bmatrix} \quad \text{and} \quad [I(z)] = \begin{bmatrix} I_1(z) \\ I_2(z) \\ \vdots \\ I_N(z) \end{bmatrix}.$$

$[Z] = [R] + j\omega[L]$ and $[Y] = [G] + j\omega[C]$. $[R]$, $[L]$, $[G]$, and $[C]$ are matrices for series resistance, series inductance, shunt conductance, and shunt capacitance per unit length, respectively. They are symmetric matrices. The values of $[R]$, $[L]$, $[G]$, and $[C]$ can be obtained by solving the TEM/quasi-TEM problem in the cross section of the multi transmission lines [18].

The voltages and currents of different signal conductors are coupled in Equations (4.37) and (4.38), that is, the matrices $[Z]$ and $[Y]$ are dense matrices. In order to solve these equations, we need to decouple them first. For this purpose, we choose nonsingular matrices $[T_V]$ and $[T_I]$ to diagonalize $[Z]$ and $[Y]$:

$$[T_V]^{-1}[Z][T_I] = \mathrm{diag}\{Z_{m,i}\} = [Z_m], \tag{4.39}$$

$$[T_I]^{-1}[Y][T_V] = \mathrm{diag}\{Y_{m,i}\} = [Y_m], \tag{4.40}$$

where $i = 1, \cdots, N$ and *diag* means the diagonal matrix. It should be noted that for the lossless conductors ($[R] = [G] = [0]$) and homogeneous surrounding medium (ε, μ), $[L][C] = \mu\varepsilon[U_N]$, where $[U_N]$ is an N by N unit matrix. In that case, $[T_V]$ and $[T_I]$ always exist and $[T_V] = [T_I] = [T]$, where the column vectors of $[T]$ are the eigenvectors of $[L]$.

By using Equations (4.39) and (4.40), the coupled Equations (4.37) and (4.38) are decoupled as

$$\frac{\partial [V_m(z)]}{\partial z} = -[Z_m][I_m(z)], \tag{4.41}$$

$$\frac{\partial[I_m(z)]}{\partial z} = -[Y_m][V_m(z)], \qquad (4.42)$$

where $[V_m] = [T_V]^{-1}[V]$ and $[I_m] = [T_I]^{-1}[I]$ are defined as the modal voltage and current, respectively.

Now the original coupled N transmission lines are decoupled into N pairs of independent modal transmission lines. Each pair of modal transmission lines can be solved separately as in Reference 18. The modal propagation constant and modal characteristic impedance for ith pair of modal transmission lines are defined as $\gamma_{m,i} = \sqrt{Z_{m,i}Y_{m,i}}$ and $Z_{m0,i} = \sqrt{Z_{m,i}/Y_{m,i}}$, respectively. $\gamma_{m,i}^2$ is also the eigenvalue of $[Z][Y]$ and $[Y][Z]$.

After solving Equations (4.41) and (4.42) for each mode, we can get the modal $ABCD$ matrix as

$$\begin{bmatrix} V_{mL} \\ I_{mL} \end{bmatrix} = \begin{bmatrix} A_m & B_m \\ C_m & D_m \end{bmatrix} \begin{bmatrix} V_{mR} \\ I_{mR} \end{bmatrix}, \qquad (4.43)$$

where

$$[A_m] = \mathrm{diag}\{\mathrm{ch}\,\theta_{m,i}\}, \qquad (4.44)$$
$$[B_m] = \mathrm{diag}\{Z_{m0,i}\mathrm{sh}\,\theta_{m,i}\}, \qquad (4.45)$$
$$[C_m] = \mathrm{diag}\{\mathrm{sh}\,\theta_{m,i}/Z_{m0,i}\}, \qquad (4.46)$$
$$[D_m] = \mathrm{diag}\{\mathrm{ch}\,\theta_{m,i}\}, \qquad (4.47)$$

with $i = 1, \cdots, N$ and $\theta_{m,i} = \gamma_{m,i}l$. l is the length of the N transmission lines. Subscripts L and R denote the left and right terminals of the transmission lines, respectively.

The modal admittance matrix is obtained as

$$\begin{bmatrix} \mathrm{diag}\left\{\dfrac{\mathrm{cth}\,\theta_{m,i}}{Z_{m0,i}}\right\} & \mathrm{diag}\left\{-\dfrac{1}{Z_{m0,i}\mathrm{sh}\,\theta_{m,i}}\right\} \\ \mathrm{diag}\left\{-\dfrac{1}{Z_{m0,i}\mathrm{sh}\,\theta_{m,i}}\right\} & \mathrm{diag}\left\{\dfrac{\mathrm{cth}\,\theta_{m,i}}{Z_{m0,i}}\right\} \end{bmatrix} \begin{bmatrix} V_{mL} \\ V_{mR} \end{bmatrix} = \begin{bmatrix} I_{mL} \\ -I_{mR} \end{bmatrix}. \qquad (4.48)$$

Since $[V_{L/R}] = [T_V][V_{mL/R}]$ and $[I_{L/R}] = [T_I][I_{mL/R}]$, we can get the ABCD matrix of the uncoupled transmission lines as

$$\begin{bmatrix} A & B \\ C & D \end{bmatrix}_{MTL} = \begin{bmatrix} T_V & 0 \\ 0 & T_I \end{bmatrix} \begin{bmatrix} A_m & B_m \\ C_m & D_m \end{bmatrix} \begin{bmatrix} T_V^{-1} & 0 \\ 0 & T_I^{-1} \end{bmatrix} = \begin{bmatrix} T_V A_m T_V^{-1} & T_V B_m T_I^{-1} \\ T_I C_m T_V^{-1} & T_I D_m T_I^{-1} \end{bmatrix}$$
(4.49)

and the admittance matrix of the uncoupled transmission lines as

$$[Y]_{MTL} = \begin{bmatrix} T_I \text{diag}\left\{ \dfrac{\text{cth}\theta_{m,i}}{Z_{m0,i}} \right\} T_V^{-1} & T_I \text{diag}\left\{ -\dfrac{1}{Z_{m0,i}\text{sh}\theta_{m,i}} \right\} T_V^{-1} \\ T_I \text{diag}\left\{ -\dfrac{1}{Z_{m0,i}\text{sh}\theta_{m,i}} \right\} T_V^{-1} & T_I \text{diag}\left\{ \dfrac{\text{cth}\theta_{m,i}}{Z_{m0,i}} \right\} T_V^{-1} \end{bmatrix},$$
(4.50)

where subscript MTL means multi transmission lines. The corresponding equivalent 2N-port network is shown in Figure 4.22.

Until now we got all the ABCD matrices for each part in Figure 4.19. The ABCD matrix of the whole PDN can be easily obtained by multiplying the ABCD matrices of the individual network. After that, it can be substituted into a SPICE-like simulator together with

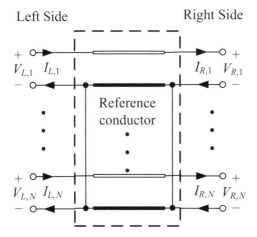

Figure 4.22 Equivalent network of the multiconductor transmission lines.

other on-board components to perform the system-level signal and PI simulation.

4.2.6 Simulation Results

In this section, two examples of PDNs are used to validate the accuracy and efficiency of the proposed IEEC method.

The first example is a PGP pair from Reference 13, which is redrawn in Figure 4.23. The signal trace includes three parts: left microstrip line, stripline, and right microstrip line. Two through-hole vias are used to connect them. Three ports are defined in Figure 4.23: Port 1 is located at an upper position and connected to the up plane and the down plane, Ports 2 and 3 are connected to the two ends of the signal trace, respectively. Port 1 is used to generate the noise between the two planes, and the coupling between Port 1 and Port 2 is analyzed. In the simulation, the thickness of the trace and planes is 0.01 mm, and the dielectric constant of the substrate is 4.6.

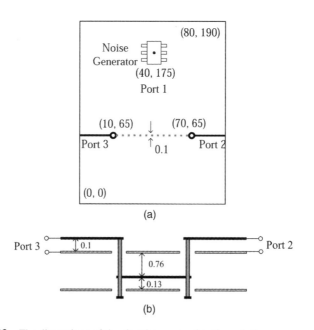

Figure 4.23 The dimensions of the signal trace passing through the power-ground planes [8]. (a) Top view and (b) side view (unit: mm)

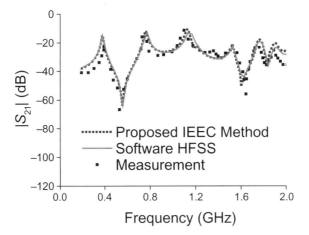

Figure 4.24 Coupling coefficient between Port 1 and Port 2 [8].

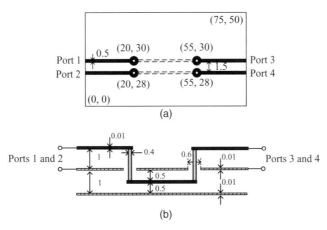

Figure 4.25 Two coupled signal traces passing through the power-ground planes [8]. (a) Top view and (b) side view. (unit: mm) The thickness of all traces and planes is 0.01 mm.

Figure 4.24 shows the magnitude of S_{21} obtained by the proposed method, the measurement result from Reference 13, and the software HFSS. Good agreement can be observed. The peaks of S_{21} are due to the strong resonances of the PGPs. At these resonant frequencies, the noise coupling between Port 1 and Port 2 is pronounced. In the practical design, the decoupling capacitors are usually used to eliminate the resonances of the PGPs.

The second example is two coupled signal traces. Their dimensions are shown in Figure 4.25. The substrate has a dielectric constant of 4.1.

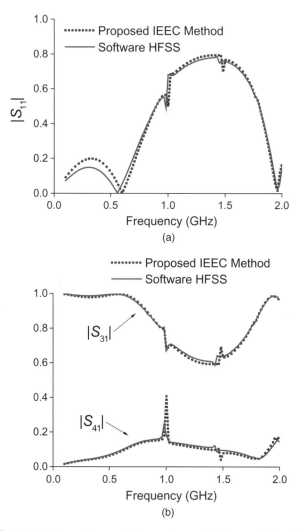

Figure 4.26 (a) Reflection, (b) transmission and crosstalk characteristics of the coupled signal traces [8].

The reflection, transmission, and crosstalk characteristics of the traces are simulated in Figure 4.26. Good agreement for all S-parameters is observed. This demonstrates the accuracy of the proposed IEEC method for the simulation of closely placed multi traces. Both the transmission line mode and parallel plate mode contribute to the cross talk. At the PGPs' inherent resonant frequency, this cross talk is amplified.

Table 4.2
Computing Times for the IEEC Method and
Three-Dimensional Simulator HFSS

Example	IEEC	HFSS
1	39.5 seconds	5 minutes
2	41 seconds	70 minutes
3	15 seconds	20 minutes

Table 4.2 lists the computing times of the proposed IEEC method and the 3D simulator HFSS. Example 1 refers to those PGPs used in Section 5.2.3. Examples 2 and 3 refer to those two examples used in this section. The computing time of the proposed IEEC method includes the solution of parallel plate mode and transmission line mode, and the combination of their equivalent circuits. It can be seen that the IEEC method greatly reduces the computing time in comparing with the HFSS. This is because the most time-consuming part of the IEEC method is only the numerical solution of the integral equation along the 1D contour C. Therefore, its number of unknowns is dramatically reduced. Whereas, for the 3D finite element method-based HFSS, a dense volume meshing is required because of the large aspect ratio of the PGPs and the tiny structures such as the through-hole and traces. This results in a lot of unknowns.

The last example is a complex PDN with three pairs of PGPs as shown in Figure 4.27. There are four copper planes: power plane 1, ground plane 1, power plane 2, and ground plane 2. The substrate is FR4 with $\varepsilon_r = 4.4$ and a loss tangent of 0.02. The thickness of the metal planes and the signal traces is 0.035 mm. All metals are copper. A circular air-hole passes through all metal planes and substrates. There are two sets of coupled signal traces. The first set of coupled traces is shown on the top of Figure 4.27a. It includes the coupled microstrip line 1 (CMS1), through-hole vias, and the coupled stripline (CSL). The second set of coupled traces is shown at the bottom of Figure 4.27a. It includes the coupled microstrip line 2 (CMS2), through-hole vias, and the coupled microstrip line 3 (CMS3). One power pin is used to connect power plane 1 and power plane 2. Another ground pin is used to connect ground plane 1 and ground plane 2.

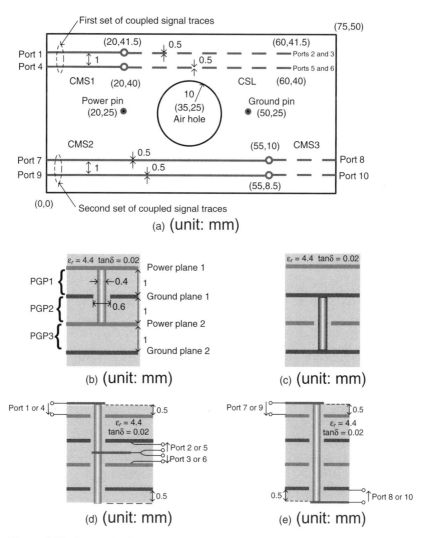

Figure 4.27 Power distribution network and two sets of coupled signal traces passing through it [9]. (a) Top view, and cross sections of (b) power pin, (c) ground pin, (d) first set of coupled traces, and (e) second set of coupled traces. (unit: mm) PGP means the power-ground plane pair.

The equivalent circuit of the entire structure is shown in Figure 4.28. Ten ports are defined at the terminals of those coupled microstrip lines and striplines. The CSLs are split into a set of up coupled stripline (up CSL) and a set of down coupled stripline (down CSL) by using the modal decoupling technology. As shown in Figure 4.27d, Ports 2 and

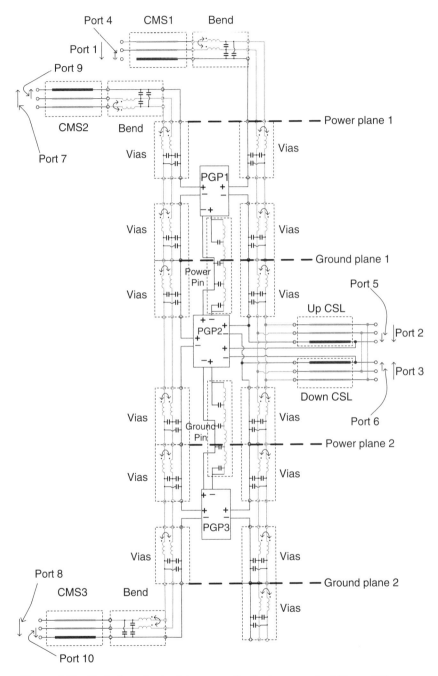

Figure 4.28   Equivalent circuit of the power distribution network of Figure 4.27.
PGP means the power-ground plane pair [9].

5 are defined between the up CSL and the ground plane 1, and Ports 3 and 6 are defined between the down CSL and the power plane 2. The equivalent networks of three PGP pairs (which are denoted as PGP1, PGP2, and PGP3 in Figs. 4.27 and 4.28) are extracted by using the integral equation. The via equivalent circuits are extracted by using the above proposed method. The bends which connect the microstrip lines and the vias are modeled as one L-C section. Their equivalent circuits are extracted by using the quasi-static method. All of these equivalent circuits are cascaded together to get the final equivalent circuit of the multilayered PDN.

The transmission ($|S_{31}|$), reflection ($|S_{11}|$), near end cross talk ($|S_{41}|$), and far end cross talk ($|S_{51}|$) of the first set of coupled traces are plotted in Figure 4.29. The transmission ($|S_{87}|$), reflection ($|S_{77}|$), near end cross talk ($|S_{97}|$), and far end cross talk ($|S_{107}|$) of the second set of coupled traces are plotted in Figure 4.30. Figure 4.31 shows the cross talk ($|S_{81}|$) between the first and second sets of coupled traces. The results from the full-wave HFSS simulation are also plotted for validation. For this complex multilayered PDN, the proposed IEEC method shows good agreement with the full-wave solver.

For the high frequency, there is a little difference between the IEEC method and the full-wave simulation. This is because the radiation from the microstrip lines and the periphery of the PGPs become strong at the high frequency. This radiation is neglected in the current proposed method. Due to this radiation, the peak in the HFSS result of Figure 4.29b is also smaller than that in the IEEC result. Radiative models of microstrip lines and PGPs can be used to reduce this difference.

4.3 3D HYBRID INTEGRAL EQUATION METHOD

4.3.1 Overview of the Algorithm

The signal and PI problem of PGPs is analyzed in the last section. In this section, we will focus on the emission and susceptibility modeling of the finite-size PGPs. The signal traces passing through PGPs produce the noise, which then propagates inside the PGPs and leaks into the surrounding area of the package through the periphery and gaps of the PGPs. On the other hand, the external interference will also induce

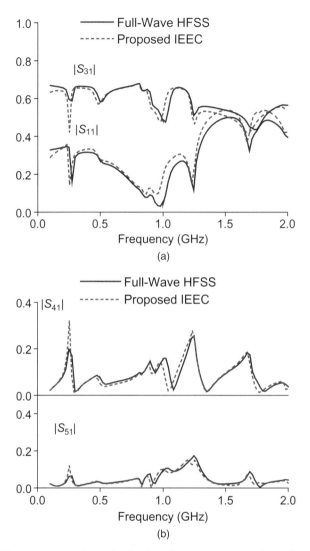

Figure 4.29 (a) Transmission and reflection, (b) near end cross talk and far end cross talk of the first set of coupled traces [9].

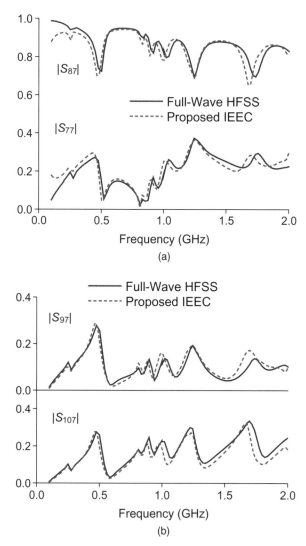

Figure 4.30 (a) Transmission and reflection, (b) near end cross talk and far end cross talk of the second set of coupled traces [9].

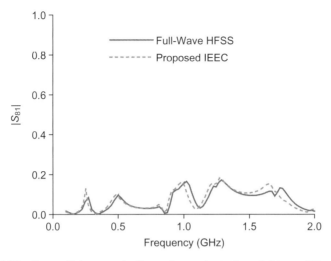

Figure 4.31   Cross talk between the first and second set of coupled traces [9].

currents on the vias inside the PGPs. It may lead to the malfunction of the circuit integrated inside the package. Above electromagnetic emission and susceptibility problems are even more pronounced, if the noise spectrum covers any of the resonance frequencies of the PGPs.

These challenges require the simultaneous simulations of emission and susceptibility. However, in most established electromagnetic solvers for PGPs, such as the Foldy–Lax multiple scattering method [2, 3] and the finite differential method [6], an assumption is made that the PGPs are continuous and extend to infinity, or that their boundaries are perfect magnetic conductors (PMCs). If the PGPs are assumed to be infinitely extended, the resonances due to the wave's reflection from the practical boundary cannot be modeled. The assumption of the PMC boundary is accurate only for the signal and PI simulation of the PGPs, but it is not suitable for the simulation of emission and susceptibility. This is because the electromagnetic waves cannot penetrate the PMC boundary; that is, the internal and external electromagnetic fields are factitiously isolated in those simulation algorithms. This is not true for real PGPs. Some researchers considered the radiation's effect on the impedance of PGPs by increasing the loss tangent of the substrate between the power and the ground planes [19]. However, the assumption of a PMC boundary is still applied, which prevents an accurate prediction of the radiation.

In this section, we propose a novel hybrid integral equation method. It is based on a PGP's model, which differs from established models in that it considers the finite size and rectangular shape of the PGPs and does not assume a PMC boundary. Since we introduce this non-PMC boundary, we can accurately calculate the external radiated field that is produced by the source inside the PGPs, as well as the electric current inside the PGPs that is induced by external sources. Therefore, this proposed method is able to solve all aspects of the EMC problems related to the PGPs, such as the coupling, emission, and susceptibility problems. Moreover, in practical designs, the PGP is often split into several "islands" to provide powers for different circuits. The gaps between these islands will introduce significant electromagnetic emissions. These gaps are also considered in our model. The proposed method is optimized by making full use of the structural features of the PGPs, so that it not only provides a much accurate solution, but also reduces the total number of unknowns.

In the proposed hybrid integral equation method, first, the whole geometry of the PGPs is decoupled into the internal subdomain (including the substrate that is sandwiched between the power and the ground planes) and the external subdomain (the free-space surrounding the PGPs). After that, the equivalent electric and magnetic currents are placed on the interface between these two subdomains. Integral equations are then created, which are based on the electric and magnetic fields produced by those equivalent currents. These integral equations are discretized to linear equations through the standard process of the method of moments (MoMs), which are then solved to get the values of the equivalent currents. Finally, the electromagnetic fields in the internal and external subdomains can be easily calculated according to the radiations from these equivalent currents.

4.3.2 Equivalent Electromagnetic Currents and Dyadic Green's Functions

A typical finite power-ground structure is shown in Figure 4.32, where the upper and lower planes are assumed to be perfect electric conductors. The planes can be split into several parts to supply powers to different circuit modules. In Figure 4.32, E^{out} denotes the potential external interference sources. E^{in} denotes the potential internal interference source, which is introduced by the signal currents. The objective

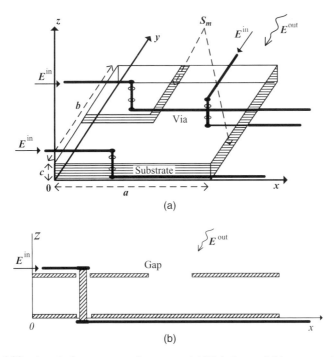

Figure 4.32   A typical power-ground structure: (a) Bird view and (b) cross section [11].

is to calculate the fields generated inside and outside of the PGPs by these sources. The detailed formulation is presented in the following.

First, the entire domain is divided into the internal subdomain (including the substrate that is sandwiched between the power and the ground planes) and the external subdomain (the free space surrounding the PGPs). Let S_m denote the periphery and gaps of the PGPs (the gray regions in Fig. 4.32a), and S be the whole closed surface of the internal subdomain (including S_m and the upper and lower planes) with $S_m \in S$. After that, by using the equivalence principle, S_m is closed by perfect electric conductors, and the equivalent magnetic currents M_S with the same values and opposite directions are placed on both sides of S_m. The equivalent electric current J_S is placed on the outside of S, and the vias are replaced by the equivalent volume electric currents J_V, as shown in Figure 4.33. Finally, each subdomain can be modeled by using the different integral equations, which are coupled through the equivalent currents M_S and J_S.

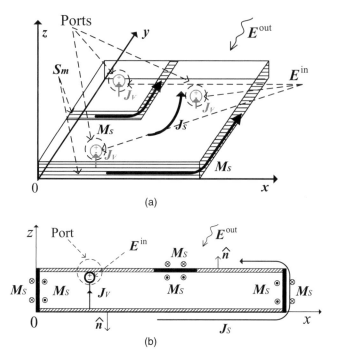

Figure 4.33 Equivalent problem of the power-ground planes: (a) Bird view and (b) cross section [11].

For the external subdomain, the integral kernel is the free-space Green's function, while for the internal subdomain, the integral kernel is the rectangular cavity's dyadic Green's function [20] with the following features:

1. There is no need for unknown electric currents to be placed on the inside of S, because their effects have already been taken care of by the dyadic Green's functions themselves.

2. When the thickness of the substrate is much smaller than the shortest wavelength of interest, only low-order modes are considered inside the PGPs. This is because the high-order modes decay very quickly along their propagation paths. In this case, the dyadic Green's functions of the internal problem are reduced to 2D functions (while the external problem is still a 3D problem), which reduces the complexity of the internal problem.

For the internal subdomain, two integral equations are derived:
Due to the continuity of the tangential magnetic field on S_{m}, we get

$$\left\lfloor L^h \left[r, M_S(r')\right] + K^h \left[r, \dot{J}_V(r')\right]\right\rfloor_t + \hat{n}/2 \times \dot{J}_S(r) = 0, \quad (4.51)$$

with the observation point $r \in S_{\mathrm{m}}$.

Due to the vanishing of the total tangential electric field on vias' surface, we get

$$\left\lfloor K^e \left[r, M_S(r')\right] - L^e \left[r, \dot{J}_V(r')\right]\right\rfloor_t = E_t^{\mathrm{in}}(r), \quad (4.52)$$

with the observation point r located on the vias.

For the external subdomain, one integral equation is derived:

Due to the vanishing of the total tangential electric field on the outside of S, we get

$$\left\lfloor K^0 \left[r, M_S(r')\right] + L^0 \left[r, \dot{J}_S(r')\right]\right\rfloor_t + \hat{n}/2 \times M_S(r) = E_t^{\mathrm{out}}(r) \quad (4.53)$$

with the observation point $r \in S$.

In above Equations (4.51–4.53), the unknown functions to be solved for are M_S, $\dot{J}_V$, and $\dot{J}_S$. $\dot{J}_S = J_S \eta_0$ and $\dot{J}_V = J_V \eta_0$ denote the scaled equivalent electric currents on the outside of S and along vias with $\eta_0 = 120\pi$, respectively. Equations (4.51–4.53) are coupled to each other, they must be solved together to get the values of the unknowns. The common unknowns in internal and external equations are $\dot{J}_S$ and M_S. They are used to couple the internal and external subdomains. They are related to the surface electromagnetic fields E and H as follows:

$$\dot{J}_S(r) = \hat{n} \times \dot{H}(r) \quad \text{and} \quad M_S(r) = E(r) \times \hat{n}, \quad (4.54)$$

where $\hat{n}$ denotes the outward unit normal vector of S, as shown in Figure 4.33b. $\dot{H} = H \eta_0$ denotes the scaled magnetic field. E_t^{in} denotes the internal excitation applied on the vias, which represents the interference that comes from the integrated circuit (IC) connected to the vias. E_t^{out} denotes the incident wave from the outside of the PGPs. The subscript t denotes the tangential component. Similar to the feed modeling commonly used in MoMs simulations of wire antennas, E_t^{in} is modeled as delta-gap voltage sources as shown in Figure 4.33. Ports are defined together with these delta-gap voltage sources, so that the PGPs with N vias are equivalent to an N-port network. Its admittance/impedance matrix can be calculated by solving $\dot{J}_V$.

The superscripts h, e, and 0 in the operators L and K denote the magnetic field in the cavity, the electric field in the cavity, and the electric field in free-space, respectively. The operators L^e, L^h, K^e, K^h, L^0, and K^0 in Equations (4.51–4.53) can be expressed in two different forms according to the different Green's functions used: one is to directly use the electric dyadic Green's functions of the first and second kinds $\bar{G}_{e1}$ and $\bar{G}_{e2}$ (this form represents the electric and magnetic field integral equations); another is to use the magnetic and electric vector potentials $\bar{G}_A$ and $\bar{G}_F$, as well as the electric and magnetic scalar potential G_Φ and G_Ψ (this form represents the mixed potential integral equation). The later one has weaker singularities than the former one. Therefore, when discretizing L^e, L^h, K^e, K^h, L^0, and K^0 into linear equations, it is easier to accurately calculate the coefficients matrix by using the potential dyadic Green's functions.

It should be noted that the PGPs are highly resonant structures. At the resonance frequency, the numerical solution becomes sensitive to the coefficients matrix's values. According to these, the potentials with a weaker singularity are used to express the operators in Equations (4.51–4.53) in the following way:

$$L^e\left[r, X(r')\right] = jk_0\mu_r \int_V \bar{G}_A\left(r,r'\right) \cdot X(r')dv'$$
$$+ j\nabla \int_V G_\Phi\left(r,r'\right)\nabla' \cdot X(r')dv' / k_0\varepsilon_r, \quad (4.55)$$

$$L^h\left[r, X(r')\right] = jk_0\varepsilon_r \int_S \bar{G}_F\left(r,r'\right) \cdot X(r')ds'$$
$$+ j\nabla \int_S G_\Psi\left(r,r'\right)\nabla'_S \cdot X(r')ds' / k_0\mu_r, \quad (4.56)$$

$$K^h\left[r, X(r')\right] = \int_V \nabla \times \bar{G}_A\left(r,r'\right) \cdot X(r')dv', \quad (4.57)$$

$$K^e\left[r, X(r')\right] = \int_S \nabla \times \bar{G}_F\left(r,r'\right) \cdot X(r')ds', \quad (4.58)$$

$$L^0\left[r, X(r')\right] = jk_0 \int_S G_0\left(r,r'\right) \cdot X(r')ds'$$
$$+ j\nabla \int_S G_0\left(r,r'\right)\nabla'_S \cdot X(r')ds' / k_0, \quad (4.59)$$

$$K^0\left[r, X(r')\right] = \int_S \nabla G_0\left(r,r'\right) \times X(r')ds', \quad (4.60)$$

where $X(r')$ is the unknown vector function ($\dot{J}_S$, $\dot{J}_V$ or M_S); $G_0(r,r') = e^{-jk_0|r-r'|}/4\pi|r-r'|$ is the free-space Green's function with the free-space wave number $k_0 = \omega\sqrt{\varepsilon_0\mu_0}$; ω is the angular frequency; ε_0 and μ_0 are the permittivity and permeability in vacuum, respectively; ε_r and μ_r are the relative permittivity and permeability of the substrate. The losses in substrate are accounted for by assuming complex ε_r and μ_r. The prime on ds' and dv' emphasizes that the integration is over r'.

The relationships between potential Green's functions are $\nabla \cdot \bar{G}_A = -\nabla' G_\Phi$ and $\nabla \cdot \bar{G}_F = -\nabla' G_\Psi$. Under the Cartesian coordinate system defined in Figure 4.32, the expressions of these potential Green's functions are written as

$$\bar{G}_A(r,r') = \sum_{n=0}^{\infty}\sum_{m=0}^{\infty}\sum_{l=0}^{\infty} C_{mnl}$$

$$\begin{bmatrix} C_xS_yS_zC_x'S_y'S_z' & 0 & 0 \\ 0 & S_xC_yS_zS_x'C_y'S_z' & 0 \\ 0 & 0 & S_xS_yC_zS_x'S_y'C_z' \end{bmatrix}, \quad (4.61)$$

$$\bar{G}_F(r,r') = \sum_{n=0}^{\infty}\sum_{m=0}^{\infty}\sum_{l=0}^{\infty} C_{mnl}$$

$$\begin{bmatrix} S_xC_yC_zS_x'C_y'C_z' & 0 & 0 \\ 0 & C_xS_yC_zC_x'S_y'C_z' & 0 \\ 0 & 0 & C_xC_yS_zC_x'C_y'S_z' \end{bmatrix}, \quad (4.62)$$

$$G_\Phi = \sum_{n=0}^{\infty}\sum_{m=0}^{\infty}\sum_{l=0}^{\infty} C_{mnl}S_xS_yS_zS_x'S_y'S_z', \quad (4.63)$$

$$G_\Psi = \sum_{n=0}^{\infty}\sum_{m=0}^{\infty}\sum_{l=0}^{\infty} C_{mnl}C_xC_yC_zC_x'C_y'C_z', \quad (4.64)$$

where $C_{mnl} = \dfrac{\delta_m\delta_n\delta_l}{abc\left(k_x^2 + k_y^2 + k_z^2 - k_0^2\varepsilon_r\mu_r\right)}$,

$$\delta_m = \begin{cases} 1, & m=0 \\ 2, & m\neq 0 \end{cases}, \delta_n = \begin{cases} 1, & n=0 \\ 2, & n\neq 0 \end{cases}, \text{and } \delta_l = \begin{cases} 1, & l=0 \\ 2, & l\neq 0 \end{cases}.$$

$k_x = m\pi/a$, $k_y = n\pi/b$, and $k_z = l\pi/c$, for $m,n,l = 0, 1, 2, \cdots$. $C_\beta = \cos(k_\beta\beta)$, $S_\beta = \sin(k_\beta\beta)$, $C_\beta' = \cos(k_\beta\beta')$, and $S_\beta' = \sin(k_\beta\beta')$, for $\beta = x/y/z$ and

$\beta' = x'/y'/z'$. The length, width, and height of the PGPs are defined to be a, b, and c, respectively, as shown in Figure 4.32a.

After the integral Equations (4.51–4.53) are obtained, we need to discretize them into the linear equations, which can be solved by the computer. To do so, first, the unknown currents $\dot{J}_S$ and M_S are expanded by using the subsectional vector basis function defined in Reference 21, while $\dot{J}_V$ is expanded by using the subsectional vector basis function which is uniform along each via, where each via is modeled as the cylinder with finite radius. After that, the basis functions of M_S are used to perform the dot products on both sides of integral Equation (4.51). Subsequently, these dot products are integrated over each basis function's domain. This is the so-called Galerkin's process. The similar Galerkin's process is applied between the basis functions of $\dot{J}_V$ and integral Equation (4.52), as well as the basis functions of $\dot{J}_S$ and the integral Equation (4.53). After this discretization, the variables r and r' in Equations (4.51–4.60) vanish, and the following linear equations can be obtained:

The discretization of the internal integral Equations (4.51) and (4.52) leads to

$$[A]\begin{bmatrix} M_S \\ \dot{J}_V \end{bmatrix} + [C][\dot{J}_S] = [E_t^{\text{in}}]. \tag{4.65}$$

The external integral Equation (4.53) is discretized to

$$[D]\begin{bmatrix} M_S \\ \dot{J}_V \end{bmatrix} + [Z][\dot{J}_S] = [E_t^{\text{out}}], \tag{4.66}$$

where $[M_S]$, $[\dot{J}_V]$, and $[\dot{J}_S]$ denote the vectors of expansion coefficients of the unknown functions. The coefficient matrices $[A]$, $[C]$, $[D]$, and $[Z]$ are obtained through the discretization of the integral operators L^e, L^h, K^e, K^h, L^0, K^0, and $\hat{n}/2 \times$ in Equations (4.51–4.53). These operators describe the electromagnetic interactions between M_S, $\dot{J}_V$, and $\dot{J}_S$. Since Galerkin's process is used, matrices $[A]$ and $[Z]$ are symmetric. It should be noted that matrices $[D]$, $[Z]$, and $[C]$ are independent of the vias' layout, so that they can be reused for the same PGPs with different vias layouts. Either Equation (4.65) or (4.66) is underdetermined; they must be solved together to get the values of the equivalent currents M_S and $\dot{J}_S$. This means that M_S and $\dot{J}_S$ are dependent on both the internal and external geometries, as opposed to merely one of them. On the other hand, we can see from Equations (4.65) and (4.66) that both internal source $[E_t^{in}]$ and external source $[E_t^{out}]$ may contribute to M_S and $\dot{J}_S$.

The linear Equations (4.65) and (4.66) are often iteratively solved for a larger number of unknowns. However, the convergence will be very slow at the resonance frequencies of the PGPs due to the ill-conditioned matrix $[A]$. We will use the same method used in Reference 22 to perform the LU factorization of $[A]$, where the symmetry of the matrix $[A]$ is used to accelerate the factorization. After that, we can decouple Equations (4.65) and (4.66), and solve the following equivalent equations:

$$\left([Z]-[D]\cdot[A]^{-1}\cdot[C]\right)\left[\dot{J}_S\right]=\left[E_t^{out}\right]-[D]\cdot[A]^{-1}\cdot\left[E_t^{in}\right], \quad (4.67)$$

$$\begin{bmatrix} M_S \\ \dot{J}_V \end{bmatrix}=[A]^{-1}\cdot\left(\left[E_t^{in}\right]-[C]\cdot\left[\dot{J}_S\right]\right). \quad (4.68)$$

Equation (4.67) is iteratively solved for $\dot{J}_S$ by using the generalized minimal residual algorithm (GMRES) method [23, 24], then M_S and $\dot{J}_V$ can be calculated from Equation (4.68) by using the previously obtained $\dot{J}_S$. Subsequently, the impedance matrix of the PGPs and the induced electric current along the vias can be calculated by using $\dot{J}_V$, while the external radiated field can be calculated by using $\dot{J}_S$ and M_S. In general, the impedance, induced current, and the radiated field are expected to be as low as possible to ensure that the high-speed electronic devices work at a normal condition.

Since the radiation from the PGPs is considered in the proposed method, the calculated impedance matrix is not an ideal lossless network matrix even for lossless substrates. This is different from most available methods where the PMC boundary assumption is applied. In the following section, the proposed method is validated through several numerical examples. Since both internal and external sources are considered in the proposed method, it can be used for both emission and susceptibility simulations.

4.3.3 Simulation Results

For validation purpose, the proposed method is compared to measurement data, as well as to results from the commercial simulators FEKO [25] and HFSS [17].

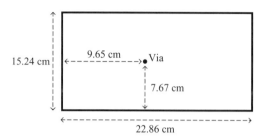

Figure 4.34 Power-ground planes with one via [11].

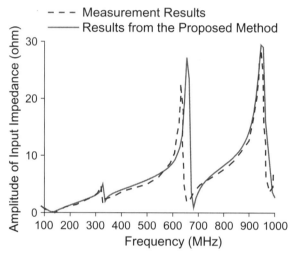

Figure 4.35 Magnitude of the input impedance of the via in Figure 4.34 [11].

A. Input and Mutual Impedances of PGPs. A single-layer power-ground structure with one via is shown in Figure 4.34. The radius of the via is 350 μm, and the substrate material is FR4 with a thickness of 0.1588 cm, $\varepsilon_r = 4.1$, and $\mu_r = 1$. In Figure 4.35, the input impedance of the via is calculated by using the proposed method and is compared to measurement results from Reference 19. Four modes are excited according to the dimension of the PGPs and the location of the via: TM_{10}, TM_{20}, TM_{30}, and TM_{02}, where TM_{30} and TM_{02} are degenerated. Good agreement can be observed.

The second example is a single-layer power-ground structure with two vias as shown in Figure 4.36. The substrate material is FR4 with

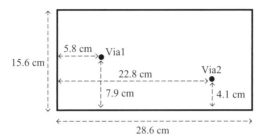

Figure 4.36   Power-ground planes with two vias [11].

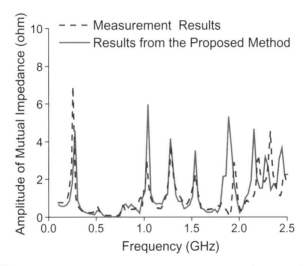

Figure 4.37   Magnitude of the mutual impedance between two vias shown in Figure 4.36 [11].

a 0.6 mm thickness, a relative permittivity of 4.2, and a 0.018 loss tangent. The radius of two vias is 100 μm. Figure 4.37 shows the calculated mutual impedance between two vias by using the proposed method along with measurement results from Reference 26. From this figure we can see that the agreement is good for the low-order modes (of which the resonance frequencies are below 1.5 GHz), while for the high-order modes, the measured resonance frequencies are slightly higher than the calculated ones. This difference is due to measurement uncertainties. At high frequencies, the permittivity and loss tangent of the substrate often differ from their nominal values at low frequencies,

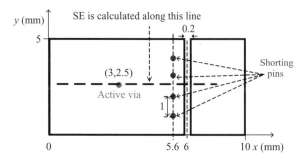

Figure 4.38   A split power-ground structure with four shorting pins, where the unit is in millimeters [11]. The SE is calculated along the line: from 1 mm, 2.5 mm, and 0.4 mm to 8 mm, 2.5 mm, and 0.4 mm.

and the measured S-parameters are more sensitive to the parasitic capacitance and inductance than those at low frequencies. These affect the high-order modes' resonance frequencies. The small shifts of the measured resonance frequencies of high-order modes are also observed in figure 3 of Reference 26.

B. Use of Shorting Pins to Reduce Radiation from Gaps. In practical package design, the PGP is often split into several parts to supply powers for different circuit modules. The gaps between different parts usually degrade the shielding performance of the PGP; that is, they strengthen the leakage radiation, compared to the unsplit PGP. The next example is used to validate the ability of the proposed method to model the interactions between internal sources and external radiated fields, as well as to examine the use of shorting pins to reduce the gaps' radiation. The dimensions of the PGPs are plotted in Figure 4.38. The substrate material is of a 0.2-mm thickness, $\varepsilon_r = 1$, and $\mu_r = 1$. The radius of the active via and shorting pins is 50 μm. The upper plane is split into two parts at $x = 6$ mm, and an active via is placed at the center of the left part. To reduce the radiated field from the gap, four shorting pins are evenly distributed along the gap as shown in Figure 4.38.

The electric shielding effectiveness (SE) is a common EMC measure used in EMC engineering to evaluate the shielding performance. In this chapter, it is defined as the ratio of the radiated electric field without the shorting pins to the radiated electric field with the presence of the shorting pins, measured at the same line, which is located 0.2 mm above the upper plane in Figure 4.38.

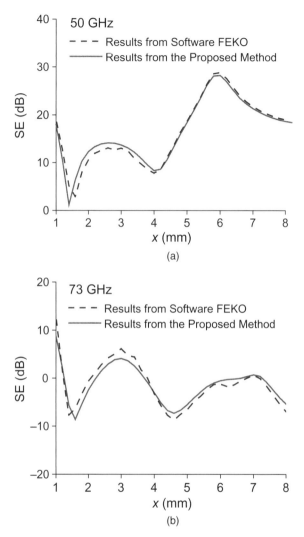

Figure 4.39 Electric shielding effectiveness of the shorting pins at (a) 50 GHz and (b) 73 GHz, along the line from 1 mm, 2.5 mm, and 0.4 mm to 8 mm, 2.5 mm, and 0.4 mm [11].

Figure 4.39a,b shows the electric SE of the shorting pins at 50 GHz and 73 GHz, respectively. The lowest resonance frequency of the left part of the PGPs is at 50 GHz according to its dimension and the location of the active via. At this resonance frequency, the radiated field from the gap is very strong. It can be observed from Figure 4.39a that by using the shorting pins, this radiated field can be reduced by 10 dB for most observation points. However, at 73 GHz, we can see from Figure 4.39b that for most observation points the SE is negative; that is, the use of the shorting pins results in an even higher radiation. This is due to a resonance between the shorting pins. They divide the whole gap into five equivalent magnetic currents, with 73 GHz being close to their resonance frequency. This suggests that for the practical package design, the distance between shorting pins should be carefully chosen according to the operating frequencies.

In order to validate the accuracy of the proposed method to model the interaction between the external radiated field and the internal source, the simulation results are compared to results obtained from the commercial software FEKO and are plotted in Figure 4.39. Again, a good agreement is observed.

C. Induced Electric Current Due to External Noise. In the above examinations, the sources are located inside the PGPs, the coupling inside the PGPs and radiated fields are calculated. These correspond to the conducted and radiated emission issues in EMC engineering. In the following examination, the source is located outside the PGPs, and the induced electric currents along vias inside the PGPs are calculated. This corresponds to another important EMC issue: susceptibility. The external source can induce electric currents along the vias, which will then propagate and produce noise inside the PGPs. These induced currents are usually small, but when their spectrums cover any resonant frequency of the PGPs, they will be amplified.

Figure 4.40 shows the considered PGPs. The upper plane is split into two parts, and each part has one via located inside. The substrate between the upper and lower planes is made of FR4 with a thickness of 0.1 mm, a relative permittivity of 4.4, and a loss tangent of 0.02. The radius of two vias is 0.1 mm. In this simulation, for the general case, the external source is assumed to be a plane wave with $-x$ polarization and $-z$ incident direction, and its amplitude is 1 V/m, as shown in Figure 4.40.

Figure 4.41a,b shows the induced electric currents along via1 and via2, respectively. The gap on the upper plane degrades the shielding of the PGPs by increasing the leakage of the electric field into the structure. The induced electric currents are amplified at the resonant frequencies of the PGPs. Figure 4.41c,d shows the calculated input admittances of via1 and via2, respectively. We can observe from Figure

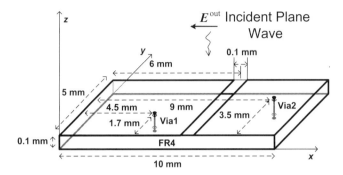

Figure 4.40   A power-ground structure and the incident plane wave [11].

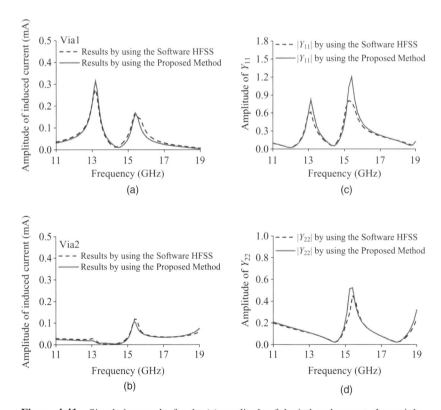

Figure 4.41   Simulation results for the (a) amplitude of the induced current along via1, (b) amplitude of the induced current along via2, (c) input admittance of via1, and (d) input admittance of via2 [11].

4.41a,c, as well as Figure 4.41b,d, that the resonant frequencies of the induced currents are the same as those of the input admittances. This is because the resonant frequencies of a power-ground structure are inherently determined by its geometry and via layout, that is, the coefficient matrices $[A]$, $[C]$, $[D]$, and $[Z]$ in Equations (4.65) and (4.66), while the excitations $[E_t^{in}]$ and $[E_t^{out}]$ at the right hand sides of Equations (4.65) and (4.66) only affect the amplitudes of these resonances.

For validation purpose, the results from the commercial simulator HFSS [17] are also plotted in Figure 4.41. Again, a good agreement can be observed. However, the proposed method outperforms HFSS in terms of CPU time (37 minutes as opposed to 2 hours and 20 minutes).

4.4 CONCLUSION

In this Chapter, 2D and 3D integral equation methods are proposed to analyze the PDN in high-speed electronic package respectively. In the 2D integral equation method, the complex 3D problem is decomposed into two simple 2D problems: the parallel plate mode problem and transmission line mode problem. Therefore, it greatly reduces the computing time and still keeps a good accuracy. The equivalent circuit is extracted from the modal analysis, which can be substituted into a circuit simulator to perform the signal integrity and PI cosimulations. For the 3D integral equation method, its major advantage its accurate simulation of all EMC problems related to the PGPs, such as the coupling, emission, and susceptibility problems.

REFERENCES

[1] T. Okoshi, *Planar Circuits for Microwave and Lightwaves*, Springer-Verlag, New York, 1985, pp. 10–42.

[2] C. J. Ong, D. Miller, L. Tsang, B. Wu, and C. C. Huang, Application of the Foldy–Lax multiple scattering method to the analysis of vias in ball grid arrays and interior layers of printed circuit boards, *Micro. Opt. Technol. Lett.*, vol. 49, no. 1, pp. 225–231, 2007.

[3] Z. Z. Oo, E. X. Liu, E. P. Li, X. C. Wei, Y. J. Zhang, M. Tan, L. W. Li, and R. Vahldieck, A semi-analytical approach for system-level electrical modeling of electronic packages with large number of vias, *IEEE Trans. Adv. Packag.*, vol. 31, no. 2, pp. 267–274, 2008.

[4] E. X. LIU, E. P. LI, Z. Z. OO, X. C. WEI, Y. J. ZHANG, and R. VAHLDIECK, Novel methods for modeling of multiple vias in multilayered parallel-plate structures, *IEEE Trans. Microwave Theory Tech.*, vol. 57, no. 7, pp. 1724–1733, 2009.

[5] A. E. RUEHLI, Equivalent circuit models for three-dimensional multiconductor systems, *IEEE Trans. Microwave Theory Tech.*, vol. 22, no. 3, pp. 216–221, 1974.

[6] A. E. ENGIN, K. BHARATH, M. SWAMINATHAN, M. CASES, B. MUTNURY, N. PHAM, D. N. de ARAUJO, and E. MATOGLU, Finite-difference modeling of noise coupling between power/ground planes in multilayered packages and boards, *56th Electronic Components and Technology Conference*, 2006, pp. 1262–1267.

[7] X. C. WEI, E. P. LI, E. X. LIU, and X. CUI, Efficient modeling of re-routed return currents in multilayered power-ground planes by using integral equation, *IEEE Trans. Electromagn. Compat.*, vol. 50, no. 3, pp. 740–743, 2008.

[8] X. C. WEI, E. P. LI, E. X. LIU, and R. VAHLDIECK, Efficient simulation of power distribution network by using integral equation and modal decoupling technology, *IEEE Trans. Microwave Theory Tech.*, vol. 56, no. 10, pp. 2277–2285, 2008.

[9] X. C. WEI and E. P. LI, Integral-equation equivalent-circuit method for modeling of noise coupling in multilayered power distribution networks, *IEEE Trans. Microwave Theory Tech.*, vol. 58, no. 3, pp. 559–565, 2010.

[10] X. C. WEI, E. P. LI, and E. X. LIU, An efficient equivalent circuit model for the EMC analysis of power/ground noise, *12th Workshop on Signal Propagation on Interconnection*, Avignon, France, May 12–15, 2008.

[11] X. C. WEI, E. P. LI, E. X. LIU, E. K. CHUA, Z. Z. OO, and R. VAHLDIECK, Emission and susceptibility modeling of finite-size power-ground planes using a hybrid integral equation method, *IEEE Trans. Adv. Packag*, vol. 31, no. 3, pp. 536–543, 2008.

[12] J. Y. FANG, Y. Z. CHEN, Z. H. WU, and D. W. XUE, Model of interaction between signal vias and metal planes in electronics packaging, *IEEE 3rd Topical Meeting on Electronic Performance of Electronic Packaging*, 1994, pp. 211–214.

[13] J. PARK, H. KIM, Y. JEONG, J. KIM, J. PAK, D. KAM, and J. KIM, Modeling and measurement of simultaneous switching noise coupling through signal via transition, *IEEE Trans. Adv. Packag*, vol. 29, no. 3, pp. 548–559, 2006.

[14] P. A. KOK and D. D. ZUTTER, Prediction of the excess capacitance of a via-hole through a multilayered board including the effect of connecting microstrips or striplines, *IEEE Trans. Microwave Theory Tech.*, vol. 42, no. 12, pp. 2270–2276, 1994.

[15] T. WANG, R. F. HARRINGTON, and J. R. MAUTZ, Quasi-static analysis of a microstrip via through a hole in a ground plane, *IEEE Trans. Microwave Theory Tech.*, vol. 36, no. 6, pp. 1008–1013, 1998.

[16] Y. J. ZHANG, J. FAN, G. SELLI, M. COCCHINI, and D. P. FRANCESCO, Analytical evaluation of via-plate capacitance for multilayer packages or PCBs, *IEEE Trans. Microwave Theory Tech.*, vol. 56, no. 9, pp. 2118–2128, 2008.

[17] Ansoft software, http://www.ansoft.com

[18] D. M. POZAR, *Microwave Engineering*, 3rd ed., John Wiley & Sons, Hoboken, NJ, 2005.

[19] R. L. CHEN, J. CHEN, T. H. HUBING, and W. M. SHI, Analytical model for the rectangular power-ground structure including radiation loss, *IEEE Trans. Electromagn. Compat.*, vol. 47, no. 1, pp. 10–16, 2005.

[20] C. TAI, *Dyadic Green's Functions in Electromagnetic Theory*, 2nd ed., IEEE Press, New York, 1994.

[21] S. M. RAO, D. R. WILTON, and A. W. GLISSON, Electromagnetic scattering by surface of arbitrary shape, *IEEE Trans. Antennas Propagat.*, vol. 30, no. 3, pp. 409–418, 1982.

[22] X. C. WEI, E. P. LI, and Y. J. ZHANG, Efficient solution to the large scattering and radiation problem using the improved finite element-fast multipole method, *IEEE Trans. Magn.*, vol. 41, no. 5, pp. 1684–1687, 2005.

[23] Y. SAAD and M. H. SCHULTZ, GMRES: A generalized minimal residual algorithm for solving nonsymmetric linear systems, *SIAM J. Sci. Stat. Comput.*, vol. 7, pp. 856–869, 1986.

[24] V. FRAYSSÉ, L. GIRAUD, S. GRATTON, and J. LANGOU, A set of GMRES routines for real and complex arithmetics on high performance computers, *CERFACS Technical Report TR/PA/03/3*, 2003. Public domain software available on http://www.cerfacs/algor/Softs

[25] FEKO Software, http://www.feko.info/index.html.

[26] C. H. CHIEN, Circuit modeling of power/ground plane structures for printed circuit boards, *Micro. Opt. Technol. Lett.*, vol. 47, no. 1, pp. 97–99, 2005.

CHAPTER 5

Systematic Microwave Network Analysis for 3D Integrated Systems

An intrinsic via circuit model is first derived through rigorous electromagnetic analysis for an irregular plate pair with multiple vias in a printed circuit board (PCB) in Section 5.1. Segmentation technique is used to divide the plate pair into a plate domain and multiple circular via domains. The via domains are assumed electrically small and not overlap each other in the interest of frequencies. Thus, they can be modeled as lumped circuits. On the other hand, the plate domain which is larger or, at least, comparable to operating wavelengths should be solved by full-wave methods. Hybrid via circuit and plate field model developed here satisfies the boundary conditions of both the via domains and the plate domain explicitly. It provides a solid theoretical foundation for conventional physics-based via circuit model which can be regarded as a low-frequency approximation of the intrinsic via circuit model.

Conventional plate-pair impedance is obtained by an average integral of the Green's functions of two-dimensional (2D) cavities over two port areas. It has been extensively used in power delivery system designs but shows no connection with signal integrity (SI) analysis.

Electrical Modeling and Design for 3D System Integration: 3D Integrated Circuits and Packaging, Signal Integrity, Power Integrity and EMC, First Edition. Er-Ping Li.
© 2012 Institute of Electrical and Electronics Engineers. Published 2012 by John Wiley & Sons, Inc.

In Section 5.2, we demonstrate that the derivation of the intrinsic via circuit model naturally leads to a new impedance definition of plate pair or power-bus, which is expressed in terms of cylindrical waves. The new plate-pair impedance has clear physics meaning and makes possible signal integrity/power integrity (SI/PI) cosimulations. Numerical and measurement examples have indicated that, while the new impedance gives almost the same results to the conventional one in a plate pair with few vias, it can correctly predict the resonant frequency shift in the case of a plate pair with a large amount of vias.

Multilayer parallel plates with a large amount of signal, ground and power vias, as well as decoupling capacitors are common structures for high-speed electronic integrated systems. In Section 5.3, the intrinsic via circuit model is extended to practical multilayer structures. For a single plate pair with many vias, all the via holes (antipads) on top and bottom plates are regarded as ports of a microwave network. The admittance matrix of the microwave network is derived by integrating the intrinsic via circuit model and the impedance matrix of the plate model which describes via couplings. A multilayer PCB structure is thus modeled as a cascaded microwave network of admittance matrices. A recursive connection scheme is presented to derive the final admittance matrix associated with the via holes or ports on the top and bottom plates of the multilayer structures. Decoupling capacitors are easily processed as they can be viewed as loads to the final admittance matrix.

In each section of this chapter, both commercial numerical solvers and measurements are used to validate the algorithm derived. Good agreements are observed between the model/algorithm developed here and numerical examples and measurements. The systematic microwave network has been proved to be an efficient method for SI/PI co-analysis or co-simulation of three-dimensional (3D) integrated systems.

5.1 INTRINSIC VIA CIRCUIT MODEL FOR MULTIPLE VIAS IN AN IRREGULAR PLATE PAIR

5.1.1 Introduction

Vias are widely used in 3D electronic package integration and multilayer high-speed PCBs to connect the signal traces on different layers or connect devices to power and ground plates [1–4]. As a kind of inevitable discontinuity, vias may cause mismatch, cross talk, mode conversion, and other SI issues in a signal link path (SLP). Moreover, vias passing through a parallel power/ground plate pair could effectively pick up the power distribution network noise, resulting in degraded signal quality. Similarly, high-speed transient currents flowing along vertical vias could also excite the parallel plates they penetrate, causing serious voltage fluctuations in power distribution network or electromagnetic interference (EMI) problems due to strong edge radiations. Therefore, modeling the electromagnetic behavior of vias in parallel plates plays a critical role in analyses of SI, PI, and EMI for multilayer PCBs and packages.

Via-plate interaction has been extensively studied by many methods. For vias crossing a single plate, either full-wave methods or quasi-static approaches are effective due to the localized field distribution near the vias [5–12]. For vias crossing a plate pair, however, quasi-static approximation and simple lumped circuits are no longer suitable as the electromagnetic fields are not simply localized near vias. Although the higher-order evanescent parallel plate modes are still refined in proximity to the vias as energy stored in the electric and magnetic fields, the propagating modes spread over the entire plate pair. This makes it more difficult to model a via crossing a plate pair than a via crossing a single plate.

An analytical method for the via-plate-pair interaction was first reported in Reference 13 for a single via and, later, extended to the coupling of two vias [14]. A magnetic frill current was assumed in a via hole in a plate (the gap region between a via and a plate), and cylindrical waves were used to describe the parallel plate modes. Boundary conditions at the vias were explicitly enforced. However, the analytical method cannot handle a plate pair with more than two vias. Moreover, the method is restricted to an infinitely large plate pair, and no edge reflection was considered.

Recently, an algorithm denoted by the Foldy–Lax multiple scattering method (FLMSM) was proposed to extend the analytical method to multiple vias in a plate pair [15–19]. The dyadic Green's function of an infinitely large plate pair or a finite circular plate pair, as well as the addition theorem of cylindrical harmonics, were used to analyze the multiple scattering among vias. All the via holes in plates were regarded as ports of a multiport microwave network whose admittance matrix was obtained from current distributions in the ports. The multiple scattering method can be regarded as an efficient semi-analytical approach. However, the method is still restricted to either an infinite or a finite circular plate pair since an analytical dyadic Green's function is not available for an irregular finite plate pair. Note that the power/ground pair is normally irregular in real-world PCB designs. This limits the engineering applications of FLMSM.

Different from the rigorous multiple scattering method, an alternative solution denoted a physics-based via circuit model was proposed in References 20–24. In this model, the via-plate-pair interaction was represented simply by a π-circuit. Each via barrel was viewed as a simple short circuit, and the displacement currents between the via and the top/bottom plates were represented by two shunt capacitors. The impedance of the plate pair, widely studied in References 25–32, was used as the return path for the signal current along via barrels. The combination of the lumped via circuits and the full-wave plate-pair impedance correctly reflects the fact that via itself is usually electrically small while the plate pair is comparable to the wavelength of interests. Moreover, the model is suitable for any irregular plate-pair, since the impedance matrix can be easily obtained from the cavity model and the segmentation technique [28, 30].

Despite its flexibility in handling edge boundaries, the physics-based via circuit model does not satisfy via boundary conditions due to its physical intuitive nature. In Reference 24, the via-plate capacitance in the physics-based via circuit model is found to be expressed only in terms of the odd-numbered parallel plate modes. All even higher-order modes are neglected. This implies that the boundary conditions at the vias are not satisfied by the simple lumped circuit in the physics-based via circuit model. In addition, the coupling among vias is described by the parallel plate impedance matrix (the TM_{z00} mode of each via) only. Therefore, when vias are so closely spaced that the coupling due to the higher-order modes cannot be neglected, the accuracy of the physics-based via circuit model will start to deteriorate.

The purpose of this section is to derive an intrinsic via circuit model from a rigorous electromagnetic analysis. The derivations will demonstrate that the new via circuit model satisfies the boundary conditions at both the vias and the plate edges. The term "physics-based via circuit model" is used herein to refer to the model proposed in References 20–24 based on physical intuition, while the term "intrinsic via circuit model" refers to the model presented in this section based on rigorous electromagnetic analysis. Later, the physics-based via circuit model

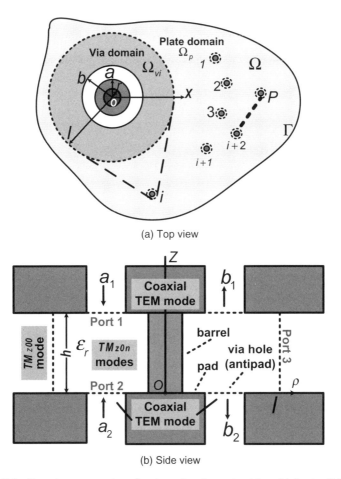

(a) Top view

(b) Side view

Figure 5.1   Domain segmentation of an irregular plate pair with multiple vias [41].

will be demonstrated to be just an approximation of the intrinsic via circuit model at low frequencies.

5.1.2 Segmentation of Vias and a Plate Pair

The top view of an irregular plate pair loaded with P vias is shown in Figure 5.1a. Bounded by the edge Γ, the region of the entire via-plate structure is denoted Ω that is decomposed into $P + 1$ separate regions: P via domains Ω_{vi}, $i = 1, 2, \cdots, P$ and one plate domain Ω_p, that is, $\Omega = \Omega_p + \sum_{i=1}^{P} \Omega_{vi}$.

As an example, the via domain for Via i, Ω_{vi}, is enlarged in Figure 5.1a,b, showing the top and side views. The barrel, pad, and via-hole radii of the via are denoted r, a, and b, respectively. The via height or the separation of the plates is h, and a dielectric layer is between the plates with a relative permittivity of ε_r. A local cylindrical coordinate system is set up with the origin at the center of the bottom plate surface in Ω_{vi} as shown in Figure 5.1b. The virtual via boundary between the via domain and the plate domain is located at $\rho = l$.

The vertical (z-directional) electric field near the i-th via can be expressed as

$$E_z = \sum_{m=-\infty}^{\infty} \sum_{n=0}^{\infty} [a_{mn} J_{mn}(\rho, \phi, z) + b_{mn} H_{mn}(\rho, \phi, z)], \tag{5.1}$$

where a_{mn} and b_{mn} are the expansion coefficients for the inward and outward TM_{zmn} parallel plate modes $J_{mn}(\rho, \phi, z)$ and $H_{mn}(\rho, \phi, z)$, respectively, which are expressed as

$$J_{mn}(\rho, \phi, z) = J_m(k_n \rho) e^{jm\phi} \cos\left(\frac{n\pi}{h} z\right), \tag{5.2}$$

$$H_{mn}(\rho, \phi, z) = H_m^{(2)}(k_n \rho) e^{jm\phi} \cos\left(\frac{n\pi}{h} z\right), \tag{5.3}$$

where $J_m(\cdot)$, $H_m^{(2)}(\cdot)$ are the mth-order Bessel and Hankel (the second kind) functions, respectively; (ρ, ϕ, z) is the cylindrical coordinates of the via; and $k_n = \sqrt{k_0^2 \varepsilon_r - (n\pi/h)^2}$ is the radial wave number. Only the TM_{zmn} parallel plate modes are considered here. The TE_{zmn} modes are negligible due to the specific excitations typical in a via geometry (a circular magnetic current in a via hole, or a vertical electric current along a via barrel).

In most practical designs, the following assumptions are satisfied:

- Only the transverse electromagnetic (TEM) mode is considered in the via holes. In other words, magnetic frill currents are assumed to excite the parallel plate pair, which have been widely used in References 13–19, 24, 33, and 34. This implies that sources in the via holes only excite the TM_{z0n}, $n \geq 0$ modes in the region between the two parallel plates, assuming vias and their corresponding via holes are concentric.

- The height h is electrically small so that only the TM_{zm0} cylindrical waves can propagate. All the higher-order modes (TM_{zmn}, $n \geq 1$) decay rapidly along the radial direction. Therefore, there is a virtual via boundary $\rho = l$ (in local coordinates) for each via beyond which all the higher-order modes can be neglected. In other words, only the TM_{zm0}, $-\infty \leq m \leq \infty$ modes are considered in the plate domain outside the via domains.

- The virtual via boundary for each via is electrically small so that the azimuthal variation of fields due to the asymmetry of the via-plate structure is negligible. This means only the TM_{z0n}, $n \geq 0$ modes need to be considered for the via-domain modeling.

These assumptions indicate that each via domain is an electrically small region bounded by a virtual circular boundary. Thus, voltages and currents are well defined and circuit ports can be specified. As shown in Figure 5.1, Ports 1 and 2 are defined as coaxial ports on the inner surface of the plates, between the vias and the plates, and across the via holes. Port 3 is defined as a radial port at the virtual circular boundary between the two parallel plates. Furthermore, these assumptions determine that only the TM_{z00} waves need to be considered in the plate domain. Therefore, properly setting the virtual circular boundaries between the via domains and the plate domain is critical for the validity of the via circuit model derived later on.

In this chapter, it is also assumed that the virtual circular boundaries do not overlap as shown in Figure 5.1a. This guarantees that the higher-order modes of one via do not illuminate another via. This assumption limits the approach presented herein to the cases where vias are not placed very closely to each other. Section 5.1.4 will provide an empirical formula to determine the virtual via boundary, or in other words the minimum spacing between two adjacent vias for the validity of the approach.

At Port 3 of via i, the port voltage and current can be defined in terms of the TM_{z00} waves as

$$V_i = -E_{zi}h, \tag{5.4}$$

$$I_i = 2\pi l H_{\phi i}. \tag{5.5}$$

Then, the plate domain in Figure 5.1a can be modeled as a P-port network to describe the coupling among the P vias using the TM_{z00} modes. The impedance matrix of this P-port network is different to the conventional impedance matrix of a rectangular plate pair widely studied in References 25–32. The new impedance definition and calculations will be introduced in Section 5.2.

The via, inside its via domain, can be viewed as a 3-port network as shown in Figure 5.1. Ports 1 and 2 can connect other layers in a multilayer PCB or package, and Port 3 connects to the network that describes the plate domain to ensure the continuity of the tangential fields at the virtual via boundary at $\rho = l$. This segmentation approach enforces the boundary conditions at both the via barrels and the plate edges, as explained later in detail.

5.1.3 An Intrinsic 3-Port Via Circuit Model

A via domain can be viewed as a via located at the center of a circular plate pair with a dimension of $\rho = l$. Then, the field distributions excited by a magnetic frill current can be derived using the same method as in Reference 24.

5.1.3.1 Field Distributions Excited by a Magnetic Frill Current

Using the equivalence principle of electromagnetics, the TEM mode at Port 1 or Port 2 of a via, shown in Figure 5.1b as an example, can be regarded as an equivalent magnetic frill current,

$$M_\phi = -\frac{V_0}{\rho' \ln(b/a)} \delta(z - z'), \tag{5.6}$$

where V_0 is the voltage across the via hole at $z' = 0$ or $z' = h$. The magnetic field distribution due to Equation (5.6) has been derived as [24],

$$H_\phi(\rho, z) = -\frac{\omega \varepsilon \pi V_0}{h \ln(b/a)} \sum_{n=0}^{\infty} G_n^S(\rho) \cos\left(\frac{n\pi}{h} z\right) \cos\left(\frac{n\pi}{h} z'\right), \tag{5.7}$$

for $l \geq \rho \geq b$, and

$$H_\phi(\rho, z) = -\frac{\omega\varepsilon\pi V_0}{h\ln(b/a)}\sum_{n=0}^{\infty}F_n^S(\rho)\cos\left(\frac{n\pi}{h}z\right)\cos\left(\frac{n\pi}{h}z'\right), \quad (5.8)$$

for $r \le \rho \le a$, and the two auxiliary functions are defined as

$$G_n^S(\rho) = \frac{\left(1-\Gamma_r^{(n)}\Gamma_l^{(n)}\right)^{-1}}{k_n(1+\delta_{n0})}\left\{\begin{matrix}[J_0(k_nb)-J_0(k_na)]\\+\Gamma_r^{(n)}\left[H_0^{(2)}(k_nb)-H_0^{(2)}(k_na)\right]\end{matrix}\right\}$$
$$\left[H_1^{(2)}(k_n\rho)+\Gamma_l^{(n)}J_1(k_n\rho)\right], \quad (5.9)$$

$$F_n^S(\rho) = \frac{\left(1-\Gamma_r^{(n)}\Gamma_l^{(n)}\right)^{-1}}{k_n(1+\delta_{n0})}\left\{\begin{matrix}\left[H_0^{(2)}(k_nb)-H_0^{(2)}(k_na)\right]\\+\Gamma_l^{(n)}\left[J_0(k_nb)-J_0(k_na)\right]\end{matrix}\right\}$$
$$\left[J_1(k_n\rho)+\Gamma_r^{(n)}H_1^{(2)}(k_n\rho)\right], \quad (5.10)$$

where $\Gamma_r^{(n)}$ and $\Gamma_l^{(n)}$ are the reflection coefficients for the nth cylindrical waves from the via barrel ($\rho = r$) and the outer radial boundary ($\rho = l$), respectively, and given as

$$\Gamma_r^{(n)} = -\frac{J_0(k_nr)}{H_0^{(2)}(k_nr)}, \quad (5.11)$$

$$\Gamma_l^{(n)} = \begin{cases}\Gamma_l\delta_{n0} & \text{Irregular} & S=I\\ -\dfrac{H_0^{(2)}(k_nl)}{J_0(k_nl)} & \text{PEC} & S=E\\ -\dfrac{H_1^{(2)}(k_nl)}{J_1(k_nl)} & \text{PMC} & S=M\\ 0 & \text{PML} & S=L,\end{cases} \quad (5.12)$$

where $J_{0(1)}(\cdot)$, $H_{0(1)}^{(2)}(\cdot)$ denote the zero (first)-order Bessel and Hankel functions of the second kind, respectively; and k_n is the radial wavenumber for the TM_{z0n} modes as

$$k_n = \sqrt{k_0^2\varepsilon_r - \left(\frac{n\pi}{h}\right)^2}. \quad (5.13)$$

As shown in Equation (5.12), the values of $\Gamma_l^{(n)}$ has four choices depending on the boundary conditions at $\rho = l$. The superscript S in the auxiliary functions G_n^S and F_n^S indicates one of the choices. PEC, PMC, and PML stand for perfect electric conductor, perfect magnetic conductor, and perfectly matched layer conditions, respectively.

According to the discussion in Section 5.1.2, $S = I$ represents the case where all the higher-order (TM_{z0n}, $n \geq 1$) modes are well confined in the via domain. Therefore, in this case, only the reflection coefficient of the TM_{z00} waves, Γ_l, is nonzero. The other boundary conditions are introduced here for the determination of the virtual via boundary $\rho = l$ in Section 5.1.4.

The electrical-field components can be obtained from Equations (5.7) and (5.8) from the magnetic field as

$$E_z(\rho, z) = \frac{j\pi V_0}{h \ln(b/a)} \sum_{n=0}^{\infty} \frac{\partial\left[\rho G_n^S(\rho)\right]}{\rho \partial \rho} \cos\left(\frac{n\pi}{h}z\right)\cos\left(\frac{n\pi}{h}z'\right), \qquad (5.14)$$

$$E_\rho(\rho, z) = \frac{-j\pi V_0}{h \ln(b/a)} \sum_{n=1}^{\infty} \frac{n\pi}{h} G_n^S(\rho) \sin\left(\frac{n\pi}{h}z\right)\cos\left(\frac{n\pi}{h}z'\right), \qquad (5.15)$$

for $l \geq \rho \geq b$ and

$$E_z(\rho, z) = \frac{j\pi V_0}{h \ln(b/a)} \sum_{n=0}^{\infty} \frac{\partial\left[\rho F_n^S(\rho)\right]}{\rho \partial \rho} \cos\left(\frac{n\pi}{h}z\right)\cos\left(\frac{n\pi}{h}z'\right), \qquad (5.16)$$

$$E_\rho(\rho, z) = \frac{-j\pi V_0}{h \ln(b/a)} \sum_{n=1}^{\infty} \frac{n\pi}{h} F_n^S(\rho) \sin\left(\frac{n\pi}{h}z\right)\cos\left(\frac{n\pi}{h}z'\right), \qquad (5.17)$$

for $r \leq \rho \leq a$. The following identity can be used to obtain the explicit expression of E_z

$$\frac{d\left[\rho B_1(\rho)\right]}{\rho d\rho} = B_0(\rho), \qquad (5.18)$$

where $B_{1(0)}$ is the first-order (zero-order) Bessel or Hankel function.

5.1.3.2 Admittance Matrix of a Two-Port Network

A via domain, shown in Figure 5.1b as an example, can be regarded as a two-port microwave network with the radial port (Port 3) terminated with the impedance of the plate domain, which is related to the reflection coefficient Γ_l as [27]

$$Z_l^+ = \frac{j\omega\mu h}{2\pi k l} \frac{H_0^{(2)}(kl) + \Gamma_l J_0(kl)}{H_1^{(2)}(kl) + \Gamma_l J_1(kl)}. \tag{5.19}$$

Therefore, an admittance matrix of the two-port network can be defined as

$$\begin{bmatrix} I_1 \\ I_2 \end{bmatrix} = \begin{bmatrix} Y_{11} & Y_{12} \\ Y_{21} & Y_{22} \end{bmatrix} \begin{bmatrix} V_1 \\ V_2 \end{bmatrix}, \tag{5.20}$$

where (V_1, I_1) and (V_2, I_2) are the voltage and current pair of Ports 1 and 2, respectively. In practice, the dielectric material between the plate pair is normally reciprocal. As a result, the two-port via network satisfies reciprocity, that is, $Y_{12} = Y_{21}$. Let $V_1 = 0$ and $V_2 = V_0$. Y_{12} and Y_{22} can then be calculated through the port currents as shown in Figure 5.2, with the top hole of the via closed by a PEC boundary and an equivalent magnetic frill current located at the bottom via hole.

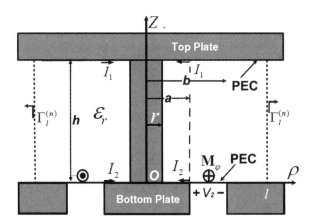

Figure 5.2 Illustration of a via domain used for extracting the admittance matrix of the via [41].

The magnetic field for $r \leq \rho \leq a$ in Equation (5.8) leads to a description of the current distribution on the via barrel and pad as [15]

$$I(\rho, z) = 2\pi\rho H_\phi(\rho, z).$$ (5.21)

Substituting Equation (5.8) into Equation (5.21) using $V_0 = V_2$ and $z' = 0$, yields

$$I(\rho, z) = -\frac{\omega\varepsilon\pi V_2}{h\ln(b/a)} \sum_{n=0}^{\infty} 2\pi\rho F_n^S(\rho)\cos\left(\frac{n\pi}{h}z\right).$$ (5.22)

As shown in Figure 5.2, the port currents can be obtained by letting $I_2 = I(a, 0)$ and $I_1 = -I(a, h)$. Then, the self-admittance Y_{22} and mutual admittance Y_{12} can be obtained from Equation (5.22) as

$$Y_{22} = -\frac{2\omega\varepsilon\pi^2 a}{h\ln(b/a)} \sum_{n=0}^{\infty} F_n^S(a),$$ (5.23)

$$Y_{12} = \frac{2\omega\varepsilon\pi^2 a}{h\ln(b/a)} \sum_{n=0}^{\infty} (-1)^n F_n^S(a).$$ (5.24)

A reciprocal two-port network can be represented by a π-type equivalent circuit as shown in Figure 5.3 [35]. Then, using Equations

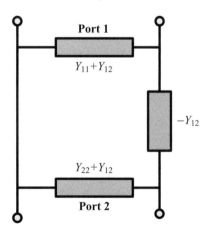

Figure 5.3 A π-type equivalent circuit for a reciprocal two-port network from its admittance matrix. The circuit is rotated by 90°, to be consistent with the ports illustrated in Figure 5.1.

(5.23) and (5.24), the following shunt and series elements for the π-type equivalent circuit model can be expressed as

$$Y_{12} + Y_{22} = -\frac{4\omega\varepsilon\pi^2 a}{h\ln(b/a)} \sum_{n=1,3,5\cdots}^{\infty} F_n^S(a), \qquad (5.25)$$

$$-Y_{12} = -\frac{2\omega\varepsilon\pi^2 a}{h\ln(b/a)} \sum_{n=0}^{\infty} (-1)^n F_n^S(a). \qquad (5.26)$$

5.1.3.3 Via Circuit Components Due to Higher-Order Modes

Shunt Capacitances. The shunt admittance $Y_{12} + Y_{22}$ has a clear physical meaning that it represents the capacitive coupling between the via (the barrel and pad) and the bottom plate. Thus, a shunt capacitance can be defined as

$$Y_{12} + Y_{22} = j\omega C_h. \qquad (5.27)$$

Therefore, substituting Equation (5.25) into Equation (5.27) yields

$$C_h = \frac{j4\varepsilon\pi^2 a}{h\ln(b/a)} \sum_{n=1,3,5\cdots}^{2N-1} F_n^S(a), \qquad (5.28)$$

where N is the mode number used to truncate the infinite summation in practical calculations.

The shunt capacitance C_h can be divided into two parts:

$$C_h = C_b + C_p, \qquad (5.29)$$

where the barrel-plate capacitance C_b and the pad-plate capacitance C_p are expressed, respectively, as

$$C_b = \frac{j4\varepsilon\pi^2 a}{h\ln(b/a)} \sum_{n=1,3,5\cdots}^{2N-1} F_n^S(r), \qquad (5.30)$$

$$C_p = \frac{j4\varepsilon\pi^2 a}{h\ln(b/a)} \sum_{n=1,3,5\cdots}^{2N-1} \left[F_n^S(a) - F_n^S(r) \right]. \qquad (5.31)$$

In the case of a via without a pad, that is, the pad radius is equal to the barrel radius ($a = r$), the pad-plate capacitance C_p vanishes, and the barrel-plate capacitance C_b in Equation (5.30) is reduced to the analytical expression of the barrel-plate capacitance given in Reference 24.

Similarly, the shunt capacitance from $Y_{12} + Y_{22}$ represents the capacitive coupling between the via and the top plate. Its value is the same as C_h if the pad and via-hole dimensions are the same in both the top and bottom plates.

Series Capacitance. The series admittance in Equation (5.26), $-Y_{12}$, is expressed in terms of both the zero-order propagating mode (TM_{z00}) and the higher-order evanescent modes (TM_{z0n}, $n \geq 1$). It can be separated into two parts

$$-Y_{12} = Y_v + j\omega C_v^1, \tag{5.32}$$

where the admittance Y_v represents the part due to the zero-order mode

$$Y_v = -\frac{2\omega\varepsilon\pi^2 a}{h\ln(b/a)} F_0^S(a), \tag{5.33}$$

and the series capacitance C_v^1 due to the higher-order modes is expressed as

$$C_v^1 = \frac{j2\varepsilon\pi^2 a}{h\ln(b/a)} \sum_{n=1}^{\infty} (-1)^n F_n^S(a). \tag{5.34}$$

Note that the series capacitance C_v^1 is not included in the previous physics-based via circuit model in References 20–22 and 24, as it cannot be attributed to a static capacitance between two separate conductors. While the fields E_ρ in Equations (5.15) and (5.17) lead to the capacitive element C_h, the higher-order parts of the fields E_z in Equations (5.14) and (5.16) are responsible for the parasitic capacitance C_v^1.

Equations (5.27) and (5.32) convert the general π-circuit shown in Figure 5.3 into the via circuit model shown in Figure 5.4. The parasitic capacitances C_h and C_v^1 characterize the energy stored in the electric field near the via due to the higher-order parallel plate modes.

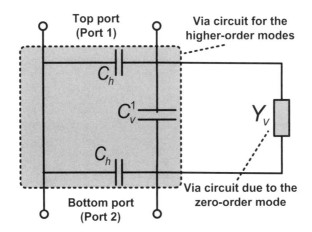

Figure 5.4   Via circuit model due to the higher-order modes with two shunt and one series parasitic capacitances [41].

5.1.3.4 Via Circuit Components Due to Zero-Order Waves

In the two-port via circuit model shown in Figure 5.4, both C_h and C_v^1 are related to the higher-order $n \geq 1$ modes, which are well confined in the via domain according to the assumptions in Section 5.1.2. Then, only the series admittance Y_v in Equation (5.33), the component due to the zero-order waves, is related to both the via and the plate domains. An alternative derivation of Y_v is provided herein based on the segmentation of the via domain and the plate domain as shown in Figure 5.5. The derivation will show clearly how Y_v can be decomposed into an intrinsic part due to the via domain and an impedance due to the plate domain.

As discussed in Section 5.1.2, the via domain I and the plate domain II are separated at $\rho = l$, as illustrated in Figure 5.5. The boundary condition on the interface requires

$$I_l^- = -I_l^+, \tag{5.35}$$

$$V_l^- = V_l^+, \tag{5.36}$$

where $I_l^\pm = 2\pi l H_\phi(l^\pm, z)$ and $V_l^\pm = -E_z(l^\pm, z)h$ stand for the left and right side port currents and voltages at the segmentation interface. These

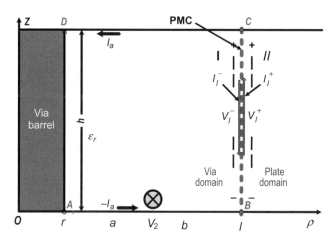

Figure 5.5 An alternative derivation of Y_v by the segmentation technique [41].

voltages and currents have been well defined similarly as in Equations (5.4) and (5.5), because of the assumption that only the zero-order waves can reach the boundary between the via and plate domains.

According to the equivalence principle, a PMC boundary can be assumed for the interface at $\rho = l$ if the equivalent currents I_l^- and I_l^+ are impressed along both sides. Then, in region I, the voltage V_l^- can be obtained as

$$V_l^- = R_m V_2 + Z_l^- I_l^-,\tag{5.37}$$

where

$$Z_l^- = -\frac{j\omega\mu h}{2\pi kl}\frac{J_0(kl) + \Gamma_r^{(0)} H_0^{(2)}(kl)}{J_1(kl) + \Gamma_r^{(0)} H_1^{(2)}(kl)},\tag{5.38}$$

and the voltage transform coefficient R_m relating V_2 to a part of V_l^-, which only results from the zero-order waves, is

$$R_m = \frac{j\pi}{\ln(b/a)}\frac{\partial\left[\rho G_0^M(\rho)\right]}{\rho\partial\rho}\bigg|_{\rho=l}.\tag{5.39}$$

Equation (5.39) is obtained from the zero-order component of Equation (5.14) at $\rho = l$ with a PMC boundary condition.

In region II, when there is only one via in the plate pair, the voltage V_l^+ is related to I_l^+ as

$$V_l^+ = Z_l^+ I_l^+, \tag{5.40}$$

where Z_l^+ is the impedance looking at Port 3 outward into the plate domain by definition, and has the same form as Equation (5.19) for a circular plate pair. For an irregular plate pair, Equation (5.40) is still valid and a generalized Equation (5.19) will be derived in Section 5.2.

Substituting Equations (5.40) and (5.37) into Equation (5.36), as well as using Equation (5.35), yields

$$I_l^- = -\frac{R_m V_2}{Z_l^+ + Z_l^-}. \tag{5.41}$$

The zero-order current component I_a flowing into the rim of the via pad at $\rho = a$ and $z = h$ is a combination of the effects of both V_2 and I_l^-. It can be obtained as

$$I_a = Y_a V_2 + R_e I_l^-. \tag{5.42}$$

The first term of the right-hand side of Equation (5.42) is derived from the zero-order current component of Equation (5.33) caused by V_2 with a PMC boundary at $\rho = l$, and

$$Y_a = -\frac{2\omega\varepsilon\pi^2 a}{h\ln(b/a)} F_0^M(a). \tag{5.43}$$

The transform coefficient R_e reflects the magnetic field at $\rho = a$ excited by the current I_l^- at $\rho = l$ as

$$R_e = \frac{a\left[J_1(ka) + \Gamma_r^{(0)} H_1^{(2)}(ka)\right]}{l\left[J_1(kl) + \Gamma_r^{(0)} H_1^{(2)}(kl)\right]}. \tag{5.44}$$

Then, substituting Equations (5.41) into (5.42), the series admittance Y_v is derived as

$$Y_v = Y_a + \frac{-R_m R_e}{Z_l^+ + Z_l^-}. \tag{5.45}$$

When all the via geometrical dimensions are electrically small, the admittance Y_a behaves like a capacitance, that is, $Y_a = j\omega C_v^0$, and the impedance Z_l^- acts as an inductance, that is, $Z_l^+ = j\omega L_v$. From Equations (5.43) and (5.38), it can be derived that

$$C_v^0 = \frac{j2\varepsilon\pi^2 a}{h\ln(b/a)} F_0^M(a), \tag{5.46}$$

and

$$L_v = -\frac{\mu h}{2\pi kl} \frac{J_0(kl) + \Gamma_r^{(0)} H_0^{(2)}(kl)}{J_1(kl) + \Gamma_r^{(0)} H_1^{(2)}(kl)}. \tag{5.47}$$

Then the series admittance Y_v in Equation (5.45) can be rewritten as

$$Y_v = j\omega C_v^0 + \frac{R_v^2}{j\omega L_v + Z_l^+}, \tag{5.48}$$

where the ideal transformation ratio R_v is defined as

$$R_v = \sqrt{-R_m R_e}. \tag{5.49}$$

The formula Equation (5.48) implies an equivalent circuit shown in Figure 5.6 [36]. A similar equivalent π-network has been proposed for radial-line/coaxial-line junction in References 37 and 38 with no circuit extraction.

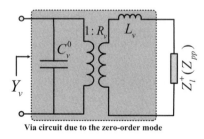

Via circuit due to the zero-order mode

Figure 5.6   Via circuit elements due to the zero-order mode.

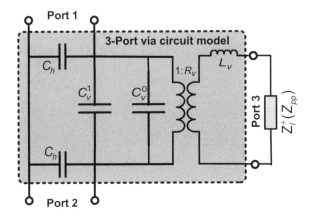

Figure 5.7 The complete 3-port via circuit model [41].

5.1.3.5 The Complete Intrinsic 3-Port Via Circuit Model

By combining the via circuits due to the higher-order modes in Figure 5.4 and the zero-order waves in Figure 5.6, a complete 3-port via circuit model and its connection to the impedance matrix of the plate domain Z_{pp} can be obtained as shown in Figure 5.7.

To demonstrate that all the parasitics C_h, C_v^1, C_v^0, L_v and R_v are intrinsic to a via domain, and also to facilitate the usage of the derived intrinsic via circuit model, their expressions are given here again explicitly as

$$C_h = \left[\frac{j2\pi a}{h} \sum_{n=1,3,5\cdots}^{2N-1} F_n^S(a) \right] C_a, \qquad (5.50)$$

$$C_v^1 = \left[\frac{j\pi a}{h} \sum_{n=1}^{2N-1} (-1)^n F_n^S(a) \right] C_a, \qquad (5.51)$$

where $C_a = 2\varepsilon_r \varepsilon_0 \pi / \ln(b/a)$ is the per-unit-length capacitance of a coaxial geometry with inner and outer radii of a and b, respectively. The auxiliary function F_n^S can be expressed in a concise form as

$$F_n^S(a) = \frac{W_{10}(k_n a, k_n r)[W_{s0}(k_n l, k_n b) - W_{s0}(k_n l, k_n a)]}{k_n(1 + \delta_{n0})W_{s0}(k_n l, k_n r)}, \qquad (5.52)$$

where

$$s = 1 \text{ for } F_n^M(a); \; s = 0 \text{ for } F_n^E(a);$$

$$F_n^L(a) = \frac{H_0^{(2)}(k_n b) - H_0^{(2)}(k_n a)}{k_n (1 + \delta_{n0}) H_0^{(2)}(k_n r)} W_{10}(k_n a, k_n r), \qquad (5.53)$$

and, $W_{mn}(x, y)$ is an auxiliary function defined as a determinant of the Bessel and Hankel functions as

$$W_{mn}(x, y) = \begin{vmatrix} J_m(x) & J_n(y) \\ H_m^{(2)}(x) & H_n^{(2)}(y) \end{vmatrix}, \qquad (5.54)$$

where $J_m(\cdot)$ and $H_n^{(2)}(\cdot)$ are the mth Bessel and nth second-order Hankel functions, respectively.

Similarly, the parameters due to the zero-order waves can be expressed in their concise forms as

$$L_v = -\frac{\mu h}{2\pi k b} \frac{W_{00}(kl, kr)}{W_{10}(kl, kr)}, \qquad (5.55)$$

$$C_v^0 = \frac{j\pi a}{2kh} \frac{W_{10}(ka, kr)}{W_{10}(kl, kr)} [W_{10}(kl, kb) - W_{10}(kl, ka)] C_a, \qquad (5.56)$$

$$R_v = \sqrt{-R_m R_e}, \qquad (5.57)$$

where

$$R_m = \frac{j\pi}{2\ln(b/a)} \frac{W_{10}(kl, kl)}{W_{10}(kl, kr)} [W_{00}(kl, kr) - W_{00}(ka, kr)], \qquad (5.58)$$

$$R_e = \frac{a}{l} \frac{W_{10}(ka, kr)}{W_{10}(kl, kr)}. \qquad (5.59)$$

It can be shown from the properties of $W_{mn}(x, y)$ provided in the Appendix that all the parasitic parameters from Equations (5.50) to (5.59) are real values despite the fact that the imaginary unit j may be included in some expressions. Moreover, at the low frequencies where $ka, kb, kr, kl \approx 1$, using the small argument approximations of $W_{10}(x, y)$ and $W_{00}(x, y)$ given in the Appendix, the static approximations of the parasitic parameters caused by the zero-order mode can be obtained as

$$L_v \simeq \frac{\mu h l}{2\pi b} \ln(l/r), \tag{5.60}$$

$$C_v^0 \simeq \frac{\varepsilon \pi}{h}\left[l^2 - \frac{b^2 - a^2}{2\ln(b/a)}\right], \tag{5.61}$$

$$R_v \simeq \sqrt{\frac{\ln(l/a)}{\ln(b/a)}}. \tag{5.62}$$

Equations (5.50) to (5.59) provide analytical expressions for all the components in the intrinsic 3-port via circuit model shown in Figure 5.7. Clearly, all of them are only related to the via structure itself, that is, are functions of r, a, b, l and ε_r only. The impact of the plate edges is described by the impedance matrix of the plate domain only, which will be addressed in Section 5.2.

It is worth giving a brief explanation on the physical meaning of each parasitic component. The shunt capacitance C_h comes from the higher-order E_ρ components in Equations (5.15) and (5.17); the series capacitance C_v^1 is due to the higher-order E_z components in Equations (5.14) and (5.16); the capacitance C_v^0 reflects the effects of the zero-order E_z component between the two radial plates AB and CD in Figure 5.5; the inductance L_v is due to the radial loop $CDAB$; and the transformer ratio R_v is to relate the vertical voltage V_l^- to the horizontal excitation V_2.

5.1.3.6 Properties of the via Parasitic Components

The properties of C_h have been extensively studied in Reference 24 for a via without a pad. Therefore, only convergence with the number of the parallel plate modes as well as the selection of the boundary condition at the virtual boundary at $\rho = l$ is discussed for both C_h and C_v^1.

An example of the convergence of the parasitic capacitances in Equations (5.50) and (5.51) with the increase of N, the number of the higher-order modes used in the calculations is shown in Figure 5.8. Here the outer boundary $\rho = l$ is selected to be 40 mm, far enough away from the via. It can be seen that only tens of the higher-order modes are required to obtain the converged values. An interesting observation is that the series capacitance C_v^1 is negative, and its physical meaning is yet to be revealed.

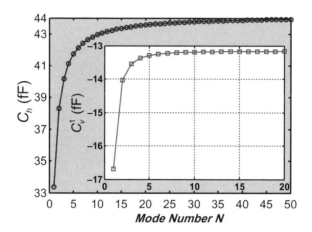

Figure 5.8 Convergence of the via capacitances versus the mode number N ($r = 0.1$, $a = 0.4$, $b = 0.8$, $h = 0.8$, $l = 40$, unit: mm; $\varepsilon_r = 3.84$; frequency = 1.0 GHz) [41].

The convergence of Equations (5.50) and (5.51) with the increase of l is demonstrated in Figure 5.9. The shunt and series capacitances C_h and C_v^1 converge to a constant value when l is approximately 1.6 times of the via-hole radius in this example, independent of the boundary condition applied at $\rho = l$. This is because the higher-order modes decay very quickly from the via due to the small plate separation h. This demonstrates the previous assertion that there is a virtual via boundary $\rho = l$ to separate the via and plate domains. The plate size, shape, or other vias outside of the via domain have negligible impact on the parasitic capacitances C_h and C_v^1.

In addition, the result shown in Figure 5.9 implies that the difference in C_h or C_v^1 values with different boundary conditions at $\rho = l$ can be used to determine where the virtual boundary should be located in order to separate the via and plate domains. Based on this idea, an empirical formula will be developed in the next section to determine the virtual boundary from geometrical dimensions.

From the discussion of the virtual boundary, it is clear that both the intrinsic and the physics-based via circuit models are only valid when the distance between any two vias is larger than $2l$. Otherwise, the multiple higher-order scattering among vias cannot be neglected. This is the limitation of this category of via circuit models.

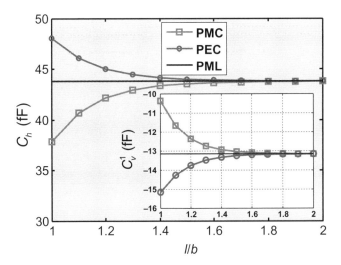

Figure 5.9   Convergence of the via capacitances versus the radius of the virtual boundary with different boundary conditions ($r = 0.1$, $a = 0.4$, $b = 0.8$, $h = 0.8$, unit: mm; $\varepsilon_r = 3.84$; mode number $N = 31$; frequency $= 1.0$ GHz) [41].

The frequency-dependent properties of the parasitics due to the zero-order mode from Equations (5.55) to (5.59) are shown in Figure 5.10, where $l = b$ is used in the calculations. At the frequencies lower than 10 GHz, these parasitic parameters remain approximately constant in this example, which are consistent with the static approximations of Equations (5.60), (5.61), and (5.62). However, beyond 10 GHz, the frequency dependence is significant, and a larger via-hole radius b results in a faster increase of their values with frequency.

5.1.4 Determination of the Virtual Via Boundary

The validity of the intrinsic via circuit model depends on the assumptions discussed in Section 5.1.2. One of the most important ones is that all the circular via domains do not overlap each other. Therefore, the virtual circular boundary can be used to quantify the limitation of the intrinsic via circuit model in practical applications.

The via-plate capacitance C_h reflects the energy stored in the higher-order electric fields adjacent to a via. So it is used herein to quantify the size of a via domain.

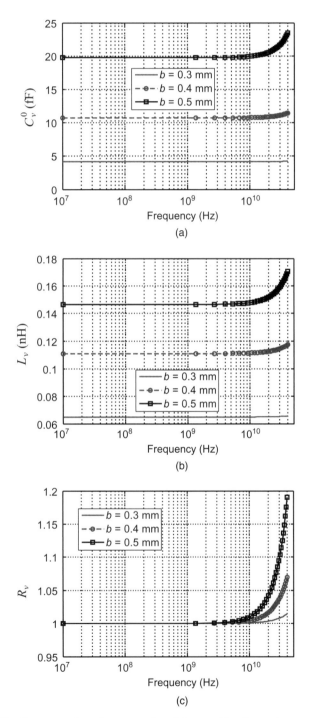

Figure 5.10 Frequency-dependent properties of the parasitics due to the zero-order waves with different via-hole radii ($r = a = 0.2$ mm, $h = 0.8$ mm, $\varepsilon_r = 4.2$) [41].

From Equation (5.50), different boundary conditions at $\rho = l$ will result in different values of C_h. However, when $\rho = l$ is far away from the via barrel, the C_h values for different boundary conditions converge to the one obtained using the PML boundary as illustrated in Figure 5.9. This implies that the higher-order modes are negligible at the boundary and boundary condition is no longer important for C_h calculation. In other words, the virtual via boundary can be determined by examining whether the C_h values using different boundary conditions are close enough. To do so, a relative difference σ is introduced to define the difference in the C_h values with the PMC and the PML boundary conditions at $\rho = l$ as

$$\sigma(l/b) = 1 - \frac{\sum_{n=1,3,5\cdots}^{2N-1} F_n^M(a)}{\sum_{n=1,3,5\cdots}^{2N-1} F_n^L(a)}, \tag{5.63}$$

where $F_n^M(a)$ and $F_n^L(a)$ are obtained by Equations (5.52) and (5.53). For higher-order modes at low frequencies, the wave number k_n in Equation (5.13) can be approximated as a purely imaginary number, that is, $k_n \simeq -jn\pi/h$. Therefore, $\sigma(l/b)$ is only related to l/b, h/b, and a/b, independent of the dielectric constant ε_r and frequency. Note that all the geometrical parameters are normalized to the via-hole radius b since l must be larger than or equal to b ($l \geq b$).

Some examples with different h/b and a/b values for the relative difference σ are shown in Figure 5.11. It can be seen that σ in logarithmic scale decays approximately linearly with l/b. Therefore, σ can be approximated as

$$\sigma(l/b) \simeq \alpha e^{-\beta(l/b)}, \tag{5.64}$$

where, for a specific via structure, α and β are two constant parameters that can be calculated by selecting two different l/b values in Equation (5.63) as

$$\beta = -\frac{\ln\sigma(l_2/b) - \ln\sigma(l_1/b)}{l_2/b - l_1/b}, \tag{5.65}$$

$$\ln\alpha = \frac{(l_1/b)\ln\sigma(l_2/b) - (l_2/b)\ln\sigma(l_1/b)}{l_1/b - l_2/b}. \tag{5.66}$$

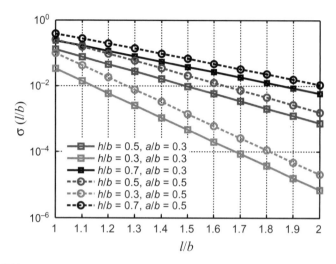

Figure 5.11 Examples of the relative difference versus the radius of the virtual boundary. ($r = 0.1$, $a = 0.4$, $b = 0.8$, unit: mm; $\varepsilon_r = 3.84$; mode number $N = 31$; frequency = 1.0 GHz) [41].

In most practical designs, $l_2/b = 1.8$ and $l_1/b = 1.2$ are suitable choices for estimating β and α. The curve-fitting formula Equation (5.64) and the accurate expression Equation (5.63) of the relative difference for several vias with different barrel-to-pad radius are compared in Figure 5.12. Note that although this is a relatively extreme case with large h/b and a/b values, good agreement is achieved between Equations (5.63) and (5.64).

Using Equation (5.64), the virtual via domain boundary at $\rho = l$ can be determined by specifying a small tolerance ε. This yields

$$l = b\left[\frac{\ln\alpha - \ln\varepsilon}{\beta}\right]. \tag{5.67}$$

For SI analyses, $0.05 \leq \sigma \leq 0.1$ is an acceptable choice. The smaller the relative difference is required, the larger the value of the virtual via boundary l.

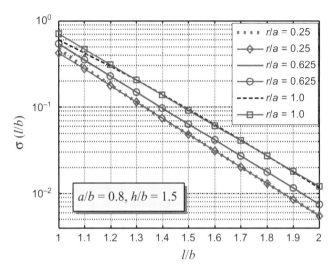

Figure 5.12   Validation of the curve-fitting formula for the relative difference (lines without symbols: from Eq. 5.63; lines with symbols: from Eq. 5.64) [41].

5.1.5 Complete Model for Multiple Vias in an Irregular Plate Pair

The intrinsic via circuit model for a single via, as shown in Figure 5.7, can be easily extended to the cases involving multiple vias. With P vias in a plate pair, the voltage and current relationship of Equation (5.40) is replaced by

$$V_i = Z_{ii}I_i + \sum_{j=1, j \neq i}^{P} Z_{ij}I_j. \tag{5.68}$$

An impedance matrix Z_{pp} is used to describe port voltages as functions of port currents in the plate domain as shown in Figure 5.1a.

Using Equation (5.68) instead of Equation (5.40), and following the same procedure in the derivation of the via circuit due to the zero-order waves, an equivalent circuit model for the ith via can be obtained as shown in Figure 5.13 using the ideal transformer theory [36]. For any arbitrarily shaped plate pair with P vias, Figure 5.13 is valid for each via, which leads to a complete circuit model shown in Figure 5.14 for multiple vias in an irregular plate pair.

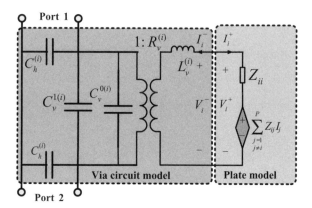

Figure 5.13 The ith via circuit model when P vias are present in a parallel plate pair [41].

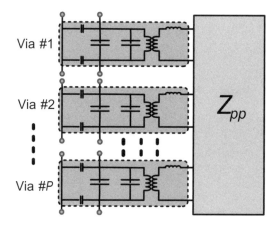

Figure 5.14 A schematic of circuit model for P vias in an irregular parallel plate pair [41].

The impedance of a plate pair used in Equation (5.40) or (5.68) is different from the conventional one defined in Reference 26 where rectangular ports and area integration of the Green's function are used to analytically evaluate the impedance matrix in a rectangular plate pair. In this application, ports in a plate pair should be circular as already pointed out in Reference 32, and Z_{pp} should be the impedance matrix for the plate pair with multiple PMC holes (with the via domains excluded). The new Z_{pp} will be addressed in detail in Section 5.2.

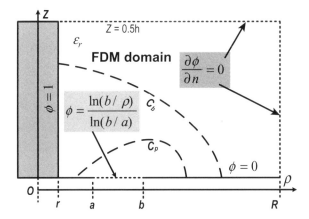

Figure 5.15   The solution domain for the shunt capacitance C_h and the boundary conditions when the finite difference method is applied [41].

5.1.6 Validation and Measurements

5.1.6.1 Evaluation of C_p, C_b, and C_h Using Finite Difference Method

In the intrinsic via circuit model, the shunt capacitance C_h, including C_p and C_b, can be validated using quasi-static approximations. Because of the radial symmetry, the potential Φ satisfies the 2D Laplace equation:

$$\frac{1}{\rho}\frac{\partial}{\partial\rho}\left(\rho\frac{\partial\Phi}{\partial\rho}\right)+\frac{\partial^2\Phi}{\partial z^2}=0. \tag{5.69}$$

The finite difference method is used to solve Equation (5.69), and the solution domain is defined in a rectangular $\rho - z$ region as shown in Figure 5.15, where the Neumann boundary condition is specified on the plane of $z = 0.5h$, as discussed in Reference 24.

The details of the finite difference method implementation have been given in Reference 39. The shunt capacitances C_p, C_b, and C_h obtained by the analytical formulas in Equations (5.28), (5.30), and (5.31) as well as by the finite difference method are compared in Figure 5.16. When other parameters are fixed, with the increase of the pad radius a, C_p increases and C_b decreases steadily; the combined capacitance C_h ($C_h = C_p + C_b$) decreases initially and then increases rapidly.

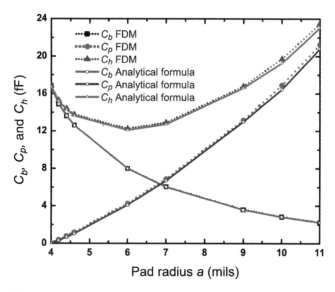

Figure 5.16 Comparisons of the barrel-plate, pad-plate and shunt capacitances using analytical formulas and the finite difference method (FDM) ($r = 0.1016$, $b = 0.4318$, $h = 0.2286$, $R = 1.27$, unit: mm; 1 mil = 0.0254 mm) [41].

The analytical evaluations and the numerical finite difference simulations agree well. The meshing grid in the finite difference method was chosen to be either 1.27×10^{-3} mm (0.05 mil) or 0.635×10^{-3} mm (0.025 mil), and convergence with mesh density was ensured.

5.1.6.2 Input Impedance of a Probe in a Circular Plate Pair

A probe can be viewed as a special via with its top coaxial port closed by PEC as shown in Figure 5.2. Consequently, an equivalent probe circuit model can be obtained by shorting the top coaxial port (Port 1) in the 3-port via circuit model given in Figure 5.7. Then, the input admittance of a centrally located probe in a circular plate pair can be obtained as

$$Y_{in}^{C} = j\omega(C_h + C_v^1 + C_v^0) + \frac{R_v^2}{j\omega L_v + Z_l^+}. \tag{5.70}$$

Another expression for the input admittance of a probe in a circular plate pair seen from Figure 5.4 is

$$Y_{in}^A = j\omega(C_h + C_v^1) + Y_v, \qquad (5.71)$$

where Y_v is obtained by Equation (5.33). Equations (5.70) and (5.71) can be regarded as two different expressions for Y_{22} in Equation (5.23), which can be further proven to be consistent with the formulas for a probe in an infinite plate pair derived in References 15, 33, and 34 when $S = L$ is selected for $F_n^S(a)$ calculations. Therefore, Y_{22} in Equation (5.23) can be viewed as a general analytical formula for the input admittance of a probe located in a circular plate pair.

The input admittance can also be obtained from the physics-based via circuit model as

$$Y_{in}^C = j\omega C_h + \frac{1}{Z_r}, \qquad (5.72)$$

where Z_r is calculated from Equation (5.19) by replacing l with r, the via barrel radius [20–22].

By numerically comparing the input admittances of a probe using Equation (5.70) of the equivalent probe circuit, Equation (5.71) of the analytical formula, and Equation (5.72) of the physics-based circuit, the equivalence of Equations (5.70) and (5.71) can be verified, and also the limitation of the physics-based via circuit model can be observed.

The input impedance results of a probe in an infinite plate pair using the three different expressions are compared in Figure 5.17. The radii of the probe barrel and the via hole are 0.254 mm (10 mil) and 0.8382 mm (33 mil), respectively. The separation of the plates is 1.4732 mm (58 mil). A Debye dielectric model, $\varepsilon_r = \varepsilon_\infty + (\varepsilon_s - \varepsilon_\infty)/(1 + j\omega\tau)$, is used for the permittivity of the dielectric material between the two plates. The relative static permittivity ε_s and the optical permittivity ε_∞ are set to be 4.3 and 4.1, respectively, and the relaxation time τ is 3.1831×10^{-11} seconds.

The input impedance calculated from Equation (5.70) of the equivalent probe circuit agrees very well with Equation (5.71) of the analytical formula for both the real and imaginary parts. The input impedance calculated from Equation (5.72) of the physics-based via circuit model, however, matches Equations (5.70) and (5.71) at low frequencies only.

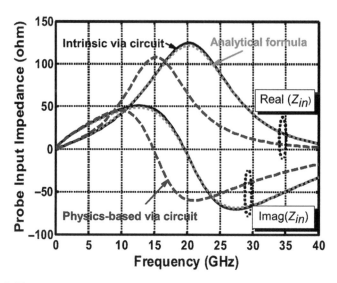

Figure 5.17   Comparisons of the input impedance results of a probe in an infinite plate pair ($r = a = 0.254$, $b = 0.8382$, $h = 1.4732$, unit: mm) [41].

Moreover, the resonant frequency predicted from Equation (5.72) is 15 GHz, as opposed to 20 GHz calculated from Equations (5.70) and (5.71).

Furthermore, Figure 5.18 compares the probe impedance for a finite PEC circular plate pair with a radius of 200 mils by the intrinsic via model of Equation (5.70), physics-based via model of Equation (5.72), analytical formula Equation (5.71), and finite difference time domain of CST Microwave Studio. It can be seen that for this example, four methods agree very well at the first resonant frequency at about 10 GHz. However, the physics-based method starts to deviate from the other three methods at the second resonant frequency. On the other hand, the intrinsic via model matches very well with the analytical formula as well as the full-wave solver in the entire frequency range up to 30 GHz.

The probe impedances of a PMC and a PEC circular plate pair are used to investigate the impact of dielectric heights on the accuracy of the physics-based via model. The results are shown in Figure 5.19. It can be seen that the physics-based via model is more accurate for the plate pair with a smaller dielectric height, h, compared with the intrinsic via circuit model.

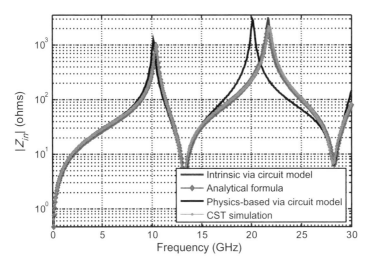

Figure 5.18 Comparisons of the input impedance results of a probe in a finite circular plate pair ($r = a = 0.127$, $b = 0.3810$, $h = 1.016$, unit: mm; $\varepsilon_\infty = 4.2$, $\varepsilon_s = 4.4$, $\tau = 1.6 \times 10^{-11}$ s; $R = 5.08$ mm with PEC boundary). (See color insert.)

The comparisons in these two figures indicate that (1) the new expression for Y_v in Equation (5.48), which is key to derive the intrinsic 3-port via circuit model, agrees very well with that in Equation (5.33); (2) the physics-based via circuit model can be viewed as an approximation of the intrinsic 3-port via circuit model at low frequencies; (3) the difference between these two via circuit models becomes larger with the increase of both operating frequencies and the plate-pair heights.

In practical applications, the physics-based circuit model is usually acceptable due to several reasons. First, only lowest resonant frequencies of a plate pair are usually critical for SI/PI analysis depending on the specific data rate and rise time; second, the height of a plate pair in practical PCBs is becoming smaller and smaller to reduce the plate pair for high-speed digital signals; and, third, to provide a current return path, shorting vias or decoupling capacitors are often used in a plate pair, which in most cases reduces the impact of the plate-pair height. Therefore, the rigorous intrinsic via circuit model derived here justifies the engineering applications of the physics-based via circuit introduced in References 20 and 21 at low frequencies in the cases with a relatively small height of the plate pair, or a plate pair with dense shorting vias/ decoupling capacitors.

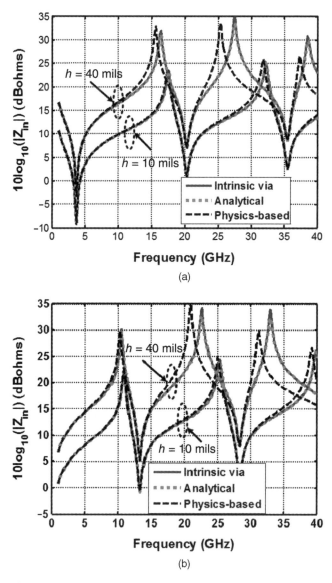

(a)

(b)

Figure 5.19 The probe impedance by three different formulas ($r = a = 0.127$, $b = 0.3810$, unit: mm; $\varepsilon_\infty = 4.2$, $\varepsilon_s = 4.4$, $\tau = 1.6 \times 10^{-11}$ s, $R = 5.08$ mm): (a) PMC cavity; (b) PEC cavity.

5.1.6.3 Measurements

Figure 5.20a illustrates a test board geometry measured to validate the intrinsic via circuit model. It contains seven plate pair with two strip-lines located in the top and bottom ones, respectively. The separations of the plate pair are 0.3048 mm (12 mils) or 0.2032 mm (8 mils). A signal via located at (4.064, 4.064) mm or (160, 160) mil, as shown in Figure 5.20a, connects these two striplines. The radii of the via barrel and via hole are 0.127 mm and 0.381 mm, respectively. For each plate pair, the fields are restricted inside a 9.144×9.144 mm^2 or 360×360 mil^2 cavity constructed by dense stitching vias connecting all the plates (an approximate PEC cavity). An additional shorting via is located at (4.064, 5.080) mm or (160, 200) mil to provide an adjacent return path for the signal via. The relative permittivity and loss tangent of the dielectric layers between the plates are 3.5 and 0.014, respectively.

The transmission property between Ports 1 and 2 of the test board geometry is simulated by the equivalent circuit shown in Figure 5.20b. Note that each stripline is split into two microstrip lines referenced to the top and bottom plates as proposed in Reference 21, in order to be connected to the intrinsic via circuit model. The length of the microstrip lines is assumed to be 6.35 mm (250 mil) in the simulation. The intrinsic via circuit models are connected to the impedance matrix of each plate pair, which is terminated by a load impedance Z_L. Two cases were studied: with and without the local shorting via at (4.064, 5.080) mm. $Z_L = 0$ is used for the case with the local shorting via while $Z_L = \infty$ for the case without the local shorting via.

The S-parameters calculated from the rigorous intrinsic via circuit model, the physics-based via circuit model, and measurements for the cases with and without the local shorting via are compared in Figures 5.21 and 5.22, respectively. The circuit models were simulated using the Advanced Design System (ADS), a circuit simulator from Agilent, Inc. The S-parameters of the test geometry were measured using an Agilent Vector Network Analyzer (VNA) E8364A in the frequency band from 45 MHz to 40 GHz. Microwave probes, model 40A G-S and S-G with a 225 micron pitch, from GGB Industries Inc., were used to minimize the effects of test fixtures. The SLOT (Short Load Open Through) calibration was conducted using a CS-14 calibration substrate, also from GGB Industries Inc.

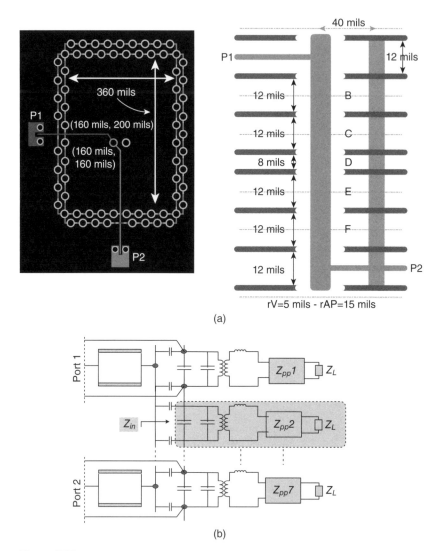

Figure 5.20 (a) Test board geometry (courtesy of the authors of Reference 21) (1 mil = 0.0254 mm); (b) equivalent circuit using the intrinsic 3-port via model for the test board geometry shown in (a).

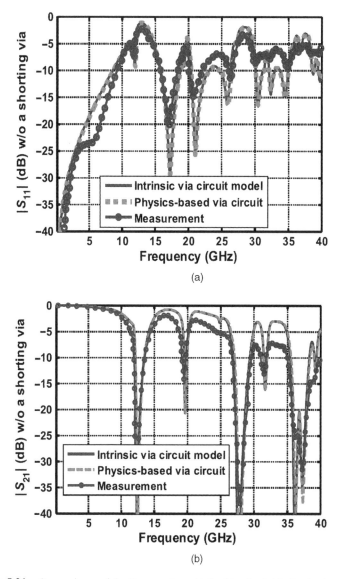

Figure 5.21   Comparisons of the *S*-parameters obtained by the intrinsic via circuit model, the physics-based via circuit model, and measurements without a local shorting via [41].

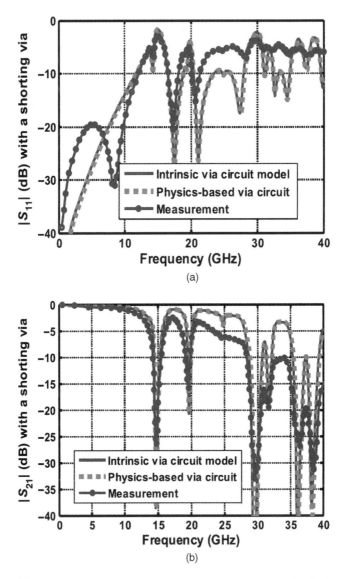

Figure 5.22 Comparisons of the S-parameters obtained by the intrinsic via circuit model, the physics-based via circuit model, and measurements when a local shorting via is present [41].

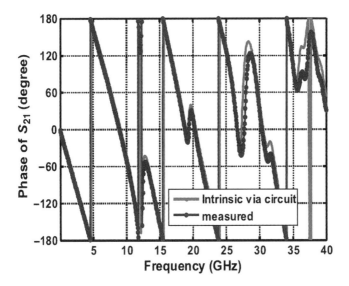

Figure 5.23 Phase comparison of the S_{21} obtained by the intrinsic via circuit model and measurements without a local shorting via [41].

There are only slight differences between the results obtained from the rigorous intrinsic via circuit model and the approximate physics-based via circuit model. These two models match the measured results very well up to 15 GHz. At higher frequencies, these two models can also predict the same trends of the S-parameters as the measurements, but with larger discrepancies. A possible reason for the discrepancies is that the trace-to-via discontinuities are neglected in the simulations.

In addition, Figure 5.23 also shows good agreement up to 30 GHz between the phase of S_{21} obtained by the intrinsic via circuit model and the measurement.

The impedance of each plate pair, Z_{in} as illustrated in Figure 5.20b, is a critical parameter to understand the simulation results of S_{21}. A smaller Z_{pp} will result in a better transmission property between the two ports.

The impedance magnitudes of the second plate pair obtained from the intrinsic and the physics-based via circuit models are compared in Figure 5.24. These two models provide almost the same series imped-ances. As mentioned earlier for the input impedance of a probe, these

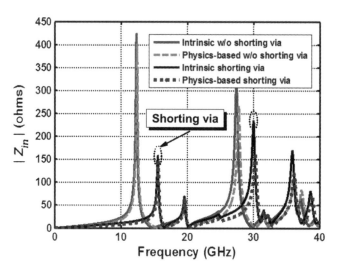

Figure 5.24   Magnitudes of the input impedance calculated from the intrinsic via circuit and the physics-based via circuit models [41].

two via circuit models result in quite similar results for a plate pair with a small height at the first several resonant frequencies. The test board geometry in this example is very small in via heights. Therefore, at the frequencies of interest (less than 40 GHz), the physics-based via circuit is still a good approximation. This explains why these two models provide almost the same S-parameters in Figures 5.21 and 5.22.

The resonant peaks in Figure 5.24 lead to the minimum $|S_{21}|$ values in both Figures 5.21b and 5.22b. This is because Z_{in} is the impedance of the return current path of the signal via. At low frequencies, the surrounding stitching vias are an effective low-impedance return current path. Thus, the signal can easily pass through several layers of the plate pair. Another observation from Figure 5.24 is that the local shorting via shifts the first resonant frequency from 12.35 GHz to 14.58 GHz, enhancing the bandwidth of the signal channel.

5.1.7 Conclusion

An intrinsic 3-port via circuit model and its connection to the impedance matrix of a plate pair is derived rigorously through electromagnetic analysis. Both boundary conditions at vias and plate edges are satisfied explicitly. This provides a solid theoretical foundation for the physics-based via circuit model that was developed ad hoc from physical intuition. Analytical formulas for the input impedance of a probe in a circular plate pair are used to verify the derived intrinsic via circuit model. It shows that the physics-based via circuit model is an acceptable approximation to the rigorous one derived here for relatively thin plate pair or for plate pair with dense shorting vias/decoupling capacitors. Furthermore, test boards with a signal via and seven plate pair with and without a local shorting via are employed to validate the intrinsic via circuit model with good agreement.

5.2 PARALLEL PLANE PAIR MODEL

5.2.1 Introduction

In last section, an equivalent circuit model is derived rigorously for a plate pair with P vias. The plate pair is divided into P via domains and one plate domain. Each via domain is characterized by an intrinsic 3-port via circuit model, whose radial port is connected to an impedance matrix, the plate-domain model. It has demonstrated that only the zero-order TM_{z00} modes need to be considered in the plate domain.

According to the segmentation technique used in Section 5.1.2, the virtual circular boundaries of via domains and the plate domain are assumed to be PMC, and radial ports with impressed currents are defined at these boundaries in both the via and plate domains so that domain connections are possible by enforcing continuity of voltages and currents. Based on these conditions, the plate domain can be modeled as an impedance matrix $Z_{pp}(P \times P)$ among the P radial ports defined at the virtual circular boundaries.

Moreover, PMC boundary conditions at the ports are consistent with the general definition of the impedance matrix of a P-port network:

$$Z_{ij} = \frac{V_i}{I_j}\bigg|_{I_s=0}, s = 1, 2, \cdots, P, \text{ and } s \neq j. \tag{5.73}$$

The port condition "$I_s = 0$, $s = 1, 2, \cdots, P$, and $s \neq j$" means the impressed current at Port s is zero. This corresponds to a zero tangential magnetic field at the port surface, that is, a PMC boundary condition, in electromagnetic analysis.

Based on the above reasons, to rigorously analyze the via-plate-pair interactions, the plate domain is not the entire solid plate pair; instead,

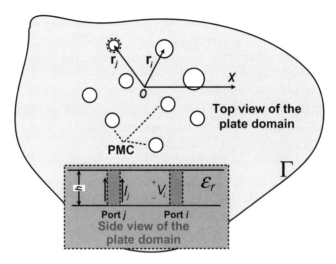

Figure 5.25 A plate domain with multiple PMC holes.

it has P PMC ports (holes) as shown in Figure 5.25 when the via domains are excluded.

Careful studies indicate that this Z_{pp}, the impedance of a plate domain, cannot be calculated from the conventional area-integral impedance definition [25, 26, 28–32], as the Green's function of a plate pair with multiple PMC holes is not available. On the other hand, although another conventional cylindrical wave definition of the power-bus impedance proposed in Reference 27 is expressed in terms of the zero-order TM_{z00} waves, it is restricted to the self-impedance of a port centered in a circular plate pair, and is not applicable for multiple ports in an arbitrarily shaped plate pair common in practical PCB or package designs.

In this section, a new Z_{pp} definition is provided so that the imped-ance matrix for multiple radial PMC ports in an irregular plate pair can be expressed in terms of the zero-order TM_{z00} waves, which is then used to connect the intrinsic via circuit models discussed in the last section. For each circular PMC hole, port voltage and current are first defined as integrations of electric and magnetic fields along the radial port. This new definition of port voltage and current reveals the equivalence of the two conventional impedance definitions. In addition, it indicates that the new Z_{pp} developed in this section is a generalized form of the conventional cylindrical wave definition proposed in Reference 27 for multiple radial ports in an irregular plate pair. Then, the outward and inward waves from the radial transmission-line ports are expressed in terms of the zero-order parallel plate waves.

A radial scattering matrix is further introduced to relate the expansion coefficients of the outward and inward waves. Furthermore, a pair of the transforms between the radial scattering matrix, denoted S_{pp}^R, and the impedance matrix Z_{pp} for a plate pair with radial transmission line ports, is introduced. Note that the radial scattering parameters are obtained from a solid plate pair as matched loads are required at the radial transmission-line ports by the definition of the scattering parameters. When Z_{pp} is calculated through the transform from S_{pp}^R, the "open" loads or the PMC boundary conditions at the ports are automatically satisfied. Therefore, the radial scattering matrix S_{pp}^R, as well as the transforms between S_{pp}^R and Z_{pp}, greatly simplifies the calculations of Z_{pp} for a plate domain with multiple PMC holes.

Analytical and numerical methods are provided to obtain the radial scattering parameters. First, an analytical solution of S_{pp}^R is derived for a circular plate pair using the addition theorem of cylindrical waves, which can be used as a validated example for numerical methods. Then, a boundary integral-equation method (BIE) is developed to calculate S_{pp}^R for an irregular plate pair which is common in practical PCB or package designs. Exciting waves are assumed to be the outward waves from the radial transmission-line port at each PMC hole. The unknown magnetic or electric currents along the edges of the plate pair satisfies an integral equation that is then solved by the method of moments (MoM). S_{pp}^R is found to be related to the impedance/admittance matrix of MoM. The impedance matrix Z_{pp} can be obtained using the transforms derived. Numerical examples and measurements are provided to validate the new definition of Z_{pp} and its calculation by the BIE method.

5.2.2 Overview of Two Conventional Z_{pp} Definitions

Consider two circular ports i and j in an irregular parallel plate pair as shown in Figure 5.24. The height, area domain, and boundary of the parallel plate pair are denoted as h, Ω, and Γ, respectively. Vector $\mathbf{r}_{i(j)}$ denotes the location of Port $i(j)$ in the global coordinates. The circumference, radius, and area of Port $i(j)$ are represented as $l_{i(j)}$, $r_{i(j)}$, and $S_{i(j)}$, respectively (Note that $r_{i(j)}$ is used to denote the radius of Port $i(j)$

instead of the length of the vector $\mathbf{r}_{i(j)}$ in this section to facilitate the following discussions).

5.2.2.1 Area-Integral Definition

The most widely used formula for the impedance of a plate pair between Ports i and j is defined as [26]

$$Z_{ij}^{(1)} = -\frac{1}{S_i S_j} \int_{S_i} \int_{S_j} G(\mathbf{r}, \mathbf{r}') dS_i dS_j, \tag{5.74}$$

where $G(\mathbf{r}, \mathbf{r}')$ is the electric field Green's function for electric currents in a parallel plate pair; the observation point $\mathbf{r} \in S_i$; and the source point $\mathbf{r}' \in S_j$. As ports are normally electrically small, rectangular ports instead of circular ones are used for analytical evaluations of Equation (5.74) for rectangular cavities.

As Green's functions for rectangular or special triangular cavities are available, the area-integral definition with the segmentation technique can provide an efficient approach for evaluating the impedance matrix of an arbitrarily shaped plate pair by decomposing the overall geometry into multiple rectangular and triangular cavities [28–30, 40].

5.2.2.2 Cylindrical Wave Definition

Another impedance definition is only used in Reference 27 for a circular parallel plate pair with a centrally located port. Its self-impedance is given in terms of cylindrical waves as

$$Z_{ii}^{(2)} = \frac{V_i}{I_i} = \frac{j\omega\mu h}{2\pi k r_i} \frac{H_0^{(2)}(kr_i) + \Gamma_R^{(0)} J_0(kr_i)}{H_1^{(2)}(kr_i) + \Gamma_R^{(0)} J_1(kr_i)}, \tag{5.75}$$

where $J_{0(1)}(\cdot)$ and $J_{0(1)}(\cdot)$ are the zero (or first)-order Bessel function and Hankel function of the second kind, respectively; and Port i is located at the center of the circular plate pair. The coefficient $\Gamma_R^{(0)}$ is used to represent the reflection of the TM_{z00} wave from the outer PMC boundary of the plate pair at $\rho = R$ as

$$\Gamma_R^{(0)} = -\frac{H_1^{(2)}(kR)}{J_1(kR)}. \tag{5.76}$$

The port voltage and current are defined on the circumference of Port i as

$$V_i = -E_{zi}h, \tag{5.77}$$

$$I_i = 2\pi r_i H_{\varphi i}. \tag{5.78}$$

Note that E_{zi} and $H_{\varphi i}$ are the total fields including both the incident and the reflected waves.

5.2.2.3 Limitations of Two Conventional Definitions

It is obvious that the area-integral definition is not suitable for the modeling of the plate domain, which is not a solid plate pair but one with multiple PMC holes at via locations. Normally, the analytical Green's function for such a plate pair with multiple PMC holes is not available. Moreover, the area-integral definition does not demonstrate a clear, relevant physical meaning for the voltages and currents associated with the TM_{z00} mode, to provide a rigorous foundation for the modeling of the plate domain in the development of the intrinsic via circuit model.

On the other hand, although the cylindrical wave definition is expressed in terms of the zero-order parallel plate waves that are relevant to the plate-domain behaviors, it is restricted to a via located at the center of a circular plate pair and only the self-impedance is given. Therefore, these two conventional impedance definitions are not adequate for modeling the normally irregular and perforated plate domain with the intrinsic via circuit model derived in the last section.

5.2.3 New Z_{pp} Definition Using the Zero-Order Parallel Plate Waves

5.2.3.1 New Port Voltage and Current Definitions

First of all, the voltage at Port i and current at Port j can be defined as line averages of the electric and magnetic fields along the circumferences of Ports i and j, respectively,

$$V_i = -\left(\frac{1}{l_i}\int_{l_i} E_z(\mathbf{r})dl_i\right)h, \tag{5.79}$$

$$I_j = \int_{l_j} H_\varphi(\mathbf{r}')dl_j, \tag{5.80}$$

where $\mathbf{r} \in l_i$ and $\mathbf{r}' \in l_j$. Obviously, Equations (5.79) and (5.80) are the generalized forms of Equations (5.77) and (5.78) developed from the zero-order cylindrical waves. It is easy to demonstrate that for a central port in a circular plate pair, the new port voltage and current definitions Equations (5.79) and (5.80) can lead to the same impedance as Equation (5.75) in the cylindrical wave definition. In other words, the new port voltage and current definitions are consistent with the conventional cylindrical wave definition.

Furthermore, assuming a constant equivalent current I along the circumference l_j of Port j as

$$J_z(\mathbf{r}') = H_\varphi(\mathbf{r}') = \frac{I}{l_j}, \tag{5.81}$$

the electric field along the circumference l_i of Port i is the radiated field by $J_z(\mathbf{r}')$ as

$$E_z(\mathbf{r}) = \int_{l_j} J_z(\mathbf{r}')G(\mathbf{r},\mathbf{r}')dl_j, \tag{5.82}$$

where $G(\mathbf{r}, \mathbf{r}')$ is the Green's function of the plate pair. Substituting Equations (5.81) and (5.82) into Equation (5.79) yields the third expression of the impedance of a plate pair as

$$Z_{ij}^{(3)} = -\frac{1}{l_i l_j}\int_{l_i}\int_{l_j} G(\mathbf{r},\mathbf{r}')dl_i dl_j. \tag{5.83}$$

Comparing Equations (5.83) to (5.74), it becomes obvious that these two expressions are in the same form except that Equation (5.83) uses the line or boundary averages, while Equation (5.74) employs the area averages. When the port sizes are electrically small, these two expressions yield almost the same results. This demonstrates that the new port voltage and current definitions are somehow consistent with the conventional area-integral definition, too.

From the above discussions, the new port voltage and current definitions Equations (5.79) and (5.80) are found reasonable and consistent with the conventional impedance definitions. The equivalence between the two conventional definitions is thus proved. Further, this implies the area-integral impedance definition can also be interpreted in terms of the zero-order parallel plate waves as in the cylindrical wave definition. However, the PMC boundaries along the ports are not enforced as the area-integral definition uses the Green's function of a solid plate pair. This means for a plate pair with only a few vias, the area-integral definition could be a good approximation as long as the total area of the PMC holes is negligible compared to the area of the entire plate pair. This observation can explain why the physics-based via circuit model using the conventional area-integral definition can still achieve reasonable results compared with measurements in References 20 and 21. For a plate pair with a large number of vias, however, the area-integral definition cannot characterize the impact of the large number of the PMC holes on the resonant frequency shifts demonstrated in the measurement provided later.

5.2.3.2 New Impedance Matrix Z_{pp} and Radial Scattering Matrix S_{pp}^R of a Plate Domain

With the new port voltage and current definitions, the impedance matrix Z_{pp} of a plate domain with multiple PMC holes is developed herein. According to the completeness of the cylindrical harmonics, the electric field distribution near Port i can be expressed in its local polar coordinates with the origin at the center of Port i as

$$E_z(r^i) = \sum_{m=-\infty}^{\infty} \left[a_i^{(m)} H_m^{(2)}(kr^i) + b_i^{(m)} J_m(kr^i) \right] e^{jm\varphi^i}, \quad (5.84)$$

where $r^i = |\mathbf{r} - \mathbf{r}_i|$ and $\varphi^i = \arg\{\mathbf{r} - \mathbf{r}_i\}$ are the local coordinates of observation point $\mathbf{r}$, and the point $\mathbf{r}_i$ is the center of Port i. Substituting Equation (5.84) into Equation (5.79), we have

$$V_i = -\left[a_i^{(0)} H_0^{(2)}(kr^i) + b_i^{(0)} J_0(kr^i) \right] h. \quad (5.85)$$

Clearly, the integration in Equation (5.79) results in the elimination of all the field components with azimuthal variations. This implies all the radial transmission-line ports are electrically small, which is one of the assumptions used in the development of the intrinsic via circuit model in Reference 41. Similarly, the current of Port i can be obtained as

$$I_i = -\frac{2\pi k r_i}{j\omega\mu}\left[a_i^{(0)}H_1^{(2)}(kr^i) + b_i^{(0)}J_1(kr^i)\right]. \tag{5.86}$$

For a P-port system, voltage and current vectors are represented as $\mathbf{V} = [V_1 \quad V_2 \quad \cdots \quad V_P]^T$ and $\mathbf{I} = [I_1 \quad I_2 \quad \cdots \quad I_P]^T$; and the wave expansion coefficient vectors are defined as $\mathbf{b} = \begin{bmatrix} b_1^{(0)} & b_2^{(0)} & \cdots & b_P^{(0)} \end{bmatrix}^T$ and $\mathbf{a} = \begin{bmatrix} a_1^{(0)} & a_2^{(0)} & \cdots & a_P^{(0)} \end{bmatrix}^T$. Then, Equations (5.85) and (5.86) are written in the matrix forms as

$$\mathbf{V} = -\mathbf{H}_0\mathbf{a} - \mathbf{J}_0\mathbf{b}, \tag{5.87}$$

$$\mathbf{I} = -\mathbf{H}_1\mathbf{a} - \mathbf{J}_1\mathbf{b}, \tag{5.88}$$

where $\mathbf{H}_0$, $\mathbf{H}_1$, $\mathbf{J}_0$, and $\mathbf{J}_1$ are all diagonal $P \times P$ matrices as

$$\mathbf{H}_0 = diag\{H_0^{(2)}(kr_i)h\}, \tag{5.89}$$

$$\mathbf{J}_0 = diag\{J_0(kr_i)h\}, \tag{5.90}$$

$$\mathbf{H}_1 = diag\left\{\frac{2\pi k r_i}{j\omega\mu}H_1^{(2)}(kr_i)\right\}, \tag{5.91}$$

$$\mathbf{J}_1 = diag\left\{\frac{2\pi k r_i}{j\omega\mu}J_1(kr_i)\right\}. \tag{5.92}$$

The plate-pair impedance matrix $\mathbf{Z}_{pp}$ relates voltage and current vectors as

$$\mathbf{V} = \mathbf{Z}_{pp}\mathbf{I}. \tag{5.93}$$

Substituting Equations (5.87) and (5.88) into Equation (5.93), yields

$$\mathbf{H}_0\mathbf{a} + \mathbf{J}_0\mathbf{b} = \mathbf{Z}_{pp}[\mathbf{H}_1\mathbf{a} + \mathbf{J}_1\mathbf{b}]. \tag{5.94}$$

On the other hand, the wave expansion coefficient vectors $\mathbf{a}$ and $\mathbf{b}$ can be viewed as the input sources and output responses of the P-port linear system of the plate domain, and they are related together as

$$\mathbf{b} = \mathbf{S}_{pp}^R\mathbf{a}, \tag{5.95}$$

where $\mathbf{S}_{pp}^R$ is a $P \times P$ radial scattering matrix for the plate pair. From the mathematical point of view, Equation (5.95) is a generalized addition theorem that relates the zero-order inward cylindrical waves in one coordinate system to the zero-order outward cylindrical waves in another one.

Using Equations (5.94) and (5.95), the transforms between the impedance matrix Z_{pp} of a plate domain with multiple PMC ports and the radial scattering matrix $\mathbf{S}_{pp}^R$ can be derived as

$$\mathbf{S}_{pp}^R = -\left[\mathbf{J}_0 - \mathbf{Z}_{pp}\mathbf{J}_1\right]^{-1}\left[\mathbf{H}_0 - \mathbf{Z}_{pp}\mathbf{H}_1\right], \tag{5.96}$$

$$\mathbf{Z}_{pp} = \left[\mathbf{H}_0 + \mathbf{J}_0\mathbf{S}_{pp}^R\right]\left[\mathbf{H}_1 + \mathbf{J}_1\mathbf{S}_{pp}^R\right]^{-1}. \tag{5.97}$$

For the completeness, the corresponding admittance matrix $\mathbf{Y}_{pp}$ for a plate pair with multiple shorting ports (shorting vias connecting two plates) can be defined as

$$\mathbf{Y}_{pp} = \left[\mathbf{H}_1 + \mathbf{J}_1\mathbf{S}_{pp}^R\right]\left[\mathbf{H}_0 + \mathbf{J}_0\mathbf{S}_{pp}^R\right]^{-1}. \tag{5.98}$$

Note that Equation (5.97) can be reduced to Equation (5.75) in the special case of only one port centered in a circular plate pair. This indicates the new Z_{pp} of Equation (5.97) is a generalized form of Equation (5.75) using the zero-order parallel plate waves for multiple ports in an arbitrary-shaped plate pair.

It is worth noting that the radial scattering parameters for a multiport network is defined as

$$S_{pp}^R(i, j) = \left.\frac{b_j^{(0)}}{a_j^{(0)}}\right|_{a_s^{(0)}=0}, \text{ when } s \neq j. \tag{5.99}$$

The condition of "$a_s^{(0)} = 0$, when $s \neq j$" indicates the boundary conditions at all the source-free ports should be the matched or PML ones. In other words, a solid plate pair instead of one with PMC holes should be used to calculate the radial scattering parameters, $\mathbf{S}_{pp}^R$. Similar to the transform between the S- and Z-parameters of a microwave network [35], the transform from $\mathbf{S}_{pp}^R$ to $\mathbf{Z}_{pp}$ in Equation (5.97) will automatically guarantee the "open" loads or the PMC boundary conditions at the source-free ports in the Z_{pp} calculations. This means Z_{pp} of a plate domain with PMC holes can be indirectly obtained from $\mathbf{S}_{pp}^{R}$ of the corresponding solid plate pair, which will be demonstrated later.

Note that the mutual impedance from the area-integral definition of Equation (5.74) is only related to the two ports, independent of all other ports. On the other hand, the new Z_{pp} of Equation (5.97) is expressed in a matrix form. This means the mutual impedance between any two ports will be affected by other ports. From the electromagnetic point of view, multiple scattering among all the PMC holes (or cylinders) will definitely influence the mutual coupling between any two ports. Simulation and measurement provided later will demonstrate that the new Z_{pp} of Equation (5.97) can reflect the impact of a large number of ports on the resonant frequency shift of the plate pair.

5.2.4 Analytical Formula for Radial Scattering Matrix S_{pp}^{R} in a Circular Plate Pair

The transform relationship between the radial scattering matrix and the impedance matrix of a plate pair with PMC ports given in Equations (5.96) and (5.97) leads to an analytical solution for a circular plate pair, which can be used as a benchmark for validating any numerical methods.

For two radial ports i and j in a circular plate pair as shown in Figure 5.26, assume that only Port j is excited and the outward wave from Port j is

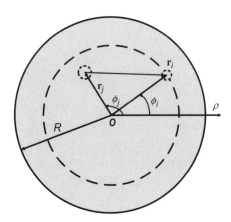

Figure 5.26 Two PMC ports i and j located in a circular plate pair.

$$E_z^{inc} = a_j^{(0)} H_0^{(2)}(k|\mathbf{r} - \mathbf{r}_j|). \tag{5.100}$$

This results in an inward wave to Port i as

$$E_z = b_j^{(0)} J_0(k|\mathbf{r} - \mathbf{r}_i|), \tag{5.101}$$

where the expansion coefficients $b_i^{(0)}$ and $a_j^{(0)}$ are related together by

$$S_{pp}^R(i, j) = b_i^{(0)} / a_j^{(0)}, \tag{5.102}$$

which is a function of port locations, plate size, and the boundary conditions of the plate edges.

With the aid of the addition theorem of cylindrical harmonics [42], the radial scattering parameter between two ports in a circular plate pair can be analytically derived as

$$S_{pp}^R(i, j) = (1 - \delta_{ij}) H_0^{(2)}(kr_{ji}) + \sum_{s=-\infty}^{\infty} J_s(kr_{io}) e^{js\varphi_i} \Gamma_R^{(s)} J_s(kr_{jo}) e^{-js\varphi_j}, \tag{5.103}$$

where $r_{ji} = |\mathbf{r}_j - \mathbf{r}_i|$ is the distance between the two port centers; and the port locations are $\mathbf{r}_j = (r_{jo}, \varphi_j)$ and $\mathbf{r}_i = (r_{io}, \varphi_i)$ in the global polar coordinates as shown in Figure 5.26. The reflection coefficient of the cylindrical waves from the edge of the circular plates at $\rho = R$ with different boundary conditions can be obtained as

$$\Gamma_R^{(s)} = \begin{cases} -\dfrac{H_s^{(2)}(kR)}{J_s(kR)} & \text{PEC at } \rho = R \\[3mm] -\dfrac{H_s^{(2)\prime}(kR)}{J_s'(kR)} & \text{PMC at } \rho = R \\[3mm] 0 & \text{PML at } \rho = R. \end{cases} \tag{5.104}$$

Note that the first term in the right-hand side of Equation (5.103) is the direct illumination from Port j to Port i, and the second summation term represents the reflected cylindrical waves from the circular edge of the plate pair.

Therefore, the impedance matrix of a circular plate pair with PMC ports can be analytically evaluated by substituting Equation (5.103) into Equation (5.97). This can be used as a benchmark to validate Z_{pp}

calculations by any numerical methods, such as the BIE method developed in the next section.

5.2.5 BIE Method to Evaluate S_{pp}^R for an Irregular Plate Pair

For any arbitrarily shaped plate pair, analytical S_{pp}^R is not available and thus, numerical methods have to be used. Various methods, including cavity model with segmentation technique [25, 28–31, 40], finite difference method [43, 44], finite element method (FEM) [45], and MoM [46], have been widely used to obtain the impedance of a solid plate pair using the conventional area-integral definition. However, it is either not possible or not efficient to apply these methods for a plate domain with PMC ports. A BIE method is introduced herein to obtain the Z_{pp} of a plate domain with PMC ports using the new impedance definition. S_{pp}^R is obtained first from the BIE method, and then Z_{pp} is calculated through its relationship with the radial scattering parameters S_{pp}^R given in Equation (5.97). This avoids the difficulties in directly calculating the impedance matrix Z_{pp} of a plate pair having multiple PMC holes (ports) as in Reference 47, where the circular port boundaries have to be meshed in addition to the plate-pair edges.

Two PMC Ports i and j in an irregular plate pair are illustrated in Figure 5.27. Assume only Port j is excited with an external source, such that the radiated electric field from Port j can be written as

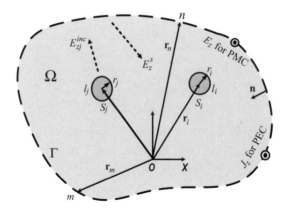

Figure 5.27 Boundary-integral equation formula to extract S_{pp}^R in an irregular plate pair.

$$E_z^{inc}(\mathbf{r}) = H_0^{(2)}\big(k|\mathbf{r}-\mathbf{r}_j|\big). \tag{5.105}$$

According to the definition of Equation (5.102), a total electric field near Port i can be expressed as

$$E_z(\mathbf{r}) = S_{pp}^R(i, j)J_0\big(k|\mathbf{r}-\mathbf{r}_i|\big). \tag{5.106}$$

Then, it can be shown that

$$S_{pp}^R(i, j) = E_z(\mathbf{r}_i). \tag{5.107}$$

This indicates $S_{pp}^R(i, j)$ actually is the total electric field, $E_z(\mathbf{r}_i)$, under the excitation at Port j as defined in Equation (5.105) with all the radial ports matched.

5.2.5.1 PMC Cavity

Usually, the separation of the parallel plates h is electrically small. Therefore, the edge boundary Γ can be approximated as a PMC boundary. According to the equivalence principle, the magnetic current $\mathbf{M}_t$ or the vertical electric field E_z is assumed to be the unknown sources for the scattered waves from the boundary Γ of the region Ω as

$$E_z^s(\mathbf{r}) = \frac{j}{4}\int_\Gamma E_z(\mathbf{r}')\nabla\big[H_0^{(2)}(k|\mathbf{r}-\mathbf{r}'|)\big]\cdot\mathbf{n}'d\Gamma, \tag{5.108}$$

where ∇ is the gradient operator of the observation point $\mathbf{r}$, and $\mathbf{n}'$ is the normal vector of the source point $\mathbf{r}'$.

The total electric field in the region Ω is the summation of both the incident and scattered waves as

$$E_z(\mathbf{r}) = E_z^{inc}(\mathbf{r}) + E_z^s(\mathbf{r}). \tag{5.109}$$

This leads to a BIE of $E_z(\mathbf{r})$, $\mathbf{r} \in \Gamma$ as

$$E_z^{inc}(\mathbf{r}) = E_z(\mathbf{r}) - \frac{j}{4}\int_\Gamma E_z(\mathbf{r}')\nabla\big[H_0^{(2)}(k|\mathbf{r}-\mathbf{r}'|)\big]\cdot\mathbf{n}'d\Gamma. \tag{5.110}$$

The normal MoM procedures are used to solve Equation (5.110) for the electric field $E_z(\mathbf{r})$ along Γ [48]. The boundary Γ is first discretized into N segments, as illustrated in Figure 5.27, and then a constant pulse function defined on each segment is used as the basis functions to expand the unknown $E_z(\mathbf{r})$ as

$$E_z(\mathbf{r}) = \sum_{n=1}^{N} e_{nj} \Pi_n(\mathbf{r}), \qquad (5.111)$$

where the pulse function $\Pi_n(\mathbf{r})$ is

$$\Pi_n(\mathbf{r}) = \begin{cases} 1 & r \in n\text{th segment} \\ 0 & \text{otherwise,} \end{cases} \qquad (5.112)$$

and e_{nj} is the expansion coefficient of the nth segment with the jth port as a source as defined in Equation (5.105). Following the same point-matching procedure described in Reference 48 results in

$$\mathbf{e}_j = \mathbf{Z}_M^{-1} \mathbf{H}_j, \qquad (5.113)$$

where the column vector $\mathbf{e}_j = \{e_{nj}\}$ contains the expansion coefficients and the column source vector $\mathbf{H}_j$ is obtained as

$$H_{nj} = H_0^{(2)}(k|\mathbf{r}_n - \mathbf{r}_j|), n = 1, 2, \cdots, N. \qquad (5.114)$$

The $N \times N$ matrix $\mathbf{Z}_M$ is the MoM matrix whose elements are obtained as

$$Z_M(m,n) = \begin{cases} j0.25k(\hat{\mathbf{r}}_{mn} \cdot \mathbf{n}) H_1^{(2)}(kr_{mn}) & m \neq n \\ 0.5w_n^{-1} & m = n, \end{cases} \qquad (5.115)$$

where $r_{mn} = |\mathbf{r}_m - \mathbf{r}_n|$ and $\mathbf{n}$ is the normal vector of the nth segment;

$$\hat{\mathbf{r}}_{mn} = \frac{\mathbf{r}_m - \mathbf{r}_n}{|\mathbf{r}_m - \mathbf{r}_n|},$$

and $\mathbf{r}_{m(n)}$ is the center of the $m(n)$ th segment; w_n is the length of the nth segment. Substituting Equations (5.108), (5.111), and (5.113) into Equation (5.109) results in the electric field at Port i as

$$E_z(\mathbf{r}_i) = H_0^{(2)}(kr_{ij}) + \sum_{n=1}^{N} M_{in} e_{nj}, \qquad (5.116)$$

where

$$M_{in} = -\frac{jk}{4}\left(\frac{\mathbf{r}_i - \mathbf{r}_n}{|\mathbf{r}_i - \mathbf{r}_n|} \cdot \mathbf{n}\right) H_1^{(2)}(k|\mathbf{r}_i - \mathbf{r}_n|). \qquad (5.117)$$

As there are P ports in the plate pair, that is, $i, j = 1, 2, \cdots, P$, repeatedly using Equations (5.113) and (5.116), the radial scattering matrix for a PMC cavity, $\mathbf{S}_{pp}^R$, according to Equation (5.107), can be expressed as

$$\mathbf{S}_{pp}^R = \mathbf{S}_{pp}^{(f)} + \mathbf{M}\mathbf{Z}_M^{-1}\mathbf{H}, \qquad (5.118)$$

where the first term of the right-hand side is a $P \times P$ matrix denoting the direct illuminations among ports as

$$S_{pp}^{(f)}(i, j) = \begin{cases} H_0^{(2)}(kr_{ij}) & i \neq j \\ 0 & i = j, \end{cases} \qquad (5.119)$$

and the second term of the right-hand side represents the reflected waves from the edge boundary Γ. The $P \times N$ matrix $\mathbf{M}$ is defined in Equation (5.117); $N \times N$ matrix $\mathbf{Z}_M$ in Equation (5.115); and $N \times P$ matrix $\mathbf{H}$ in Equation (5.114).

Note that Equation (5.118) relates the radial scattering matrix of a plate pair $\mathbf{S}_{pp}^R$ to the corresponding MoM matrix $\mathbf{Z}_M$. Furthermore, the impedance matrix of a plate domain with PMC holes can be readily obtained from Equation (5.118) using Equation (5.97). As the integral Equation (5.110) is valid for any irregular plate boundary Γ, the boundary-integral-equation method proposed here is flexible and efficient for Z_{pp} calculation from $\mathbf{S}_{pp}^R$ of Equation (5.118).

5.2.5.2 PEC Cavities

For the completeness, the derivations of the BIE method for the radial scattering matrix among ports in a PEC cavity are also provided here. For a cavity enclosed by a PEC boundary Γ, the scattered wave is expressed as

$$E_z^s(\mathbf{r}) = -\frac{k\eta_0}{4}\int_{\Gamma} J_z(\mathbf{r}')H_0^{(2)}(k|\mathbf{r} - \mathbf{r}'|)d\Gamma, \qquad (5.120)$$

where η_0 is the wave impedance of free space. On the PEC boundary Γ, the equivalent current J_z needs to satisfy the following BIE:

$$E_z^{inc}(\mathbf{r}) = \frac{k\eta_0}{4}\int_\Gamma J_z(\mathbf{r}')H_0^{(2)}(k|\mathbf{r}-\mathbf{r}'|)d\Gamma. \tag{5.121}$$

With the similar derivations to the PMC cavity, the boundary Γ is discretized, and the pulse functions are used as basis functions for J_z. The radial scattering matrix of the plate pair can be obtained as

$$\mathbf{S}_{pp}^R = \mathbf{S}_{pp}^{(f)} - \mathbf{H}\mathbf{Z}_E^{-1}\mathbf{H}, \tag{5.122}$$

where the source matrix $\mathbf{H}$ is the same as that for the PMC cavity; and $\mathbf{H}^T$ is the transpose matrix of $\mathbf{H}$; the MoM matrix $\mathbf{Z}_E$ can be obtained by [48]

$$Z_E(m,n) = \begin{cases} H_0^{(2)}(kr_{mn}) & m \neq n \\ 1 - j\dfrac{2}{\pi}\left[\ln\left(\dfrac{\gamma kw_n}{4}\right) - 1\right] & m = n, \end{cases} \tag{5.123}$$

where $\gamma = 1.781072418$ is the Euler constant.

5.2.6 Numerical Examples and Measurements

First, a circular plate pair with two radial ports is studied. The radius of the plate pair is 10 cm. The dielectric layer between the two parallel plates has a thickness of 0.508 mm, a relative permittivity of 4.2, and a loss tangent of 0.02. The ports are located at $(0, 0)$ and $(7.0711$ cm, $\pi/4)$ in the polar coordinates as shown in Figure 5.26. The port radius is 0.5 mm.

The real and imaginary parts of the scattering parameter $S_{pp}^R(1, 2)$ calculated by Equation (5.103) are shown in Figure 5.28a where $s \in [-11, 11]$ is used. Note that unlike the general scattering parameters of a passive microwave network, the amplitude of S_{pp}^R may be larger than one. This is due to the different functions or modes used in expressing the inward and outward waves (Bessel functions for the inward waves and Hankel functions for the outward ones here). Moreover, S_{pp}^R also reflects the resonant properties of the plate pair. Compared with the $Z_{pp}(1, 2)$ curve shown in Figure 5.28b, $S_{pp}^R(1, 2)$ gives exactly the same resonant frequencies at 893.6 MHz and 1634 MHz.

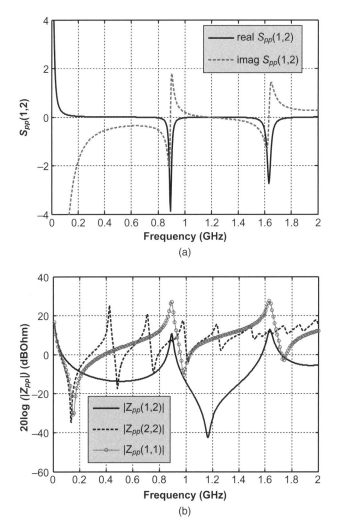

Figure 5.28 (a) Real and imaginary parts of $S_{pp}^{R}(1,2)$; (b) self and mutual impedances of two ports in a circular plate pair as shown in Figure 5.26.

The self and mutual impedances shown in Figure 5.28b are obtained by Equation (5.97). Clearly, the impedance of Port 1, which is located in the center of the circular plate pair, has less resonant frequencies than that of Port 2, due to its symmetric location in the geometry from which many modes cannot be excited.

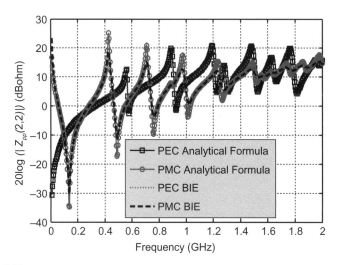

Figure 5.29   Comparison of the self impedance results of Port 2 from the analytical method and the boundary integral-equation method for the geometry shown in Figure 5.26. (See color insert.)

The impedances of the two ports in the circular plate pair are also obtained by calculating the radial scattering parameters using the BIE method introduced earlier and then using the transform Equation (5.97). These impedance results are compared with those obtained using the analytical method by Equation (5.103) and then Equation (5.97). As shown in Figure 5.29 where only the $Z_{pp}(2, 2)$ results are plotted for brevity, good agreement is observed. The resonant frequencies agree very well with only slight differences in amplitudes. In the BIE simulation, 20 segments per wavelength for the highest frequency of interest (2 GHz) are used to discretize the circular plate boundary.

The top view of another test geometry with two PMC ports in an irregular plate pair is shown in Figure 5.30. The dimensions of the polygon are shown in the figure. The dielectric constant of the dielectric layer between the two parallel plates is 4.2 and the loss tangent is 0.02. The diameter of the ports is 0.6096 mm (24 mil). The separation of two copper plates is chosen to be 0.254 mm (10 mil), and the electric conductivity of the copper plates is $5.8 \times 10^7 \, S \cdot m^{-1}$. The impedance results using HFSS (a FEM tool from Ansoft, Inc.), the cavity model with segmentation technique using the conventional area-integral definition, and the BIE method with the transform Equation (5.97) are

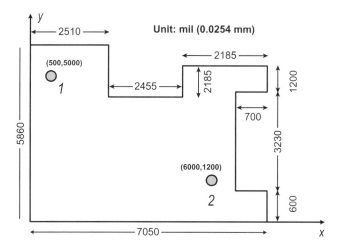

Figure 5.30   An irregular plate pair with two ports (unit: mil; 1 mil = 0.0254 mm).

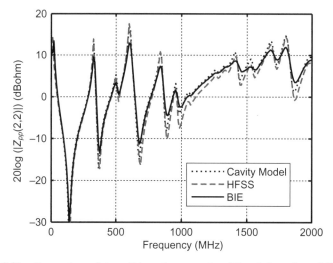

Figure 5.31   Comparison of the self-impedance results of Port 2 from three different methods for the test geometry shown in Figure 5.30.

compared in Figure 5.31. It can be seen that the $Z_{pp}(2, 2)$ obtained by the BIE method agrees well with the other two methods especially at the first several resonant frequencies.

The cavity model with segmentation technique needs to divide the plate pair into many rectangular or special triangular cavities [30]. The connections of these cavities will quickly slow down simulation speed with the increase of geometrical complexity. On the other hand, the BIE method only requires meshing plate boundaries into segments, very suitable for arbitrary shapes. It is worth pointing out that the conventional area-integral impedance definition is used in the cavity model. As shown in the results, when the number of the PMC ports is small, the difference between the conventional and the new definitions is also small.

To manifest the difference between the new Z_{pp} definition Equation (5.97) and the conventional area-integral definition, a plate pair with a large number of ports is investigated as shown in Figure 5.32a. This is a 160×100 mm^2 rectangular plate pair with 310 radial ports. Four side walls are PMC boundaries. There are 10 rows of holes (ports) and each row has 31 with a radius of 1.128 mm. The first hole is located at (5, 5) mm and the via-hole array has 10 mm and 5 mm separations along the x and y directions, respectively. The plate pair height is 0.254 mm (10 mils), filled with a dielectric material with a permittivity of 4.2 and a tangent loss of 0.02. Only the input impedance of Port 1 at (50, 10) mm is investigated.

Figure 5.32b compares the self-impedance of Port 1 obtained by Equation (5.74) of the conventional area-integral definition and that by Equation (5.97) of the new Z_{pp} using BIE. From Equation (5.74), it is clear that the self-impedance Z_{11} by the conventional impedance definition is not affected by other ports. However, the new Z_{pp} defined through the radial scattering matrix S_{pp}^R considers the PMC boundaries at all the ports. It can be seen that at low frequencies, Z_{pp} by the new definition is about 0.7 dB higher than that by the conventional area-integral definition. This can be explained by the fact that the 310 PMC holes decrease the area of the plate domain and thus reduce the parallel plate capacitance at low frequencies from 2.3414 nF to 2.1601 nF. This results in about 0.7 dB increase of the self-impedance Z_{11} at those frequencies. Therefore, the new definition of Z_{pp} can correctly capture the physics at low frequencies while the conventional area-integral definition cannot. Further, the new Z_{pp} by BIE also predicts that the large number of PMC holes can shift the resonant frequencies to higher as shown in Figure 5.32b. This phenomenon has not been reported before. The following measurements will demonstrate this phenomenon of resonant frequency shift due to the large number of PMC holes.

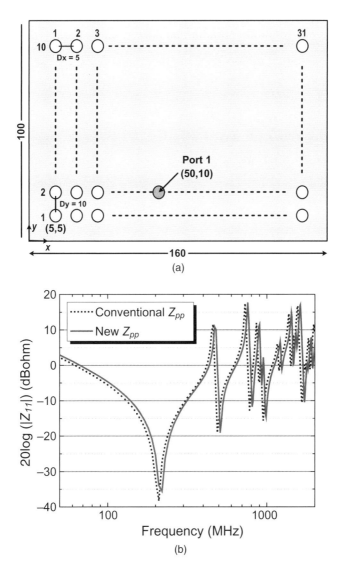

Figure 5.32 (a) A rectangular plate pair with 310 ports (unit: mm); (b) comparison of self-impedance of Port 1 by the conventional area-integral definition of Z_{pp} and the new definition of Equation (5.97) by BIE.

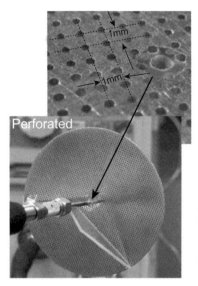

Circular 3-in. Disk
Resonator
Plate-pair: *R*=3 in., *h*=58 mil,
Dielectric constant=4.2,
Loss tangent=0.02
Probe: *a*=10, *b*=33, unit: mil

Figure 5.33 A perforated disk resonator with a test port at the center and 4500 unplated holes in a 1 mm by 1 mm grid (1 mil = 0.0254 mm).

The new Z_{pp}, as well as the BIE method, is further validated with the measurements of some disk resonator geometries as shown in Figure 5.33. These disk resonators are circular with a diameter of 76.2 mm (3 in.). A test port is located in the center of the disks with a via radius of 0.254 mm (10 mils) and a via-hole radius of 0.8382 mm (33 mils). The dielectric material between the two parallel conductive plates is 1.4732 mm (58 mils) thick, and its relative permittivity and loss tangent are of 4.2 and 0.02, respectively. The test port is further attached to a semirigid cable with an SMA connector. Two disk resonator structures are studied here: a solid one with only the test port, and a perforated one with 4500 unplated holes, with radius of 0.254 mm (10 mils), in a 1 mm × 1 mm grid in addition to the test port. The disk resonators are measured using a vector network analyzer, and are also modeled using the BIE method. The effect of the semi-rigid cable and the SMA connector attached to the test port is carefully measured and included in the modeling as an extra test fixture. The boundaries of the unplated holes in the perforated disk resonator, as well as the external boundary of the disks, are assumed to be PMC due to the small thickness of the dielectric layer between the parallel plates. The magnitude and phase results of the input impedance at the test port are shown in Figure 5.34a,b, respectively, where the measured results are compared to the corresponding simulated results using the BIE method.

It is clear from Figure 5.34 that the impedance results from the solid and the perforated disk resonators are different enough, when the number of the PMC holes is so large that the conventional impedance definitions are not suitable any more. Further, the simulated results using the BIE method show good agreement with the measured one, especially at low frequencies. More discrepancies are observed for the perforated disk resonator at high frequencies. This could be partially contributed to the manufacturing tolerances in drilling the holes (locations and sizes), and partially contributed to the fact that the higher order scattering among the holes are not included in the new Z_{pp} as discussed in Reference 41 as one of the important assumptions. Also, with the increase of frequency, the PMC hole model is eventually not a good approximation for the drilling holes. The full-wave simulations using HFSS are not possible for the perforated disk resonator with the available computer resource, due to the complexity of the geometry. Hence, they are not included in the comparisons in Figure 5.34.

From the above examples, it is clear that the new Z_{pp} based on the zero-order parallel plate waves is more rigorous for a plate pair with PMC ports. More importantly, it provides a rigorous foundation to connect the plate domain modeled as Z_{pp} with the via domains described as the intrinsic 3-port lumped circuit models. Finally, it should be mentioned that, at very low frequencies, dense mesh segments (a large N in Eq. 5.111) would lead to a singular $\mathbf{Z}_M$ matrix that could cause inaccurate results in the BIE method. This is a well-known low-frequency problem for integral-equation approaches [49], which needs further investigation.

5.2.7 Conclusion

The impedance matrix in a plate domain with multiple PMC ports is defined in this section using the zero-order parallel plate waves, where the conventional impedance definitions are not suitable anymore. Cylindrical waves are used to express the electric and magnetic fields and this leads to the introduction of the radial scattering parameters, S_{pp}^R, in a plate pair. S_{pp}^R can be regarded as a generalized addition theorem to relate the zero-order cylindrical harmonics in different polar

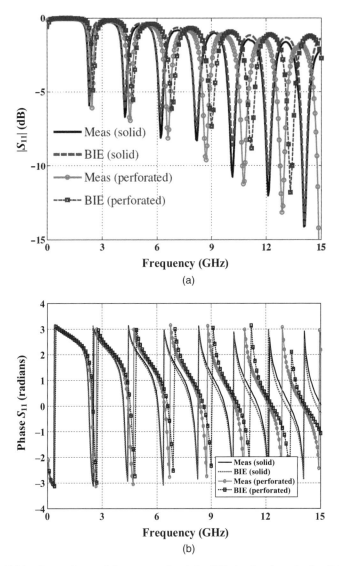

Figure 5.34   Comparisons of the measured and the BIE simulated results for disk resonators: (a) magnitudes of the input impedance; (b) phases of the input impedance.

coordinates. The transforms between Z_{pp} and S_{pp}^R are derived for a plate pair with PMC ports. An analytical formula of S_{pp}^R is derived for a circular plate pair and can be used as a benchmark to validate numerical methods used in Z_{pp} calculations. BIE method is introduced to calculate the S_{pp}^R for both PMC and PEC cavities. The introduction of S_{pp}^R and the transforms between S_{pp}^R and Z_{pp} greatly simplifies the calculations of Z_{pp} for plate pairs with multiple PMC holes. The effectiveness of the BIE method is illustrated by comparing with the analytical formula for a circular plate pair as well as with other numerical methods such as the cavity model with segmentation technique and the FEM (HFSS) for an irregular plate pair. The advantage of BIE in terms of boundary discretization makes it flexible and powerful in obtaining Z_{pp} for arbitrarily shaped plates common in PCBs and packages. The measurements of solid and perforated disk resonators further validate the new Z_{pp} definition and demonstrate the limitations of the conventional definitions for a plate pair with many PMC holes.

Note that the expressions of Z_{pp} and S_{pp}^R in terms of cylindrical waves reveal the physical meaning of Z_{pp} as the ratios of the zero-order radial transmission-line modes. This new definition of Z_{pp} makes the connections natural between the plate and via domains. Combined with that from Reference 41, a rigorous via circuit model has thus been developed for multiple vias in high-speed multilayer PCBs and packages, which can be useful for SI/PI analyses as well as EMC modeling.

5.3 CASCADED MULTIPORT NETWORK ANALYSIS OF MULTILAYER STRUCTURE WITH MULTIPLE VIAS

5.3.1 Introduction

The trend of lowering power supply voltage levels and increasing signal switching speeds requires accurate yet efficient simulation methods for the SI/PI analysis of PCB.

A multilayer PCB can be divided electrically into two functional parts: SLP and power delivery network (PDN). An SLP may consist of traces and vias to set up a communication channel between drivers and receivers which usually are pins of integrated circuits (ICs). On the other hand, a PDN usually contains a stack-up of power/ground plates, a large amount of power/ground vias, and many decoupling capacitors to provide an extremely low impedance path from voltage regulator module (VRM) to power/ground pins of IC chips mounted on PCBs. Normally, a PI analysis for a PDN system is to extract an impedance matrix of a pair of power/ground plates, namely Z_{pp} herein [26]. Lower Z_{pp} means lower simultaneous switching noise (SSN) coupling and thus better PDN performance. Various numerical methods such as the cavity model with segmentation method, finite difference method, FEM, integral equation method, partial element equivalent circuit (PEEC), and so on, have been implemented to calculate the Z_{pp} [26, 28–32, 50–52]. Decoupling capacitors are extensively used as a part of PDN systems and the guidelines of their locations and values have been summarized in References 53–55.

Besides the power delivery function, a PDN system also serves as the return current path for the SLP. While power or ground plates provide good reference planes for impedance, well-controlled transmission lines including microstrip and strip lines, a pair of parallel plates may also support the propagation of switching noise excited by vertical currents along vias crossing the plate pair. In this case, the entire plate pair structure acts as the return path of via currents. This has been clearly demonstrated in the derivations of the intrinsic via circuit model and its connection to Z_{pp} [41, 56]. Therefore, SI analysis for an SLP in a multilayer PCB cannot be separated from the PI analysis of the corresponding PDN system. From this point of view, via-plate interaction plays the central and critical roles in SI/PI cosimulation in the design of high-speed multilayer PCBs.

The multilayer PDN including vias effects has been studied by transmission line matrix method (TLM) in Reference 57 where the plates as well as vias are all described by distributed RLGC circuit elements. Recently, a semianalytical solution called FLMSM has been developed in the analysis of multiple vias in an infinite or finite circular plate pair [15–19].

This section can be viewed as an extension of References 41 and 56 from a single plate pair to multilayer plate pair with multiple vias. Section 5.3.2 explains the geometric structure and the main ideas of the microwave network method. Section 5.3.3 provides a systematic microwave network method for a multilayer plate pair with multiple vias. First, an admittance matrix of a single via is derived from the 3-port intrinsic via circuit model, then gives the derivation of the admittance matrix of a single plate pair with many vias. As some via holes are shorted to plates or terminated by decoupling capacitors, a loaded admittance matrix is discussed consequently. Finally, an efficient recursive algorithm is proposed to combine the cascaded admittance network of the multilayer plate pair. Numerical examples and measurements are provided to validate the method in Section 5.3.4 which is followed by a brief conclusion.

5.3.2 Multilayer PCB with Vias and Decoupling Capacitors

Figure 5.35a illustrates a stack-up of power/ground plates with multiple vias and decoupling capacitors. The power/ground plates could be irregular shapes at different layers. While signal vias do not connect with any plates, power (ground) vias are short to power (ground) plates.

In the intrinsic via circuit model, all vias are assumed to be electrically small and only transverse electromagnetic fields exist in via holes [41]. Thus, port voltage and current can be defined in each via hole. As shown in Figure 5.35a, via holes are assumed to be the only energy leakage path from one plate pair to another. This leads to an equivalent microwave network representation of the stack-up as shown in Figure 5.35b.

It can be seen that each plate pair is modeled as a multiport microwave network characterized by an admittance matrix with some ports terminated by different impedance loads. (Z-matrix or S-parameters can also be used.) For clarity, only the network for the first plate pair is shown in detail in Figure 5.35b. Note that there are three different impedance loads included here: short circuit for shorting ports, a small capacitance C_O for "nearly open ports," and lumped circuits for decoupling capacitors. Although actual C_O is not given here, from our experience, C_O could be set to be a very small value in actual simulations. This means "nearly open ports" in Figure 5.35a can be regarded as open circuits in the frequency of interests.

The main objective of this section is to present an efficient algorithm to obtain the transmission properties of signal vias in a multilayer PCB with multiple vias and decoupling capacitors, for example, the extraction of transmission coefficients S_{12}, S_{34} and the crosstalk parameters S_{13}, S_{14} and so on, as shown in Figure 5.35 as an example. These parameters are critical for SI/PI analysis for practical multilayer PCB designs.

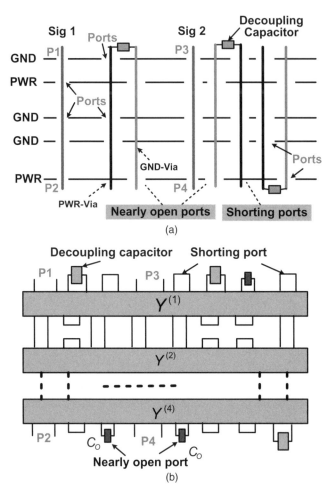

Figure 5.35 (a) Schematic of a multilayer printed circuit board with various vias and decoupling capacitors; (b) microwave network representation of the stack-up of power/ground plates.

5.3.3 Systematic Microwave Network Method

In this section, the admittance matrix of a via is first derived from the intrinsic via circuit model. Then all vias in a plate pair construct a via network which is connected with the plate network (plate domain) described as Z_{pp}. Combination of the via network and plate network leads to an admittance matrix for a single plate pair. Load impedance for some ports are processed before a recursive algorithm is developed to integrate cascaded admittance matrices of two plate pair.

5.3.3.1 Admittance Matrix of a Single Via

The geometry of a via is illustrated in Figure 5.36a. The radii of the via barrel, pad, and antipad (via hole) are denoted as r, a, and b. The via height or the separation of plates is h and between two plates is a dielectric material with relative permittivity of ε_r. In Reference 41, a 3-port intrinsic via circuit model has been derived as shown in Figure 5.36b to describe the mode conversion among two coaxial ports (Port 1 and Port 2) and one radial port (Port 3). The via-plate capacitance, C_h, is for the displacement current between the via barrel (pad) and top/bottom plates; the C_v^1 is due to the vertical electric component of the higher-order evanescent modes; and the C_v^0, L_v, and R_v are parasitic capacitance, inductance, and transformer ratio due to the zero-order mode.

The parasitic elements in this 3-port via circuit are summarized as [41]

$$C_h = \frac{j4\varepsilon\pi^2 a}{h\ln(b/a)} \sum_{n=1,3,5,\dots}^{\infty} F_n^L(a), \tag{5.124}$$

$$C_v^1 = \frac{j4\varepsilon\pi^2 a}{h\ln(b/a)} \sum_{n=1}^{\infty} (-1)^n F_n^L(a), \tag{5.125}$$

where

$$F_n^L(a) = \left[\frac{H_0^{(2)}(k_n b) - H_0^{(2)}(k_n a)}{k_n H_0^{(2)}(k_n r)} \right] W_{10}(k_n a, k_n r). \tag{5.126}$$

The radial wavenumber $k_n = \sqrt{k_0^2 \varepsilon_r - (n\pi/h)}$ (k_0 is the free wavenumber) and $W_{10}(k_n a, k_n r)$ can be obtained from a general auxiliary function $W_{mn}(x, y)$ defined as a determinant of the Bessel and Hankel functions

$$W_{mn}(x, y) = \begin{vmatrix} J_m(x) & J_n(y) \\ H_m^{(2)}(x) & H_n^{(2)}(y) \end{vmatrix}, \tag{5.127}$$

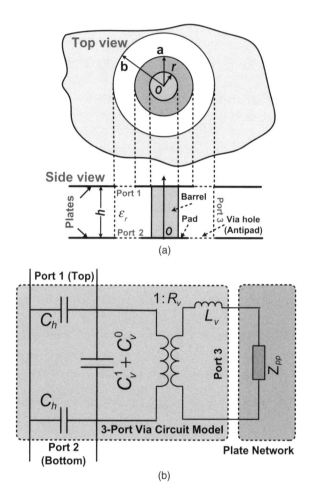

Figure 5.36 (a) Top and side views of a via geometric structure; (b) 3-port intrinsic via circuit model.

where $J_m(x)$ and $H_n^{(2)}(y)$ are the mth Bessel and nth second-order Hankel functions, respectively. The parasitic inductance L_v, capacitance C_v^0, and the transformer ratio R_v are

$$L_v = -\frac{\mu h}{2\pi kb} \frac{W_{00}(kb, kr)}{W_{10}(kb, kr)}, \tag{5.128}$$

$$C_v^0 = \frac{j\varepsilon\pi^2 a}{kh\ln(b/a)} \frac{W_{10}(ka, kr)}{W_{10}(kb, kr)}[W_{10}(kb, kb) - W_{10}(kb, ka)], \tag{5.129}$$

$$R_v = \sqrt{-R_m R_e}, \tag{5.130}$$

where

$$R_m = \frac{j\pi}{2\ln(b/a)} \frac{W_{10}(kb, kb)}{W_{10}(kb, kr)}[W_{00}(kb, kr) - W_{00}(ka, kr)], \tag{5.131}$$

$$R_e = \frac{aW_{10}(ka, kr)}{bW_{10}(kb, kr)}, \tag{5.132}$$

where $k = k_0\sqrt{\varepsilon_r}$. Note that the radial port (Port 3) boundary is set to be at the radius of the via hole ($\rho = b$) instead of a virtual boundary outside of the via-hole radius. It is feasible if the intrinsic via circuit model is valid according to the assumptions in Reference 41 (vias are small electrically and the separation of vias is large enough).

Due to the geometric symmetry, for each via (the ith via as an example), a 3-port admittance matrix can be derived from the intrinsic via circuits as

$$\begin{bmatrix} I_t^{(i)} \\ I_b^{(i)} \\ I_h^{(i)} \end{bmatrix} = \begin{bmatrix} Y_{tt}^{(i)} & Y_{bt}^{(i)} & Y_{ht}^{(i)} \\ Y_{bt}^{(i)} & Y_{tt}^{(i)} & -Y_{ht}^{(i)} \\ Y_{ht}^{(i)} & -Y_{ht}^{(i)} & Y_{hh}^{(i)} \end{bmatrix} \begin{bmatrix} V_t^{(i)} \\ V_b^{(i)} \\ V_h^{(i)} \end{bmatrix}, \tag{5.133}$$

where

$$Y_{tt}^{(i)} = j\omega(C_h + C_v^1 + C_v^0) + \frac{R_v^2}{j\omega L_v}, \tag{5.134}$$

$$Y_{bt}^{(i)} = -j\omega(C_v^1 + C_v^0) - \frac{R_v^2}{j\omega L_v}, \tag{5.135}$$

$$Y_{ht}^{(i)} = \frac{R_v}{j\omega L_v}, \tag{5.136}$$

$$Y_{tt}^{(i)} = \frac{1}{j\omega L_v}. \tag{5.137}$$

The subscript t, b, and h denote top coaxial port (Port 1), bottom coaxial port (Port 2), and horizontal or parallel radial port (Port 3) as shown in Figure 5.36, respectively.

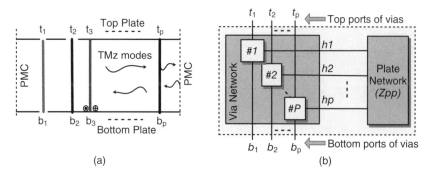

Figure 5.37 (a) A single plate pair with multiple vias; (b) microwave network model for a single plate pair with P vias.

5.3.3.2 Admittance Matrix of a Single Plate Pair with Multiple Vias

A single plate pair with multiple vias can be viewed as a special cavity bounded by top/bottom plates of PECs and side walls of PMCs as shown in Figure 5.37a. The special cavity is loaded with multiple via barrels inside and excited by coaxial ports on top/bottom plats. According to the equivalent principle in electromagnetic theory, the voltage crossing a via hole can be replaced by a magnetic frill current which excites parallel plate TM_z modes, and further induces port currents in all via holes. Therefore, an admittance matrix can be used to characterize the transmission and coupling properties among coaxial ports. Different from the FLMSM [17], the intrinsic via circuit model is used here to obtain the admittance matrix of an irregular plate pair with multiple vias.

In References 41 and 56, multiple 3-port via circuits are connected to the impedance matrix Z_{pp}, the plate model to represent a plate pair with multiple vias. Here, to facilitate the discussion, all 3-port via circuits are combined together as a $3P$-port via network linking to the P-port plate network as shown in Figure 5.37b. All vias are first regarded as signal vias and shorting coaxial ports of vias will be processed later. Clearly, the plate network Z_{pp} can be viewed as a loading impedance matrix to the via network which is linking to other layers of plate pair through its top or bottom P ports.

The admittance matrix of the via network can be derived easily from the 3-port admittance matrix of each via as

$$\begin{bmatrix} \mathbf{I}_t \\ \mathbf{I}_b \\ \mathbf{I}_h \end{bmatrix} = \begin{bmatrix} \mathbf{Y}_{tt} & \mathbf{Y}_{tb} & \mathbf{Y}_{th} \\ \mathbf{Y}_{bt} & \mathbf{Y}_{bb} & \mathbf{Y}_{bh} \\ \mathbf{Y}_{ht} & \mathbf{Y}_{hb} & \mathbf{Y}_{hh} \end{bmatrix} \begin{bmatrix} \mathbf{V}_t \\ \mathbf{V}_b \\ \mathbf{V}_h \end{bmatrix}, \tag{5.138}$$

where $\mathbf{I}_{t,b,h}$ and $\mathbf{V}_{t,b,h}$ are port current and voltage vectors of all top, bottom, and horizontal ports,

$$(\mathbf{V})\mathbf{I}_t = \left[(V)I_t^{(1)}, \quad (V)I_t^{(2)}, \quad \cdots \quad (V)I_t^{(P)}, \right]^T, \tag{5.139}$$

$$(\mathbf{V})\mathbf{I}_b = \left[(V)I_b^{(1)}, \quad (V)I_b^{(2)}, \quad \cdots \quad (V)I_b^{(P)}, \right]^T, \tag{5.140}$$

$$(\mathbf{V})\mathbf{I}_h = \left[(V)I_h^{(1)}, \quad (V)I_h^{(2)}, \quad \cdots \quad (V)I_h^{(P)}, \right]^T, \tag{5.141}$$

and $\mathbf{Y}_{\alpha\beta}$, α, $\beta = t, b, h$, is a diagonal submatrix formed from corresponding admittance elements of each via in Equation (5.133) as

$$\mathbf{Y}_{\alpha\beta} = diag\{Y_{\alpha\beta}^{(i)}\}, i = 1, 2, \cdots, P. \tag{5.142}$$

Note that in the via network, there is no via couplings as they have been included in the plate network. From Figure 5.37b, the horizontal ports are related together by the impedance matrix Z_{pp} as

$$\mathbf{V}_h = -\mathbf{Z}_{pp}\mathbf{I}_h. \tag{5.143}$$

Using Equations (5.138) and (5.143), we have the admittance matrix lth layer of plate pair as

$$\begin{bmatrix} \mathbf{I}_t \\ \mathbf{I}_b \end{bmatrix} = \begin{bmatrix} \mathbf{Y}_{tt}^{(l)} & \mathbf{Y}_{tb}^{(l)} \\ \mathbf{Y}_{bt}^{(l)} & \mathbf{Y}_{bb}^{(l)} \end{bmatrix} \begin{bmatrix} \mathbf{V}_t \\ \mathbf{V}_b \end{bmatrix}, \tag{5.144}$$

where submatrices can be obtained by

$$\mathbf{Y}_{\alpha\beta}^{(l)} = \mathbf{Y}_{\alpha\beta} - \mathbf{Y}_{\alpha h}\mathbf{Z}_{pp}(\mathbf{I} + \mathbf{Y}_{hh}\mathbf{Z}_{pp})^{-1}\mathbf{Y}_{h\beta}, \tag{5.145}$$

where $\mathbf{I}$ is a unit matrix and α, $\beta = t, b, h$, respectively.

5.3.3.3 Loaded Admittance Matrix of a Single Plate Pair

Equation (5.145) gives the admittance matrix for all the top and bottom ports of a plate pair. In reality, however, some ports are shorted to top/

bottom plates as shown in Figure 5.37a. On the other hand, a lot of ports may be loaded by decoupling capacitors illustrated in Figure 5.35 to provide low impedance return paths for signal vias or to eliminate resonant modes of plate pair.

By renumbering the "open" top/bottom ports and loaded ports, Equation (5.145) can be reorganized as

$$
\begin{bmatrix} \tilde{\mathbf{I}}_t^{(l)} \\ \tilde{\mathbf{I}}_b^{(l)} \\ \tilde{\mathbf{I}}_L^{(l)} \end{bmatrix} = \begin{bmatrix} \tilde{\mathbf{Y}}_{tt} & \tilde{\mathbf{Y}}_{tb} & \tilde{\mathbf{Y}}_{tL} \\ \tilde{\mathbf{Y}}_{bt} & \tilde{\mathbf{Y}}_{bb} & \tilde{\mathbf{Y}}_{bL} \\ \tilde{\mathbf{Y}}_{Lt} & \tilde{\mathbf{Y}}_{Lb} & \tilde{\mathbf{Y}}_{LL} \end{bmatrix} \begin{bmatrix} \tilde{\mathbf{V}}_t^{(l)} \\ \tilde{\mathbf{V}}_b^{(l)} \\ \tilde{\mathbf{V}}_L^{(l)} \end{bmatrix},
\tag{5.146}
$$

where $\tilde{\mathbf{I}}_{t,b}^{(l)}$ and $\tilde{\mathbf{V}}_{t,b}^{(l)}$ represent port current and port voltage vectors of "open" top and bottom ports on plates while $\tilde{\mathbf{I}}_L^{(l)}$ and $\tilde{\mathbf{V}}_L^{(l)}$ denote the current and voltage vectors of loaded ports. They are related to each other as

$$
\tilde{\mathbf{V}}_L = -\mathbf{Z}^L \tilde{\mathbf{I}}_L,
\tag{5.147}
$$

where $\mathbf{Z}^L$ is a diagonal matrix obtained as

$$
\mathbf{Z}^L(i,i) = \begin{cases} 0 & \text{short ports} \\[2mm] \dfrac{1}{j\omega C_O} & \text{nearly open ports} \\[3mm] R_i + j\omega L_i + \dfrac{1}{j\omega C_i} & \text{decoupling capacitors,} \end{cases}
\tag{5.148}
$$

where C_i is the capacitance of a decoupling capacitor, and R_i and L_i are the equivalent series resistance (ESR) and equivalent series inductance (ESL), respectively [30].

Substituting Equation (5.147) into Equation (5.146) yields the admittance matrix of the lth loaded plate pair as

$$
\begin{bmatrix} \tilde{\mathbf{I}}_t^{(l)} \\ \tilde{\mathbf{I}}_b^{(l)} \end{bmatrix} = \begin{bmatrix} \tilde{\mathbf{Y}}_{tt}^{(l)} & \tilde{\mathbf{Y}}_{tb}^{(l)} \\ \tilde{\mathbf{Y}}_{bt}^{(l)} & \tilde{\mathbf{Y}}_{bb}^{(l)} \end{bmatrix} \begin{bmatrix} \tilde{\mathbf{V}}_t^{(l)} \\ \tilde{\mathbf{V}}_b^{(l)} \end{bmatrix},
\tag{5.149}
$$

where

$$
\tilde{\mathbf{Y}}_{\alpha\beta}^{(l)} = \tilde{\mathbf{Y}}_{\alpha\beta} - \tilde{\mathbf{Y}}_{\alpha h} \mathbf{Z}^L (1 + \tilde{\mathbf{Y}}_{hh} \mathbf{Z}^L)^{-1} \tilde{\mathbf{Y}}_{h\beta}.
\tag{5.150}
$$

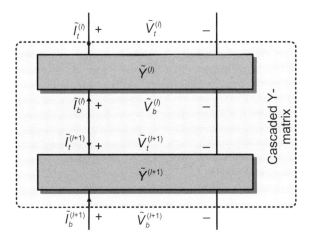

Figure 5.38   Combination of two cascaded Y-matrix network.

5.3.3.4 Recursive Algorithm for Cascaded Admittance Matrix

As discussed in Section 5.3.2, a multilayer PCB as shown in Figure 5.35 can be viewed as a cascaded microwave network. The network of each layer of plate pair is described as an admittance matrix which has been derived as Equation (5.149). For example, Figure 5.38 shows two cascaded networks of $\tilde{\mathbf{Y}}^{(l)}$ and $\tilde{\mathbf{Y}}^{(l+1)}$. The loaded admittance matrix of the $l + 1$th layer network is expressed as

$$\begin{bmatrix} \tilde{\mathbf{I}}_t^{(l+1)} \\ \tilde{\mathbf{I}}_b^{(l+1)} \end{bmatrix} = \begin{bmatrix} \tilde{\mathbf{Y}}_{tt}^{(l+1)} & \tilde{\mathbf{Y}}_{tb}^{(l+1)} \\ \tilde{\mathbf{Y}}_{bt}^{(l+1)} & \tilde{\mathbf{Y}}_{bb}^{(l+1)} \end{bmatrix} \begin{bmatrix} \tilde{\mathbf{V}}_t^{(l+1)} \\ \tilde{\mathbf{V}}_b^{(l+1)} \end{bmatrix}. \tag{5.151}$$

It can be seen that the bottom ports of the lth layer network share the same via holes with the top ports of the $l + 1$th layer network. So we have the following equations to satisfy the boundary conditions in the corresponding via holes:

$$\tilde{\mathbf{V}}_t^{(l+1)} = \tilde{\mathbf{V}}_b^{(l)}, \tag{5.152}$$

$$\tilde{\mathbf{I}}_t^{(l+1)} = -\tilde{\mathbf{I}}_b^{(l)}. \tag{5.153}$$

Using Equations (5.149), (5.151), (5.152), and (5.153) leads to the following combined admittance matrix

$$\left[\begin{array}{c} \tilde{\mathbf{I}}_t^{(l)} \\ \tilde{\mathbf{I}}_b^{(l+1)} \end{array} \right] = \mathbf{Y}_{l,l+1}^C \left[\begin{array}{c} \tilde{\mathbf{V}}_t^{(l)} \\ \tilde{\mathbf{V}}_b^{(l+1)} \end{array} \right], \tag{5.154}$$

where

$$\mathbf{Y}_{l,l+1}^C = \left[\begin{array}{cc} \tilde{\mathbf{Y}}_{tt}^{(l)} - \tilde{\mathbf{Y}}_{tb}^{(l)} \mathbf{Q} \tilde{\mathbf{Y}}_{bt}^{(l)} & -\tilde{\mathbf{Y}}_{tb}^{(l)} \mathbf{Q} \tilde{\mathbf{Y}}_{tb}^{(l+1)} \\ -\tilde{\mathbf{Y}}_{bt}^{(l+1)} \mathbf{Q} \tilde{\mathbf{Y}}_{bt}^{(l)} & \tilde{\mathbf{Y}}_{bb}^{(l+1)} - \tilde{\mathbf{Y}}_{bt}^{(l+1)} \mathbf{Q} \tilde{\mathbf{Y}}_{tb}^{(l+1)} \end{array} \right], \tag{5.155}$$

where the auxiliary matrix $\mathbf{Q}$ is defined as

$$\mathbf{Q} = \left[\tilde{\mathbf{Y}}_{bb}^{(l)} + \tilde{\mathbf{Y}}_{tt}^{(l+1)} \right]^{-1}. \tag{5.156}$$

Thus, Equation (5.155) provides a recursive algorithm to combine the cascaded admittance matrix of a multilayer PCB. The main advantage of the recursive algorithm is that the computer memory requirement is almost irrelevant to the layer number of a PCB.

5.3.4 Validations and Discussion

The proposed microwave network method is validated with both numerical simulations and measurements. Figure 5.39a shows the measurement setup and a test board with rectangular plate pair. The length and width of the plate pair are 75 mm and 50 mm, respectively. Between the two plates is a dielectric layer with the thickness of 1.0 mm, a relative permittivity of 4.1, and a loss tangent of 0.015. Two subminiature version A (SMA) probes are located at (20, 20) mm and (40, 30) mm, respectively. The diameters of the inner and outer conductors of the SMA probes are 1.27 mm and 4.06 mm, respectively.

A vector network analyzer, Agilent 8753D, is used to measure the S-parameter among the two SMA ports. The standard SLOT procedure is adopted for calibration. In total, 801 sampling points are measured from 0.1 GHz to 5.0 GHz.

Figure 5.39b illustrates the circuit model for the measurement setup. The intrinsic via circuit is used to model the transitions between SMAs and the parallel plane pair as a probe is regarded as a special via by shorting one coaxial port to the corresponding plate. For this probe, the capacitances, C_h and C_v^1, are 96.3 fF and -32.8 fF, respectively. On the other hand, according to Equations (5.124–5.132), C_v^0,

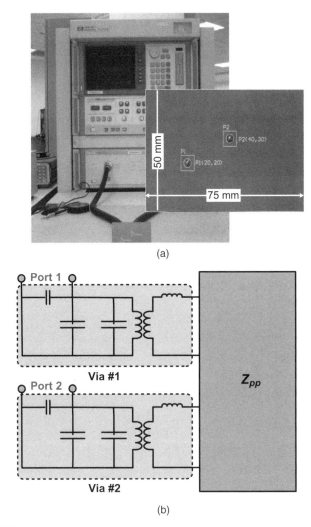

Figure 5.39   (a) Measurement setup for two SMA ports in a rectangular plate pair; (b) equivalent intrinsic via circuit model for the measurement setup.

L_v, and R_v are frequency dependent and vary from 1.15 pF, 0.23 nH and 1.0 at 0.1 GHz, to 1.49 pF, 0.29 nH and 1.7 at 5.0 GHz, respectively.

Both reflection and transmission coefficients, S_{11} and S_{21}, are compared in Figure 5.40 for the present via circuit model, the FEM of HFSS, and the measurements. It can be seen that the microwave network method agrees very well with the full-wave numerical solver,

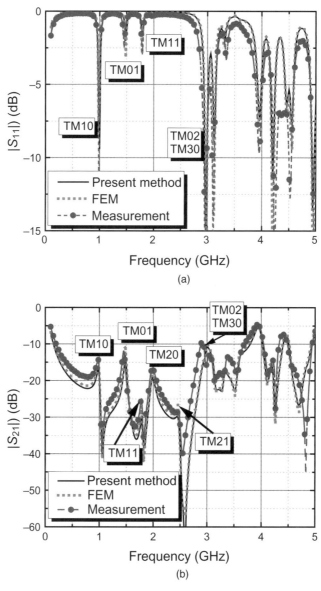

Figure 5.40 Comparison of 2-port S-parameters obtained by the present via circuit method, finite element method (HFSS, Ansoft Inc.) and the measurements.

HFSS. Although there are observable differences, the results obtained by the present via circuit model demonstrate same trends and resonances to the measurement results. The first several valleys in S_{11} and the peaks of S_{21} are mainly due to the resonant modes of the rectangular cavity as depicted in Figure 5.40.

The second example here is a plate pair excited by an SMA probe with a decoupling capacitor located nearby. The structure of the example is given in Reference 55 (fig. 3). The plate pair is a rectangular one with a length of 100 mm, a width of 50 mm, and a thickness of 1.124 mm (the separation of plates). The dielectric material is FR-4 with a relative permittivity of 4.7 and a loss tangent of 0.02. The probe is located at (30, 20) mm, and the decoupling capacitor is at (35, 20) mm (the origin of the coordinates is at the left corner of the plate pair). The capacitance of the decoupling capacitor is chosen to be 8.14 nF with the effective series inductance and resistance (ESL/ESR) of 1.57 nH and 666 mΩ, respectively.

Figure 5.41 compares the simulation results by the present method and the measurements given in Reference 55 (fig. 4). It can be seen that the decoupling capacitor effectively lowers the input impedance at

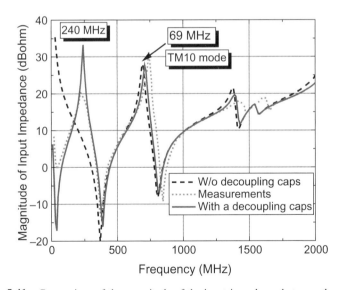

Figure 5.41 Comparison of the magnitude of the input impedance between the present method and measurements in Reference 55 (fig. 4) (short dash: without decoupling capacitors; short dot: measurements; solid: with a closely spaced capacitor).

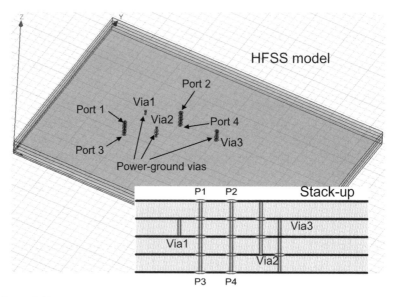

Figure 5.42 A multilayer structure with two signal vias, five parallel plates, and five power/ground vias.

low frequencies. However, it has little impact at high frequencies due to the large impedance of the ESL. The simulation results by the present method can correctly predict the shift of resonant frequencies due to the decoupling capacitor although the amplitude of input impedance has a little difference from the measurement.

Another example is a 5-layer stack-up with two signal vias and three power/ground vias as shown in Figure 5.42. The length and width of the rectangular plate pair are of 75 mm and 50 mm, respectively. The separation of plates is 1 mm. A dielectric material, with a relative permittivity of 4.1 and a loss tangent of 0.023, is sandwiched between two plates. Two signal vias are located at (20, 20) mm for Port 1-Port 3 and (30, 30) mm for Port 2-Port 4, respectively. Power/ground vias or Via1, Via2, and Via3 are located at (22, 27), (27.5, 22.5), and (40, 30) mm, and their connections to plates are illustrated in the stack-up shown in Figure 5.42. The radii of via barrel and antipad (via hole) are 0.25 mm and 0.5 mm.

Comparisons of the present method, the FEM of HFSS and SIwave (Ansoft Inc.) for the structure of Figure 5.42 are given in Figures 5.43

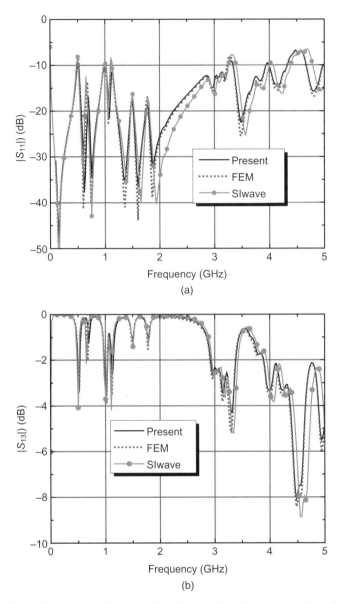

Figure 5.43 Comparison of the results for Figure 5.42 by the present method, FEM, and SIwave (HFSS, Ansoft Inc.): (a) reflection coefficient S_{11}; (b) transmission coefficient S_{13}.

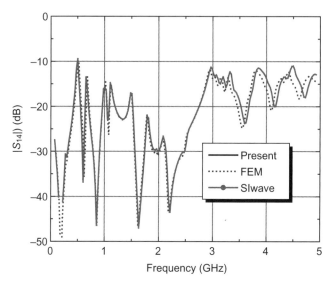

Figure 5.44   Comparison of forward crosstalk parameter S_{14} for Figure 5.41 by the present method and FEM (HFSS, Ansoft Inc.).

and 5.44, respectively. It can be seen that the results by the present method match very well with those by FEM of HFSS for the reflection coefficient S_{11}, transmission coefficient S_{13}, and forward crosstalk parameter S_{14}. The present microwave network method is more efficient than FEM in the aspect of computer resources. It takes only 42 seconds of CPU time against 1820 seconds by using FEM of HFSS. For the memory resources, the present method used 0.7 Mbytes, only about 1% of 69.3 Mbytes for FEM simulation. This example demonstrates that the present microwave network method not only can achieve almost the same accuracy as a full-wave numerical solver, but also is more efficient in both CPU time and memory costs. This makes the present method a feasible solver for SI or PI simulation of a multilayer parallel structure with a lot of vias. Note that the results by SIwave begin to deviate from the results obtained by the present method and FEM of HFSS as illustrated in Figures 5.43 and 5.44 at the high frequency end. This implies that the present method is more accurate than SIwave in comparing with the full-wave solver, HFSS.

To demonstrate the capability of the microwave network method, a two-layer square plate pair with a 31×31 via array is studied as

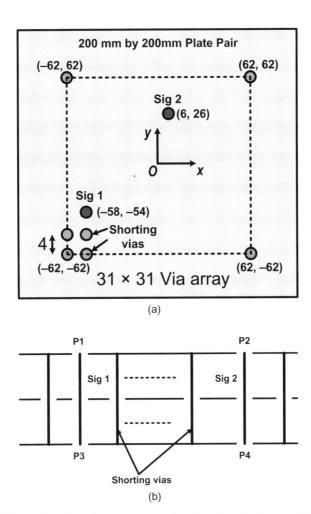

Figure 5.45 (a) Top view of a square plate pair with a 31 × 31 via array; (b) side view of the two layers of plate pair (unit: mm).

shown in Figure 5.45. The side length of the plate pair is 200 mm, and the separation of the two plates is 1 mm, which is filled by a dielectric material with relative permittivity of 4.1 and a loss tangent of 0.015. The origin of a coordinate system is set at the center of the plate pair. The location of the via array starts at (−62, −62) mm, and the periodicity of the array is 4 mm along both x-axis and y-axis. Among these 961 vias, only two signal vias are located at (−58, −54) mm and (6, 26) mm.

Their barrel and antipad radii are 0.25 mm and 0.5 mm, respectively. All other 959 vias are shorting vias which connect the top and bottom plates as shown in Figure 5.45b. The via-hole radius of these shorting vias on the middle plate is 0.5 mm, and their barrel radius is 0.25 mm.

Figure 5.46a presents the reflection and crosstalk parameters, S_{11} and S_{14} of the structure shown in Figure 5.45, obtained by the present method and SIwave, respectively. It is worth noting that the general full-wave solvers, such as HFSS, can no longer handle this complicated example.

It can be seen that the cross talk between two signal vias is still quite weak up to 5 GHz although it becomes stronger with the increase of frequencies. This is due to the fact that there are many shorting vias surrounding each signal via, which act as "shielding walls." From a circuit point of view, these shorting vias provide a very low impedance path for the return current. Thus, the structure is expected to achieve very good transmission properties as shown in Figure 5.46b. Moreover, both the cross talk and transmission coefficients by the two different solvers agree as demonstrated in Figure 5.46. (Note that although transmission coefficient S_{13} seems quite different between two methods, the absolute differences are less than 0.3 dB.)

Figure 5.46a shows that the crosstalk parameter S_{14} by the present method matches well with that obtained by SIwave. Although the two methods have the same trend for the reflection coefficient S_{11} with frequency, the difference between the two methods can be up to nearly 20 dB. This may be caused by the different calculation processes and port definitions in these two methods.

As the signal via 1 is surrounded by several shorting vias as shown in Figure 5.45, it can be approximately viewed as a coaxial cable with an 8 mm square outer conductor. Such a coaxial cable with only a length of 2 mm is expected to have excellent transmission properties with very weak reflections. Therefore, the results obtained by the present method for S_{11} and S_{14} are quite reasonable.

5.3.5 Conclusion

Based on the intrinsic via circuit model and the impedance matrix of a plate pair, a multilayer PCB can be modeled as a cascaded microwave network whose ports are defined on the via holes of plates. A recursive algorithm is developed to combine the admittance matrix layer by

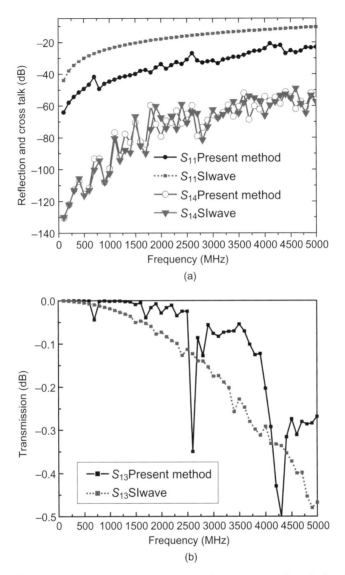

Figure 5.46 Comparison of *S*-parameters by the microwave network method and SIwave (Ansoft Inc.): (a) reflection and crosstalk parameters; (b) transmission coefficient.

layer. This can minimize the memory requirements for a board with many layers of power/ground plates. Decoupling capacitors, which are extensively used in the design of PDN, are naturally regarded as imped- ance loads to the cascaded microwave network. The proposed method has been validated by a full-wave FEM and the measurements. All of these examples demonstrate that the microwave network method is as accurate as a full-wave solver and at the same time, it is faster and more efficient in either CPU time or memory costs. Thus, the method is very suitable for SI/PI analysis for 3D package integration and multilayered PCBs with multiple vias and decoupling capacitors.

APPENDIX: PROPERTIES OF THE AUXILIARY FUNCTION $W_{mn}(x, y)$

1. For real values of x and y, $W_{mn}(x, y)$ is imaginary as

$$W_{mn}(x, y) = -j \begin{vmatrix} J_m(x) & J_n(y) \\ Y_m(x) & Y_n(y) \end{vmatrix}, \qquad (A.5.1)$$

   where $Y_m(x)$ is a Neumann function or mth order Bessel function of the second kind.

2. For real values of x and y, $W_{mn}(jx, jy)$ can be expressed as

$$W_{mn}(jx, jy) = \frac{2}{\pi} j^{n-m+1} \begin{vmatrix} I_m(x) & I_n(y) \\ K_m(x) & K_n(y) \end{vmatrix}, \qquad (A.5.2)$$

   where $I_m(x)$ and $K_m(x)$ are the mth order modified Bessel func- tions of the first and second kind, respectively.

3. Using small argument approximations of the Bessel and Hankel functions ($x \to 0$ and $y \to 0$), it can be shown that

$$W_{10}(x, y) \simeq -j \frac{2}{\pi} \left[\frac{1}{x} + \frac{x}{2} \ln y - \frac{y^2}{4x} \right], \qquad (A.5.3)$$

$$W_{00}(x, y) \simeq j \frac{2}{\pi} \ln \frac{x}{y}. \qquad (A.5.4)$$

REFERENCES

[1] H. W. JOHNSON and M. GRAHAM, *High Speed Digital Design: A Handbook of Black Magic*, Prentice-Hall, Englewood Cliffs, NJ, 1993.

[2] S. H. HALL, G. W. HALL, and J. A. McCALL, *High-Speed Digital System Design: A Handbook of Interconnect Theory and Design Practices*, John Wiley & Sons, New York, 2000.

[3] E. LAERMANS, J. GEEST, D. ZUTTER, F. OLYSLAGER, S. SERCU, and D. MORLION, Modeling complex via hole structure, *IEEE Trans. Adv. Packag.*, vol. 25, no. 2, pp. 206–214, 2002.

[4] R. ABHARI, G. V. ELEFTHERIADES, and E. V. DEVENTER-PERKINS, Physics-based CAD models for the anaysis of vias in parallel-plate environments, *IEEE Trans. Microwave Theory Tech.*, vol. 49, no. 10, pp. 1697–1707, 2001.

[5] S. MAEDA, T. KASHIWA, and I. FUKAI, Full wave analysis of propagation characteristics of a through hole using the finite-difference time-domain method, *IEEE Trans. Microwave Theory Tech.*, vol. 39, no. 12, pp. 2154–2159, 1991.

[6] T. WANG, R. F. HARRINGTON, and J. R. MAUTZ, Quasi-static analysis of a microstrip via through a hole in a ground plane, *IEEE Trans. Microwave Theory Tech.*, vol. 36, no. 6, pp. 1008–1013, 1988.

[7] P. KOK and D. D. ZUTTER, Capacitance of a circular symmetric model of a via hole including finite ground plane thickness, *IEEE Trans. Microwave Theory Tech.*, vol. 39, no. 7, pp. 1229–1234, 1991.

[8] P. A. KOK and D. D. ZUTTER, Prediction of the excess capacitance of a via-hole through a multilayered board including the effect of connecting microstrips or striplines, *IEEE Trans. Microwave Theory Tech.*, vol. 42, no. 12, pp. 2270–2276, 1994.

[9] K. S. OH, J. E. SCHUTT-AINE, R. MITTRA, and W. BU, Computation of the equivalent capacitance of a via in a multilayered board using the closed-form Green's function, *IEEE Trans. Microwave Theory Tech.*, vol. 44, no. 2, pp. 347–349, 1996.

[10] A. W. MATHIS, A. F. PETERSON, and C. M. BUTLER, Rigorous and simplified models for the capacitance of a circularly symmetric via, *IEEE Trans. Microwave Theory Tech.*, vol. 45, no. 10, pp. 1875–1878, 1997.

[11] F. TEFIKU and E. YAMASHITA, Efficient method for the capacitance calculation of circularly symmetric via in multilayered media, *IEEE Microw. Guid. Wave Lett.*, vol. 5, no. 9, pp. 305–307, 1995.

[12] S.-G. HSU and R.-B. WU, Full-wave characterization of a through hole via in multilayered packaging, *IEEE Trans. Microwave Theory Tech.*, vol. 43, no. 5, pp. 1073–1081, 1995.

[13] Q. GU, Y. E. YANG, and M. A. TASSOUDJI, Modeling and analysis of vias in multilayered integrated circuits, *IEEE Trans. Microwave Theory Tech.*, vol. 41, no. 2, pp. 206–214, 1993.

[14] Q. GU, A. TASSOUDJI, S. Y. POH, R. T. SHIN, and J. A. KONG, Coupled noise analysis for adjacent vias in multilayered digital circuits, *IEEE Trans. Circ. Syst.*, vol. 41, no. 12, pp. 796–804, 1994.

[15] H. CHEN, Q. LIN, L. TSANG, C.-C. HUANG, and V. JANDHYALA, Analysis of a large number of vias and differential signaling in multilayered structures, *IEEE Trans. Microwave Theory Tech.*, vol. 51, no. 3, pp. 818–829, 2003.

[16] L. TSANG and D. MILLER, Coupling of vias in electronic packaging and printed circuit board structures with finite ground plane, *IEEE Trans. Adv. Packag.*, vol. 26, no. 4, pp. 375–384, 2003.

[17] C. C. HUANG, L. TSANG, C. H. CHAN, and K. H. DING, Multiple scattering among vias in planar waveguides using preconditioned SMCG method, *IEEE Trans. Microwave Theory Tech.*, vol. 52, no. 1, pp. 20–28, 2004.

[18] C.-J. ONG, D. MILLER, L. TSANG, B. WU, and C.-C. HUANG, Application of the Foldy-Lax multiple scattering method to the analysis of vias in ball grid arrays and interior layers of printed circuit boards, *Microw. Opt. Technol. Lett.*, vol. 49, no. 1, pp. 225–231, 2007.

[19] X. GU and M. B. RITTER, Application of Foldy-Lax multiple scattering method to via analysis in multi-layered printed circuit board, *Proc. IEC DesignCon Conf.*, Santa Clara, CA, February 4–February 7, 2008.

[20] C. SCHUSTER, Y. KWARK, G. SELLI, and P. MUTHANA, Developing a "physical" model for vias, in *Proc. IEC DesignCon Conf.*, Santa Clara, CA, February 6–9, 2006, pp. 1–24.

[21] G. SELLI, C. SCHUSTER, Y. H. KWARK, M. B. RITTER, and J. L. DREWNIAK, Developing a physical via model for vias—Part II: Coupled and ground return vias, in *Proc. IEC DesignCon Conf.*, Santa Clara, CA, January 29–February 1, 2007, pp. 1–22.

[22] G. SELLI, C. SCHUSTER, and Y. KWARK, Model-to-hardware correlation of physics-based via models with the parallel-plate impedance included, *Proceedings IEEE Symposium on Electromagnetic Compatibility*, Portland, OR, August 14–18, 2006, pp. 781–785.

[23] M. PAJOVIC, J. XU, and D. MILOJKOVIC, Analysis of via capacitance in arbitrary multilayer PCBs, *IEEE Trans. Electromagn. Compat.*, vol. 49, no. 3, pp. 722–726, 2007.

[24] Y. ZHANG, J. FAN, G. SELLI, M. COCCHINI, and F. D. PAULIS, Analytical evaluation of via-plate capacitance for multilayer printed circuit boards and packages, *IEEE Trans. Microwave Theory Tech.*, vol. 56, no. 9, pp. 2118–2128, 2008.

[25] Y. T. LO, D. SOLOMON, and W. F. RICHARDS, Theory and experiment on microstrip antennas, *IEEE Trans. Antennas Propagat.*, vol. AP-27, no. 2, pp. 137–145, 1979.

[26] G.-T. LEI, R. W. TECHENTIN, P. R. HAYES, D. J. SCHWAB, and B. K. GILBERT, Wave model solution to the ground/power plane noise problem, *IEEE Trans. Instrum. Meas.*, vol. 44, no. 2, pp. 300–303, 1995.

[27] M. XU and T. H. HUBING, The development of a closed-form expression for the input impedance of power-return plane structures, *IEEE Trans. Electromagn. Compat.*, vol. 45, no. 3, pp. 478–485, 2008.

[28] Z. L. WANG, O. WADA, Y. TOYOTA, and R. KOGA, Convergence acceleration and accuracy improvement in power bus impedance calculation with a fast algorithm using cavity modes, *IEEE Trans. Electromag. Compat.*, vol. 47, no. 1, pp. 2–9, 2005.

[29] Y. JOEONG, A. C. LU, L. L. WAI, W. FAN, B. K. LOK, H. PARK, and J. KIM, Hybrid analytical modeling method for split power bus in multilayered package, *IEEE Trans. Electromagn. Compat.*, vol. 48, no. 1, pp. 82–94, 2006.

[30] C. WANG, J. MAO, G. SELLI, S. LUAN, L. ZHANG, J. FAN, D. J. POMMERENKE, R. E. DUBROFF, and J. L. DREWNIAK, An efficient approach for power delivery network design with closed-form expressions for parasitic interconnect inductances, *IEEE Trans. Adv. Packag*, vol. 29, no. 2, pp. 320–334, 2006.

[31] J. TRINKLE and A. CANTONI, Impedance expressions for unloaded and loaded power ground planes, *IEEE Trans. Electromagn. Compat.*, vol. 50, no. 2, pp. 390–398, 2008.

[32] K.-B. WU, G.-H. SHIUE, W.-D. GUO, C.-M. LIN, and R.-B. WU, Delaunay-Voronoi modeling of power-ground planes with source correction, *IEEE Trans. Adv. Packag.*, vol. 31, no. 2, pp. 303–310, 2008.

[33] J.-X. ZHENG and D. C. CHANG, End-correction network of a coaxial probe for microstrip patch antennas, *IEEE Trans. Antennas Propagat.*, vol. 39, no. 1, pp. 115–118, 1991.

[34] H. XU, D. R. JACKSON, and J. T. WILLIAMS, Comparison of models for the probe inductance for a parallel plate waveguide and a microstrip patch, *IEEE Trans. Antennas Propagat.*, vol. 53, no. 10, pp. 3229–3235, 2005.

[35] D. M. POZAR, *Microwave Engineering*, John Wiley & Sons, New York, 2005.

[36] K. ALEXANDER and M. N. O. SADIKU, *Fundamentals of Electric Circuits*, 2nd ed., McGraw Hill Higher Education, New York, 2002, pp. 573–577.

[37] A. G. WILLIAMSON, Equivalent circuit for radial-line/coaxial-line junction, *Electron. Lett.*, vol. 17, no. 8, pp. 300–301, 1987.

[38] G. WILLIAMSON, Radial-line/coaxial-line step junction, *IEEE Trans. Microwave Theory Tech.*, vol. MTT-33, no. 1, pp. 56–59, 1985.

[39] Y. ZHANG, E. LI, A. R. CHADA, and J. FAN, Calculation of the via-plate capacitance of a via with pad using finite difference method for signal/power integrity analysis, *2009 International Symposium on Electromagnetic Compatibility*, July 20–24, Kyoto, Japan, 2009.

[40] P. LIU and Z.-F. LI, An efficient method for calculating bounces in the irregular power/ground plane structure with holes in high-speed PCBs, *IEEE Trans. Electromagn. Compat.*, vol. 47, pp. 889–898, 2005.

[41] Y.-J. ZHANG and J. FAN, An intrinsic via circuit model for multiple vias in an irregular plate pair through rigorous electromagnetic analysis, *IEEE Trans. Microwave Theory Tech.*, vol. 58, no. 8, pp. 2251–2265, 2010.

[42] W. C. CHEW, *Waves and Fields in Inhomogeneous Media, Appendix D*, Van Nostrand Reinhold, New York, 1990.

[43] T.-K. WANG, S.-T. CHEN, C.-W. TSAI, S.-M. WU, J. L. DREWNIAK, and T.-L. WU, Modeling noise coupling between package and PCB power/ground planes with an efficient 2-D FDTD/lumped element method, *IEEE Trans. Adv. Packag*, vol. 30, no. 4, pp. 864–871, 2007.

[44] A. E. ENGIN, K. BHARATH, and M. SWAMINATHAN, Multilayered finite-difference method (MFDM) for modeling of package and printed circuit board planes, *IEEE Trans. Electromagn. Compat.*, vol. 49, no. 2, pp. 441–447, 2007.

[45] C. Guo and T. H. Hubing, Circuit models for power bus structures on printed circuit boards using a hybrid FEM-SPICE method, *IEEE Trans. Adv. Packag.*, vol. 29, no. 3, pp. 441–447, 2006.

[46] W. Shi and J. Fang, New efficient method of modeling electronics packages with layered power/ground planes, *IEEE Trans. Adv. Packag*, vol. 25, no. 3, pp. 417–423, 2002.

[47] X. C. Wei, E. P. Li, E. X. Liu, and R. Vahldieck, Efficient simulation of power distribution network by using integral-equation and model-decoupling technology, *IEEE Trans. Microwave Theory Tech.*, vol. 56, no. 10, pp. 2277–2285, 2008.

[48] E. Peterson, S. L. Ray, and R. Mittra, *Computational Methods for Electromagnetics*, Oxford University Press, Oxford, 1998.

[49] J.-S. Zhao and W. C. Chew, Integral equation solution of Maxwell's equations from zero frequency to microwave frequencies, *IEEE Trans. Antennas Propagat.*, vol. 48, no. 10, pp. 1635–1645, 2000.

[50] O. M. Ramahi, V. Subramanian, and B. Archambeault, A simple finite-difference frequency-domain (FDFD) algorithm for analysis of switching noise in printed circuit boards and packages, *IEEE Trans. Adv. Packag.*, vol. 26, no. 2, pp. 191–198, 2003.

[51] H. Wang, Y. Ji, and T. H. Hubing, Finite-element modeling of coaxial cable feeds and vias in power-bus structures, *IEEE Trans. Electromagn. Compat.*, vol. 44, no. 4, pp. 569–574, 2002.

[52] B. Achambeault and A. E. Ruehli, Analysis of power/ground-plane EMI decoupling performance using the partial-element equivalent circuit technique, *IEEE Trans. Electromagn. Compat.*, vol. 43, no. 4, pp. 437–445, 2001.

[53] J. Knighten, B. Archambeault, J. Fan, G. Selli, S. Connor, and J. Drewniak, PDN design strategies: I. Ceramic SMT decoupling capacitors—what values should I choose? *IEEE EMC Soc. Newsl.*, vol. 207, pp. 46–53, 2005.

[54] J. Knighten, B. Archambeault, J. Fan, G. Selli, L. Xue, S. Connor, and J. Drewniak, PDN design strategies: II. Ceramic SMT decoupling capacitors— Does location matter? *IEEE EMC Soc. Newsl.*, vol. 208, pp. 56–67, 2006.

[55] J. Fan, J. Drewniak, J. Knighten, N. Smith, A. Orlandi, T. Van Doren, T. H. Hubing, and R. E. DuBroff, Quantifying SMT decoupling capacitor placement in DC power-bus design for multilayer PCBs, *IEEE Trans. Electromagn. Compat.*, vol. 43, no. 4, pp. 588–599, 2001.

[56] Y. Zhang, G. Feng, and J. Fan, A novel impedance definition of a parallel plate pair for an intrinsic via circuit model, *IEEE Trans. Microwave Theory Tech.*, vol. 58, no. 12, pp. 3780–3789, 2010.

[57] J.-H. Kim and M. Swaminathan, Modeling of multilayered power distribution planes using transmission matrix method, *IEEE Trans. Adv. Packag*, vol. 25, no. 2, pp. 189–199, 2008.

CHAPTER 6

Modeling of Through-Silicon Vias (TSV) in 3D Integration

This chapter briefly reviews the through-silicon via (TSV) technology for three-dimensional (3D) integration. A wideband scalable model for electrical simulation of TSVs is then presented in detail, which is followed by a discussion of the metal-oxide-semiconductor (MOS) capacitance effects of TSVs.

6.1 INTRODUCTION

Rapid growth and convergence of digital computing and wireless communication have been driving the semiconductor technology in today's nanometer regime to continue its evolution following the scaling law, that is, Moore's law at the semiconductor transistor device and on-chip level, and "More than Moore" at the intra-chip interconnect and packaging level [1, 2]. The TSV technology [3, 4], which is explored by many researchers for 3D integration [5], represents such a latest example of scaling. A TSV is essentially a coated metal via residing in a silicon substrate for vertical interconnection (see Fig. 6.1). It has the potential

Electrical Modeling and Design for 3D System Integration: 3D Integrated Circuits and Packaging, Signal Integrity, Power Integrity and EMC, First Edition. Er-Ping Li.
© 2012 Institute of Electrical and Electronics Engineers. Published 2012 by John Wiley & Sons, Inc.

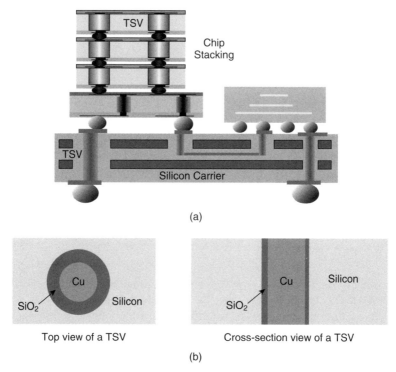

(a)

Top view of a TSV   Cross-section view of a TSV

(b)

Figure 6.1 (a) Illustration of TSVs used for stacking chips and forming a silicon carrier; (b) top and cross-section views of a single TSV.

benefits of improving the electrical performance including speed, bandwidth, and functionality, and reducing power consumption by shortening the interconnection path in the 3D integrated circuits (ICs). However, the development of TSVs faces several challenges, such as processing, heat dissipation, stress, reliability, and design for test and packaging [6].

6.1.1 Overview of Process and Fabrication of TSV

The TSV technique is hailed as one of the enabling techniques for next-generation 3D IC chip and packaging integration [5, 7]. Figure 6.1 illustrates that TSVs can be used either for stacking chips vertically, or to form a silicon carrier [4, 8] (also called a silicon interposer).

Although the focus of this chapter is electrical modeling of TSVs, it is beneficial to understand briefly the TSV fabrication process.

General Electric (GE) was among the first companies to fabricate and model TSVs [3]. History repeats itself—it was almost two decades later that a paper published in 2005 by IBM [4] on the TSV substrate carrier rekindled the research interests in this topic and marked the beginning of numerous new TSV studies.

TSVs can be processed and fabricated before, during, or after the IC fabrication process, either by chip stacking, or solely as a silicon interposer or carrier without embedded active devices. Nevertheless, they possess a number of well-defined common features: a hole that must be etched in the silicon substrate; an isolation layer to separate the TSV electrically from the silicon substrate; a barrier layer to prevent diffusion of metals into the silicon and increase metal adhesion, and the via-hole filling by a conductive material. Whereas more detailed description can be found in Reference 1, we summarize here the following key points about TSV fabrication:

1. Via-hole forming: The TSV via hole is formed by a blind-via approach, in which the hole is only etched down to a certain depth or to an etch-stop layer. This approach is compatible with the mainstream semiconductor process in use by industry. TSV etching widely uses the plasma etching technique, that is, the deep reactive ion etching (DRIE) (e.g., the Bosch recipe), to produce high aspect ratio TSVs, although laser drilling was initially tried. Fast etching speed for mass production and high quality of the via hole are critical aspects that make TSV etching techniques challenging.

2. TSV liner process: The TSV liner electrically isolates the TSV metal from the silicon substrate. The liner shall fulfill the requirement of low leakage current, large breakdown voltage, and low capacitance for signal transmission. Deposition of the liner must be conducted at a temperature that does not damage the devices, for example, below 200°C for postprocessing of a dynamic random-access memory (DRAM) device. Oxide or nitride layers are usually deposited by chemical or physical vapor deposition (CVD or PVD). Polymer isolation layers are also possible for wafer-level via-last TSVs.

3. TSV barrier layer: A barrier layer is required to prevent the TSV metal from migrating into the silicon. It also improves adhesion between the TSV metal and the liner. Titanium nitride (TiN) or

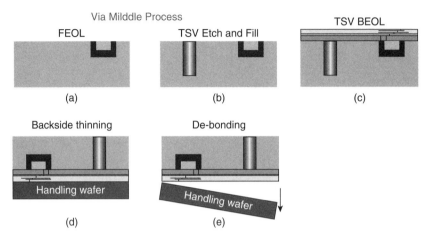

Figure 6.2 Main process steps in TSV fabrication by a via-middle approach [29].

tantalum (Ta) by CVD or PVD, which are commonly used for advanced logic devices, are still the best candidates for the TSV barrier layer.

4. TSV metal filling process: The materials for filling the TSV via hole could be copper (Cu) which is applied by electrochemical plating (ECP) or a CVD process, tungsten (W) which is applied by a CVD process, and poly-Silicon for the via-first approach. The Cu TSV process includes Cu seed deposition, Cu-via fill by ECP, and CMP (Chemical Mechanical Polishing) removal of Cu-overburden. Copper is likely to prevail in mainstream applications due to its high conductivity and low cost.

The TSV fabrication process also involves wafer handling, bonding and debonding, stacking, and so on. Figure 6.2 shows the main steps required for fabricating via-middle TSVs. Via-first and via-last fabrication approaches are also being evaluated by researchers. An example is given in Figure 6.3, in which TSVs in a via-last approach are produced by a fast electroplating process.

Note that according to the ITRS 2009 report (Table INTC3 and INTC4 in Reference 1), the dimensions for the TSV at global interconnect level in five years' time are: minimum diameter from 8 μm to 2 μm; minimum pitch from 16 μm to 4 μm; minimum depth within

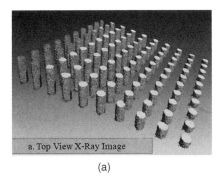
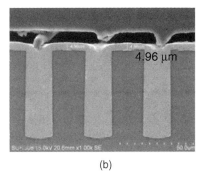

(a)   (b)

Figure 6.3   TSVs in a via-last approach are produced by a fast electroplating process: (a) top view X-ray image of TSVs; and (b) TSVs after metal filling [30].

20–50 μm; maximum aspect ratio around 5:1 to 20:1. For the intermediate interconnect level, the above numbers are: 2 μm–0.8 μm (minimum diameter); 4–1.6 μm (minimum pitch); 6–10 μm (depth); and 5:1 to 20:1 (maximum aspect ratio).

6.1.2 Modeling of TSV

Modeling is an essential part in designing and developing TSVs. TSV modeling involves several domains, such as electrical and electromagnetic modeling, thermal modeling, mechanical modeling, and so on. Therefore, TSVs require a multiphysics modeling, although the focus of this chapter is on electrical and electromagnetic modeling of TSVs.

6.1.2.1 Multiphysics Modeling of TSV

Multiphysics modeling and simulation has been used by the electrical, electromagnetic community, and other engineering communities for quite some time. Initial attempts at multiphysics modeling have been made over the past years; it is beyond the scope of this chapter to review those attempts. However, we wish to point out that the TSV presents itself as an excellent application for such a multiphysics study involving (1) semiconductor device physics and electromagnetic/optical properties of TSVs; (2) thermal-electrical properties of TSVs; (3) mechanical-electrical properties (e.g., impact of mechanical stress of

TSVs on the performance of metal-oxide-semiconductor [MOS] transistors); (4) thermal-mechanical properties, and many others.

In general, two levels of studies are highly demanded: (1) studies of simple structures to understand the nature of TSVs, for which an analytical approach is often preferred, and (2) system-level simulation and validation of TSV-based components and systems, which often involve extensive numerical approaches. This chapter reports investigations at the first level of study. Note that our previous study on multiple via modeling, presented in Reference 9 and the previous chapter of this book, shall be useful for the second level of study.

6.1.2.2 Electrical-Electromagnetic Modeling of TSV

From the electrical and electromagnetic point of view, TSV is nothing more than a metal via residing in a silicon substrate for vertical interconnection. TSVs shorten the interconnection path and thus reduce the resistance-capacitance (RC) delay.

In the electrical and electromagnetic regime, two categories of simulations have been carried out regarding silicon as a substrate. If the silicon substrate has low Ohmic loss, the electromagnetic response can be considered as linear. An effective conductivity or loss tangent could thus be used in the electromagnetic simulation. However, if the silicon behaves as a semiconductor, a self-consistent coupling of Maxwell's equations with the semiconductor physics equations must be considered. We employ in this chapter the second approach, although not in a strictly self-consistent manner.

In the following sections we will first present a wideband equivalent-circuit model for electrical modeling of TSVs. Then the MOS capacitance effect of TSVs is discussed.

6.2 EQUIVALENT CIRCUIT MODEL FOR TSV

This section presents a recently developed wideband equivalent-circuit model for electrical modeling of TSVs in 3D stacked ICs and packaging. Rigorous closed-form formulae for the resistance and inductance of TSVs are derived from the magneto-quasi-static theory with a Fourier–Bessel expansion approach, whereas analytical formulae from static solutions are used to compute the capacitance and conductance.

The equivalent-circuit model can capture the important parasitic effects of TSVs, including the skin effect, the proximity effect, the lossy effect of silicon and the semiconductor effect. Therefore, it yields accurate results comparable to those with 3D full-wave solvers.

6.2.1 Overview

Many papers have been published on electrical models of TSVs. We roughly classify the literature on electrical models of TSVs into two general approaches, that is, the "forest" approach and the "wood" approach. The "forest" approach refers to the one used in those papers that build models for many TSVs (TSV array or bundles), involve numerical modeling approaches requiring large-matrix solution, and study electrical properties of TSVs at the system level [10–13]. In contrast, the "wood" approach refers to those that build models for only a couple of TSVs arranged in some basic configurations, such as single TSV, GS, GSG, GSSG (G-ground, S-signal), involve analytical or empirical formulae with no large-matrix solution, and study the electrical properties of basic configurations of TSVs.

This chapter is focused on the latter approach, that is, the electrical modeling of TSVs with a physics-based wideband equivalent-circuit model. Nearly two decades ago, J. P. Quine et al. applied the even- and odd-mode approach to study the insertion loss and cross talk of vias in a shielded silicon board [3]. Among the many recent efforts in developing equivalent-circuit models, J. Kim and coauthors proposed in Reference 14 a π-type equivalent-circuit model for TSVs arranged in a ground-signal-ground configuration. The values of the circuit elements were obtained by parameter fitting through optimization, and not tied to the geometrical and physical parameters of TSVs. The above model was again investigated in Reference 15, and closed-form formulae were used for the circuit elements in the equivalent-circuit model. However, the resistance-inductance (RL) model, which was not rigorously derived for a two-TSV system, was not very accurate as shown in Reference 15. Readers may refer to Reference 15 for more discussion of some open literature on electrical models of TSVs.

We developed a π-type equivalent circuit model with rigorous closed-form formulae. By applying the magneto-quasi-static theory to a two-TSV system, we formulate a rigorous RL model. Combining the RL model with the capacitance-conductance model from static solution, we present a wideband equivalent-circuit model for the electrical modeling of TSVs. The wideband equivalent-circuit model captures the important parasitic effects of TSVs, including the skin effect, the proximity effect, the lossy effect of silicon and the semiconductor effect. It thus yields accurate results comparable to those with 3D full-wave solvers.

6.2.2 Problem Statement: Two-TSV Configuration

Two- or three-TSV systems are the basic units used in a 3D IC for signal transmission. It is important to study those basic units to understand fully the TSV characteristics. Because the equivalent-circuit model for a three-TSV system can be directly derived from that of a two-TSV system, this work focuses primarily on the equivalent-circuit model of a two-TSV system. Note that our previous work in Reference 9 on multiple via modeling can be extended to perform system-level simulation of 3D ICs with multiple TSVs. Figure 6.4 shows the cross section of a two-TSV system—one is a signal via and the other a ground via, which are separated by a center-to-center distance c. Each TSV comprises a central conductor core of the radius a for guiding signal and a silicon dioxide cladding layer with the thickness of $(b - a)$ for direct current isolation. d denotes the radius of a depletion region of a TSV due to the semiconductor effect discussed in Section 6.3. The following symbols are used for material properties: σ_c is the conductiv-

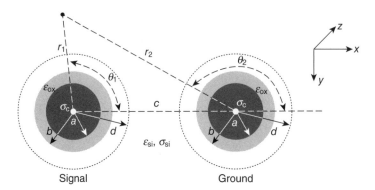

Figure 6.4 Cross-section view of a two-TSV (signal-ground) system in global Cartesian and local cylindrical coordinate systems [31].

ity of the TSV metal, ε_{ox} is the permittivity of the silicon dioxide, and ε_{si} is the permittivity and σ_{si} is the conductivity of the silicon substrate. One global Cartesian coordinate system (x, y, z) and two local cylindrical coordinate systems (r_1, θ_1, z) and (r_2, θ_2, z) are defined as shown in Figure 6.4.

6.2.3 Wideband Pi-Type Equivalent-Circuit Model

Note that the length of TSVs, which is around 50 to 200 μm, is shorter than one-twentieth of the guided wavelength ($\lambda_g = 8.7$ mm at 10 GHz) in silicon substrate. Therefore, the previously mentioned two-TSV system can be modeled with a π-type equivalent circuit as shown in Figure 6.5a. R and L in the π-type circuit denote the per-unit-length (p.u.l.) resistance and inductance of the TSV. C_g is the p.u.l. capacitance due to the oxide layer with or without consideration of the MOS effect

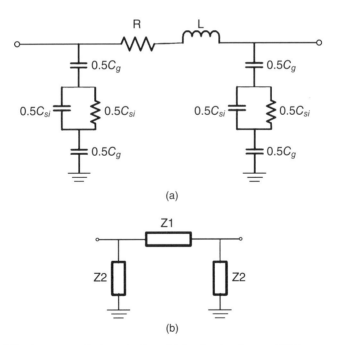

(a)

(b)

Figure 6.5 A π-type equivalent-circuit model for the signal-ground TSV system: (a) its topology featuring physics-based equivalent-circuit elements expressed in rigorous closed-form formulae, and (b) its simplified network representation [31].

of TSVs. C_{si} and G_{si} represent the p.u.l. capacitance and conductance regarding the silicon substrate and its finite conductivity, respectively.

The two-wire system has been well studied in the microwave regime with closed-form p.u.l. resistance, inductance, capacitance, and conductance (RLCG) model. The two-TSV system is essentially such a system but with coated metal wires. The closed-form formulae for the *RLCG* at high frequency without consideration of the MOS capacitance effect are given by adapting the formulae in References 16 and 17 for a two-wire structure:

$$C_g = C_{ox} = \frac{2\pi\varepsilon_{ox}}{\ln(b/a)}, \quad C_{si} = \frac{\pi\varepsilon_{si}}{\ln\left[(c/2b) + \sqrt{(c/2b)^2 - 1}\right]},$$

$$G_{si} = \frac{\sigma_{si}C_{si}}{\varepsilon_{si}} = \frac{\pi\sigma_{si}}{\ln\left[(c/2b) + \sqrt{(c/2b)^2 - 1}\right]}.$$

$$(6.1)$$

$$L = \frac{\mu_0}{\pi} \ln\left[(c/2a) + \sqrt{(c/2a)^2 - 1}\right],$$

$$R = \frac{R_s}{\pi a\sqrt{1 - (2a/c)^2}} = \sqrt{\frac{\pi f\mu_0}{\sigma_c}} \frac{1}{\pi a\sqrt{1 - (2a/c)^2}}.$$

$$(6.2)$$

where f is the frequency and μ_0 is the free space permeability. R_s is the skin-effect surface resistance of the TSV conductor.

The skin effect forces the current in the TSV conductor to flow around its outer edge, whereas the proximity effect drives the current to the inner edges of the signal-ground TSV system. Both effects increase the resistance and decrease the internal inductance of the two-TSV system at high frequencies. However, the analytical model with Equations (6.1) and (6.2) cannot fully capture those two effects, and thus is not a wideband model. The scattering parameters of a two-TSV system obtained by the analytical model do not match those with the 3D full-wave solver (see Fig. 6.7). Experimental results in Reference 18 confirmed that the R and L for on-chip interconnects strongly depend on frequency, and the C and G are weakly dependent on the frequency. We thus propose a wideband π-type equivalent-circuit model with a hybrid approach—the C and G are computed by Equation (6.1) to account for the dielectric and lossy silicon effect, and the R and L use a set of rigorous closed-form formulae derived in the following section to account both for the skin and the proximity effects of the two-TSV system.

6.2.4 Rigorous Closed-Form Formulae for Resistance and Inductance

The Fourier–Bessel expansion approach presented in Reference 19 is used to develop a set of rigorous closed-form formulae for the resistance and inductance of the two-TSV system shown in Figure 6.4. Both the skin and proximity effects are considered by the resultant equivalent-circuit model.

A time convention of $e^{j\omega t}$ and wave propagation along the positive z-direction with $e^{-j\gamma z}$ are used and suppressed throughout the chapter. The electromagnetic fields for the transverse magnetic mode regarding the TSVs can be expressed solely by the longitudinal electric field E_z [17] in the local cylindrical coordinate system shown in Figure 6.4:

$$
\begin{cases}
E_r = \dfrac{-\gamma}{k_r^2}\dfrac{\partial E_z}{\partial r}; & E_\theta = \dfrac{-\gamma}{k_r^2 \cdot r}\dfrac{\partial E_z}{\partial \theta} \\[2ex]
H_r = \dfrac{j\omega\varepsilon_{eff}}{k_r^2 \cdot r}\dfrac{\partial E_z}{\partial \theta}; & H_\theta = \dfrac{-j\omega\varepsilon_{eff}}{k_r^2}\dfrac{\partial E_z}{\partial r},
\end{cases}
\tag{6.3}
$$

where γ is the wave propagation constant and k_r is the cutoff wavenumber. They are related to each other by $k_r^2 = \omega^2\mu\varepsilon_{eff} + \gamma^2$ with $\varepsilon_{eff} = \varepsilon + \sigma/j\omega$. Equation (6.3) is simplified for the magnetic fields in the conductor because the propagation constant γ is usually a very small number [19]. Finally, the magnetic fields in the conductor ($r < a$) are given by

$$
H_r = \frac{-1}{j\omega\mu r}\frac{\partial E_z}{\partial \theta}; \quad H_\theta = \frac{1}{j\omega\mu}\frac{\partial E_z}{\partial r}.
\tag{6.4}
$$

In addition, the conduction current in a good conductor dominates over the displacement current. Thus, the wavenumber for conductors is $k_c = j\sqrt{j\omega\mu\sigma_c}$.

The longitudinal electric field E_z is governed by the following cylindrical wave equation:

$$\frac{1}{r}\frac{\partial}{\partial r}\left(r\frac{\partial E_z}{\partial r}\right) - j\omega\mu\sigma_c E_z = 0. \tag{6.5}$$

The E_z in the signal TSV conductor ($r < a$) can be expressed by the following Fourier–Bessel expansion:

$$E_z = \sum_{n=0}^{\infty} A_n J_n(k_c r_1)\cos n\theta_1 \tag{6.6}$$

and in the ground TSV conductor by

$$E_z = -\sum_{n=0}^{\infty} A_n J_{-n}(k_c r_2)\cos n\theta_2. \tag{6.7}$$

The magnetic fields in TSV conductors ($r < a$) are obtained from Equation (6.4)

$$
\begin{aligned}
H_r &= \frac{1}{j\omega\mu r}\sum_{n=0}^{\infty} n A_n J_n(k_c r)\sin n\theta \\
H_\theta &= \frac{k_c}{j\omega\mu}\sum_{n=0}^{\infty} A_n J'(k_c r)\cos n\theta.
\end{aligned}
\tag{6.8}
$$

Note that the following property of the Bessel function is used in deriving Equation (6.8):

$$\frac{dJ_n(kr)}{dr} = \frac{n}{r}J_n(kr) - kJ_{n+1}(kr). \tag{6.9}$$

The electromagnetic fields in the dielectric region involve Hankel functions and can be formulated under the small argument approximation in the Cartesian coordinate system as

$$H_x = \sum_{n=1}^{\infty} B_n \left\{ \frac{\sin n\theta_1}{r_1^n} + (-1)^n \frac{\sin n\theta_2}{r_2^n} \right\}$$

$$H_y = -\sum_{n=1}^{\infty} B_n \left\{ \frac{\cos n\theta_1}{r_1^n} + (-1)^n \frac{\cos n\theta_2}{r_2^n} \right\}.$$

(6.10)

With coordinate transformation [19], the magnetic fields in the dielectric close to the surface of the signal TSV conductor are obtained by

$$H_\theta = -\frac{B_1}{a} - \sum_{n=1}^{\infty} \cos n\theta \left(\frac{B_{n+1}}{a^{n+1}} + \frac{(-1)^n}{(n-1)!} \left(\frac{a}{c} \right)^{n-1} \xi_{n-1} \right)$$

$$H_r = \sum_{n=1}^{\infty} \sin n\theta \left(\frac{B_{n+1}}{a^{n+1}} + \frac{(-1)^{n+1}}{(n-1)!} \left(\frac{a}{c} \right)^{n-1} \xi_{n-1} \right),$$

(6.11)

where $\xi_n = \sum_{m=0}^{\infty} (-1)^m \frac{(n+m)}{m!} \frac{B_{m+1}}{c^{m+1}}$.

If the total current flowing in the signal TSV is I, the coefficient A_0 can be determined from the Ampere's law $\oint_l H_\theta|_{r=a} \cdot d\mathbf{l} = I$. The remaining unknown coefficients A_n and B_n are determined by the boundary conditions involving magnetic fields at the surface of the conductor. Due to the symmetry of the two-TSV system shown in Figure 6.4, the boundary conditions enforced at the surface of the signal TSV are automatically satisfied at that of the ground TSV. If the conductors used in the TSVs are nonmagnetic, then the coefficients are given by

$$A_0 = \frac{-j\omega\mu I}{2\pi k_c a J_1(k_c a)},$$

(6.12)

$$\alpha_{n(n>0)} = A_{n(n>0)} / A_0$$

$$= \frac{(-1)^n 2(\kappa\varsigma)^n J_1(k_c a)}{J_{n-1}(k_c a)} \cdot \left(1 - \frac{2n\kappa^2}{\varsigma^{n-1}} \frac{J_1(k_c a)}{k_c a \cdot J_0(k_c a)} \right),$$

(6.13)

$$B_1 = \frac{-I}{2\pi}, \tag{6.14}$$

$$\begin{aligned}
\beta_{n-1(n>1)} &= B_{n-1(n>1)} \big/ (-a^n I) \\
&= \frac{-2\kappa^{n-1} J_n(k_c a)}{J_{n-2}(k_c a)} \left[\varsigma^{n-1} - (n-1)\kappa^2 \varsigma \left(1 + \frac{J_2(k_c a)}{J_0(k_c a)} \right) \right], \tag{6.15}
\end{aligned}$$

where $\varsigma = \left(1 - \sqrt{1 - (2\kappa)^2} \right) \big/ (2\kappa^2)$, and $\kappa = a/c$.

The p.u.l. resistance and inductance of the two-TSV system are calculated by applying the Faraday's law in the integral form to a surface between the two TSV conductors (see Fig. 6.6). By comparing the resultant equation to the telegraph equation, we obtain the following

$$Z_{tsv} = \frac{2 E_z \big|_{\theta_1 = 0}}{I}, \quad L_{tsv} = -\int_a^{c-a} \frac{Hy}{I} \cdot dx. \tag{6.16}$$

With Equations (6.6), (6.10), and (6.12–6.15), the p.u.l. resistance and inductance of the two-TSV system is obtained from Equation (6.16):

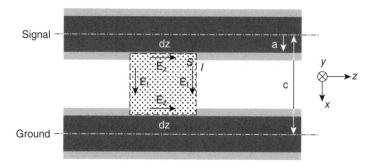

Figure 6.6 Longitudinal view of the two-TSV system: a flat surface element S and its associated rectangular contour ℓ are used with the Faraday's law to determine the per-unit-length resistance and inductance [31].

$$Z_{tsv} = \frac{-j\omega\mu_0}{\pi k_c a \cdot J_1(k_c a)}\left[J_0(k_c a) + \sum_{n=1}^{\infty} \alpha_n J_n(k_c a)\right], \qquad (6.17)$$

$$L_{tsv} = \frac{\mu_0}{\pi}\ln\left(\frac{1-\kappa}{\kappa}\right) + \frac{\mu_0}{4\pi}\sum_{n=2}^{\infty}\frac{2\beta_{n-1}}{n-1}\left[1 - \left(\frac{\kappa}{1-\kappa}\right)^{n-1}\right]. \qquad (6.18)$$

The above Z_{tsv} and L_{tsv} are in general complex numbers. Z_{tsv} contains the internal self-inductance of TSVs, and L_{tsv} includes the external inductance of the two-TSV system. The real-valued p.u.l. R and L of the two-TSV system are obtained from

$$R + j\omega L = Z_{tsv} + j\omega L_{tsv}. \qquad (6.19)$$

6.2.5 Scattering Parameters of Two-TSV System

The scattering parameters of the two-port network formed by the two-TSV system are calculated from the ABCD network parameters

$$\begin{bmatrix} A & B \\ C & D \end{bmatrix} = \begin{bmatrix} 1 + \dfrac{Z_1}{Z_2} & Z_1 \\ \dfrac{2}{Z_2} + \dfrac{Z_1}{Z_2^2} & 1 + \dfrac{Z_1}{Z_2} \end{bmatrix}, \qquad (6.20)$$

where Z_1 and Z_2 in Figure 6.5b are given by

$$Z_1 = (R + j\omega L)\cdot h$$
$$Z_2 = 2\left(\frac{1}{G_{si} + j\omega C_{si}} + \frac{1}{j\omega 0.5C_g}\right)\Big/ h, \qquad (6.21)$$

where h is the height of the TSV.

The two-TSV system makes a reciprocal and symmetrical two-port network. Therefore, its scattering parameters can be expressed as the following with a reference impedance Z_0:

$$\begin{cases} S11 = S22 = (B/Z_0 - CZ_0)/\Delta \\ S12 = S21 = 2/\Delta, \end{cases} \qquad (6.22)$$

where $\Delta = 2A + B/Z_0 + CZ_0$.

Table 6.1
List of Dimensions of the Simulated Two-
TSV Systems

(μm)	a	b	c	h
Case-A	2.5	2.8	10	40
Case-B	15	16	60	200

6.2.6 Results and Discussion

The wideband equivalent-circuit model with Equations (6.1) and (6.19) is validated against the full-wave finite element method (FEM) simulation with the HFSS software [20]. Two numerical examples are given. The dimensions of the TSVs are tabulated in Table 6.1. Their material properties are $\sigma_c = 5.8 \times 10^7$ S/m, $\varepsilon_{ox} = 4\varepsilon_0$, $\varepsilon_{si} = 11.9\varepsilon_0$, ($\varepsilon_0 = 8.854187 \times 10^{-12}$ F/m), and $\sigma_{si} = 10$ S/m. For Case-A and Case-B, the magnitude and the phase of the scattering parameters with the wideband equivalent-circuit model agree well with the full-wave FEM simulation results as shown in Figures 6.7 and 6.8, respectively. However, the simulation time required by the equivalent-circuit model using five terms in Equations (6.17) and (6.18) is less than 1 minute with a Matlab program, in contrast to 15 minutes required by the HFSS software on the same computer. Figure 6.7 also shows the comparison of the results with the analytical model (Eqs. 6.1 and 6.2), the wideband model (Eqs. 6.1 and 6.19) of this work, and the FEM simulation with the HFSS software. The comparison confirms that the model in this work is accurate over a wideband of frequency.

For Case-A, we also compare the resistance and inductance results calculated by Equation 6.19 against those by Equation 6.2. Figure 6.9 reveals that the resistance model by Equation 6.19 is an accurate wideband model—the resistance of the two-TSV system approaches the DC resistance value at low frequency, and the skin-effect surface resistance value computed by Equation (6.2) at high frequency. Equation (6.2) is only valid for high frequency. The frequency-dependent inductance computed by Equation 6.19 consists of the internal inductance and the external inductance, which decreases with the increase in frequency. Note that the frequency-independent inductance by Equation (6.2) is based on static solution and only has a fixed value (see Fig. 6.9).

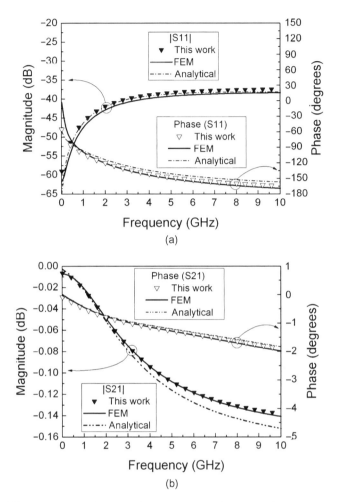

Figure 6.7 Comparison of the magnitude and phase of the scattering parameters (S11 and S21) for Case-A obtained by this work, the full-wave FEM solution with HFSS software, and the analytical model [31].

Although calculation of the resistance and inductance from Equation (6.19) involves infinite series, the convergence of the series is fast for all the simulation cases, and only about 5–10 terms are needed to reach converged results. Figure 6.10 shows that the resistance and inductance for Case-A converge to steady values only with five terms used in Equations (6.17) and (6.18).

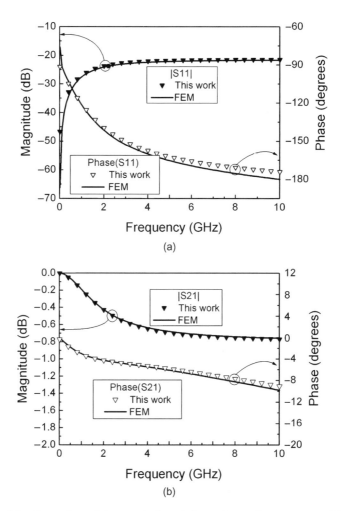

Figure 6.8 Comparison of the magnitude and phase of scattering parameters (S11 and S21) for Case-B by this work against the full-wave FEM simulation [31].

In conclusion, the π-type model with Equation (6.1) and (6.19) provides accurate and fast solution, which is verified up to 10 GHz in the above two examples, for the wideband electrical modeling of TSVs. Note that if the length of TSVs is comparable to the diameter of the TSVs, the above equivalent-circuit model can be amended to maintain accuracy by adding circuit elements to consider the fringing fields at the two ends of a TSV pair.

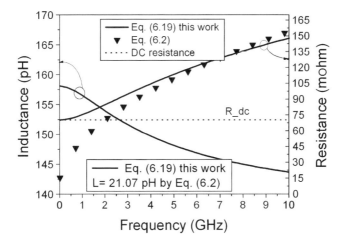

Figure 6.9   Comparison of the resistance-inductance model for TSVs, which are defined by Equations (6.19) and (6.2) [31].

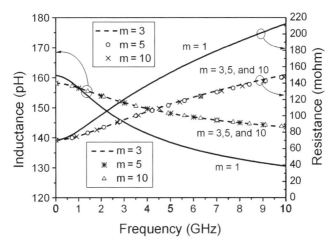

Figure 6.10   The results of the resistance and inductance for Case-A converge with five terms (m = 5) used in Equations (6.17) and (6.18) [31].

As mentioned earlier, the equivalent circuit model considers both the skin effect and proximity effect. The skin effect forces the current in the TSV conductor to flow around its outer edge, whereas the proximity effect drives the currents to the inner edges of the signal-ground TSV system, as confirmed by the FEM simulation (see Fig. 6.11). Both

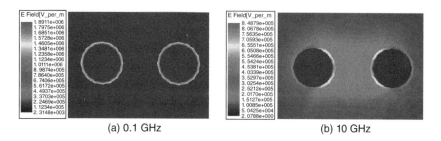

(a) 0.1 GHz   (b) 10 GHz

Figure 6.11 Skin and proximity effects of TSVs, demonstrated by the electromagnetic field plots with HFSS software [29]. (See color insert.)

effects will change the resistance and internal inductance of the TSVs. It may be inferred from Figure 6.11 that the equivalent circuit model is able to capture both effects.

Finally, we briefly discuss the concept of "eddy currents." Eddy currents, also called Foucault currents [21], are currents induced in conductors, or in other words, due to finite conductivity of the materials. Two types of eddy currents exist for TSVs—one is related to the TSV metal, the other is due to the finite conductivity of silicon.

For TSV metal, the skin effect is caused by self-induced eddy currents in conductors [21, 22], whereas the proximity effect is caused by externally induced eddy currents [17]. Both skin and proximity effects are linked to eddy currents, and should be modeled concurrently for the TSV pair. The diffusion equation is used to model the effect of eddy currents in a material [22, 23]. It is based on a magneto-quasi-static approximation of the Maxwell equations, that is, by neglecting the term $\partial D/\partial t$ in the equation $\nabla \times H = J + \partial D/\partial t$. Equation (6.5) is the diffusion equation that we used to derive R and L.

Besides the above eddy currents, which exhibit as skin and proximity effects for TSV metal, the other eddy current is related to the finite conductivity of silicon substrate. Note that the conductivity of the silicon substrate is usually 10 S/m (its resistivity is 10 Ohm·cm), which is 6–7 order smaller than that of copper (around 5.69–5.80e7 S/m). The eddy current due to silicon is small, and can be accurately modeled as lossy transmission lines with CG elements [17, 22, 23]. The full-wave validation has confirmed the accuracy of the *RLCG* model presented in this section, although the accuracy of the model reduces for TSVs with a large pitch.

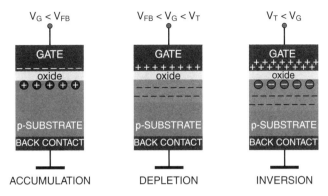

Figure 6.12 Different modes of operation (accumulation, depletion, and inversion) and its charge distribution in an n-type Metal-Oxide-Semiconductor structure (p-type substrate), where V_{tsv}, V_{FB}, and V_T denote the TSV bias voltage, flatband voltage, and threshold voltage, respectively [24].

6.3 MOS CAPACITANCE EFFECT OF TSV

6.3.1 MOS Capacitance Effect

A planar MOS capacitor shown in Figure 6.12 can operate in three different modes—accumulation, depletion, and inversion [24, 25]. Figure 6.13 shows the capacitance-voltage (C-V) curve of a MOS capacitor under (a) accumulation, (b) depletion, and (c)-(e) inversion. Similar to a planar MOS capacitor, a TSV assumes a cylindrical MOS capacitance configuration and has a bias voltage-dependent MOS capacitance. The TSV as a MOS capacitor can also operate in three different modes—the accumulation, depletion, and inversion modes. Some experimental evidence was reported in Reference 26.

6.3.2 Bias Voltage-Dependent MOS Capacitance of TSVs

The discussion in the preceding sections does not consider the MOS effect of TSVs. A TSV assumes a MOS capacitance structure and has a bias voltage-dependent MOS capacitance. A TSV perceived as an ideal cylindrical MOS capacitor was studied in Reference 27. We apply the same method used for a planar MOS capacitor as in Reference 28

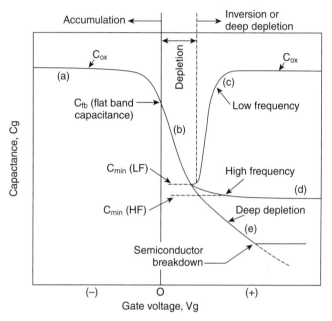

Figure 6.13 Capacitance-voltage (C-V) curve of a MOS capacitor under (a) accumulation, (b) depletion, and (c)–(e) inversion [28].

to study the TSV cylindrical MOS capacitor in the presence of interface trapped charges and oxide charges.

The TSV MOS capacitor can be modeled by the following Poisson's equation [25]:

$$\frac{1}{r}\frac{d}{dr}\left(r\frac{d\phi(r)}{dr}\right) = \frac{qN_a}{\varepsilon_{si}}. \tag{6.23}$$

Note the similarity and difference between Equations (6.5) and (6.23). The general solution to Equation (6.23) is

$$\phi(r) = \frac{qN_a}{4\varepsilon_{si}}r^2 + \alpha_1 \ln r + \alpha_2. \tag{6.24}$$

The unknown constants in Equation (6.24) are determined by the following boundary conditions:

$$\begin{cases} \phi(r)|_{r=d} = 0 \\ E|_{r=d} = \dfrac{d\phi(r)}{dr}\bigg|_{r=d} = 0. \end{cases} \qquad (6.25)$$

Finally, we obtain the following solution of the electric potential:

$$\phi(r) = \frac{qN_a}{2\varepsilon_{si}} \left(d^2 \ln \frac{d}{r} - \frac{d^2 - r^2}{2} \right), \qquad (6.26)$$

where N_a is the substrate doping concentration. The flatband voltage of a TSV MOS capacitor depends on the work function difference between the metal and the silicon, and the presence of interface-trapped charges Q_{it} and oxide charges with a surface density of $\rho_{ox}(r)$:

$$V_{FB} = \phi_{ms} - \frac{qQ_{it}}{C_{ox}} - \frac{1}{C_{ox}} \int_a^b \frac{\rho_{ox}(r) \cdot 2\pi r}{b - a} dr. \qquad (6.27)$$

$$\phi_{ms} = \phi_m - \phi_s. \qquad (6.28)$$

$$\phi_s = \chi_{si} + \frac{E_g}{2q} + V_t \ln \frac{N_a}{n_i}. \qquad (6.29)$$

An equivalent interface charge Q_{eq} is often used in circuit simulation models to represent the effect of both the interface trapped and oxide charges. Therefore, the total band bending due to ϕ_{ms} and Q_{eq} is

$$V_{FB} = \phi_{ms} - \frac{qQ_{eq}(2\pi b)}{C_{ox}}. \qquad (6.30)$$

It is noted that work function is the voltage required to extract an electron from the Fermi energy to the vacuum level. This voltage is between 3 and 5 eV for most metals. The actual value of the work function of a metal deposited onto silicon dioxide is not exactly the same as that of the metal in vacuum [24].

The threshold voltage is defined as the gate voltage for which the electron density at the Si-SiO$_2$ surface equals N_a. This corresponds to the situation where the total potential across the surface equals twice the bulk potential ϕ_F:

$$\phi_F = V_t \ln \frac{N_a}{n_i}, \tag{6.31}$$

where $V_t = KT/q$ is the thermal voltage, the value of which is 0.026 V at 300 K (around room temperature of 27°C), and N_a is the substrate doping concentration. K is the Boltzmann's constant, that is, 1.3807×10^{-23}, and T denotes the absolute temperature.

The threshold voltage is thus obtained as

$$V_T = V_{FB} + 2V_t \ln \frac{N_a}{n_i} + \frac{qN_a \left[\pi(d_m^2 - b^2)\right]}{C_{ox}}, \tag{6.32}$$

where n_i is the intrinsic carrier concentration ($n_i = 1.02 \times 10^{10}$ cm^{-3} at 300 K), and the maximum radius of the depletion region is determined by the following transcendental equation:

$$\frac{qN_a}{2\varepsilon_{si}} \left(d_m^2 \ln \frac{d_m}{b} - \frac{d_m^2 - b^2}{2} \right) = 2V_t \ln \frac{N_a}{n_i}. \tag{6.33}$$

The radius d of the depleted region ranges from b to d_m and is given by

$$V_{tsv} = V_{FB} + \frac{qN_a}{2\varepsilon_{si}} \left(d^2 \ln \frac{d}{b} - \frac{d^2 - b^2}{2} \right) + \frac{qN_a \left[\pi(d^2 - b^2)\right]}{C_{ox}}. \tag{6.34}$$

Therefore, the final p.u.l. capacitance of the TSV MOS capacitor for different modes of operation is given by

$$C_g = \begin{cases} C_{ox} = \dfrac{2\pi\varepsilon_{ox}}{\ln(b/a)} & V_{tsv} < V_{FB} \text{ (accumulation)} \\[2ex] \left(\dfrac{1}{C_{ox}} + \dfrac{1}{C_d}\right)^{-1} & V_{FB} < V_{tsv} < V_T \text{ (depletion)}, \end{cases} \tag{6.35}$$

and the depletion capacitance is $C_d = 2\pi\varepsilon_{si}/\ln(d/b)$.

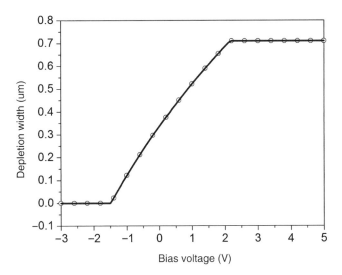

Figure 6.14   The depletion width changes with the bias voltage.

6.3.3 Results and Analysis

First, we studied the depletion region and the MOS capacitance of a TSV. The diameter of the TSV is 5 μm, and the oxide thickness is 500 nm. The doping concentration of the p-type silicon substrate is $N_a = 1.5 \times 10^{15}\,cm^{-3}$. The intrinsic carrier concentration is taken as $n_i = 1.02 \times 10^{10}\,cm^{-3}$. The equivalent interface charge is assumed as $Q_{eq} = 5.0 \times 10^{10}\,cm^{-2}$. The work function for the barrier metal is 4.25 eV. Figure 6.14 shows the results of the width of the depletion region subject to different bias voltages. Figure 6.15 shows the corresponding p.u.l. capacitance of the TSV.

Second, the two-TSV system for Case-A in the previous section is studied again to demonstrate the effect of the MOS capacitance on the scattering parameters. If we assume a fully depleted region for the TSVs, the radius of the depletion region achieves maximum, and the depletion capacitance C_d becomes minimum. Due to the addition of the depletion capacitance in series with the oxide capacitance C_{ox}, the total effective capacitance of the TSV pair is reduced, so does the RC delay. As a consequence, the signal integrity of the TSV pair with the depletion capacitance is improved, that is, the transmission coefficient—|S21| becomes slightly larger (Fig. 6.16). Note that in practice,

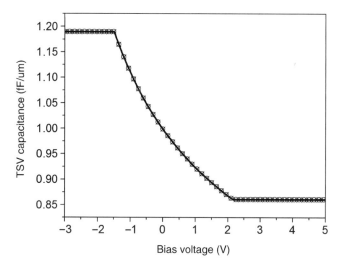

Figure 6.15 The per-unit-length capacitance of the TSV due to MOS effect.

it is hard to determine the exact frequency range for which the TSV operates at a specific mode. By assuming a fully depleted region under the depletion mode of operation of TSVs, we are able to infer from Figure 6.16 that despite the modes of operation of TSVs, their scattering parameters will be bounded by those with and without consideration of such a MOS capacitance.

6.4 CONCLUSION

First, a wideband equivalent-circuit model is developed for electrical modeling of TSVs. The model can capture the important parasitic effects including the skin effect, the proximity effect, and the MOS capacitance effect of TSVs. Numerical examples show that the wideband model is accurate compared to commercial 3D full-wave solvers. The above wideband model is applicable not only to TSV structures, but also to other via structures. The wideband model for the signal-ground via structure can be extended to simulate other configuration of via structures, such as ground-signal-ground via structures.

Second, we briefly discuss the future work for TSV studies. In the via-last process the TSVs pierce through the low-k dielectric

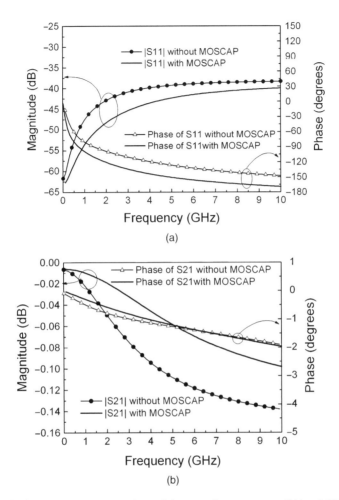

Figure 6.16 The magnitude and phase of the scattering parameters (S11 and S21) for Case-A with and without consideration of TSV MOS capacitance (assuming a fully depleted region) in the wideband equivalent-circuit model of this work [31]. Parameters used to calculate the depletion radius are $N_a = 1.5 \times 10^{15}\,\mathrm{cm}^{-3}$, $Q_{eq} = 1.0 \times 10^{11}\,\mathrm{cm}^{-2}$, $\phi_{ms} = -0.2\mathrm{V}$, and the maximum radius is $d_m = 3.512\,\mu\mathrm{m}$.

back-end-of-lines (BEOLs) in a chip. Therefore, the model developed in this chapter may need to be modified to accommodate it. More work may be needed to study the possibility of using TSVs for THz and quasi-optical signal transmission. Extensive studies on the MOS effect of TSVs may still be required, as for planar MOS capacitors [25], which include carrier transport effects, quantum-mechanical effects to calculate charges at the insulator-semiconductor interface, and dielectric breakdown to study the current carrying capability of TSVs for high power chips. Last, we argue that equivalent circuits could be used to draw together different physics to perform multiphysics modeling of TSVs.

REFERENCES

[1] International Technology Roadmap for Semiconductors (ITRS), http://public.itrs.net/.

[2] E.-P. LI, X. WEI, A. C. CANGELLARIS, E.-X. LIU, Y. ZHANG, M. D'AMORE, J. KIM, and T. SUDO, Progress review of electromagnetic compatibility analysis technologies for packages, printed circuit boards, and novel interconnects, *IEEE Trans. Electromagn. Compat.*, vol. 52, no. 2, pp. 248–265, 2010 (invited paper).

[3] J. P. QUINE, H. F. WEBSTER, H. H. II GLASCOCK, and R. O. CARLSON, Characterization of via connections in silicon circuit boards, *IEEE Trans. Microwave Theory Tech.*, vol. 36, no. 1, pp. 21–27, 1988.

[4] J. U. KNICKERBOCKER, P. S. ANDRY, L. P. BUCHWALTER, A. DEUTSCH, R. R. HORTON, K. A. JENKINS, Y. H. KWARK, G. MCVICKER, and C. S. PATEL, Development of next-generation system-on-package (SOP) technology based on silicon carriers with fine-pitch chip interconnection, *IBM J. Res. Dev.*, vol. 49, no. 4.5, pp. 725–753, 2005.

[5] P. GARROU, C. BOWER, and P. RAMM, eds., *Handbook of 3D Integration*, 2nd ed., Wiley-VCH, Weinheim, 2008.

[6] G. V. D. PLAS, P. LIMAYE, A. MERCHA, H. OPRINS, C. TORREGIANI, S. THIJS, D. LINTEN, M. STUCCHI, K. GURUPRASAD, D. VELENIS, D. SHINICHI, V. CHERMAN, B. VANDEVELDE, V. SIMONS, and I. D. WOLF, Design issues and considerations for low-cost 3D TSV IC technology, in *Proc. International* Solid State Circuits Conf. *(ISSCC)*, San Francisco, CA, February 2010, pp. 148–150.

[7] Y. XIE, J. CONG, and S. SAPATNEKAR, eds., *Three-Dimensional Integrated Circuit Design*, Springer, New York, 2009.

[8] N. KHAN, V. S. RAO, S. LIM, H. S. WE, V. LEE, X. ZHANG, Y. RUI, L. EBIN, N. RANGANATHAN, T. C. CHAI, V. KRIPESH, and J. LAU, Development of 3D silicon module with TSV for system in packaging, *IEEE Trans. Compon. Packag. Technol.*, vol. 33, no. 1, pp. 3–9, 2010.

[9] E.-X. LIU, E.-P. LI, Z. Z. OO, X. WEI, Y. ZHANG, and R. VAHLDIECK, Novel methods for modeling of multiple vias in multilayered parallel-plate structures, *IEEE Trans. Microwave Theory Tech.*, vol. 57, no. 7, pp. 1724–1733, 2009.

[10] K. J. HAN and M. SWAMINATHAN, Polarization mode basis functions for modeling insulator-coated through-silicon via (TSV) interconnections, in *IEEE Workshop on Signal Propagation on Interconnects,* May 12–15, 2009, 2009, pp. 1–4.

[11] R. WEERASEKERA, M. GRANGE, D. PAMUNUWA, H. TENHUNEN, and Z. LI-RONG, Compact modelling of through-silicon vias (TSVs) in three-dimensional (3-D) integrated circuits, in *Proc. IEEE International conf. on 3D System Integration*, 28–30 Sept. 2009, 2009, pp. 1–8.

[12] X. GU, B. WU, M. RITTER, and L. TSANG, Efficient full-wave modeling of high density TSVs for 3D integration, in *Proc. Electronic Components and Technology Conference (ECTC)*, Las Vegas, June 2010, pp. 663–666.

[13] Z. GUO and G. PAN, On simplified fast modal analysis for through silicon vias in layered media based upon full-wave solutions, *IEEE Trans. Adv. Packag.*, vol. 33, no. 2, pp. 517–523, 2010.

[14] C. RYU, J. LEE, H. LEE, K. LEE, T. OH, and J. KIM, High frequency electrical model of through wafer via for 3-D stacked chip packaging, in *1st Electronics System integration Technology Conference*, Dresden, September 2006, pp. 215–220.

[15] C. XU, H. LI, R. SUAYA, and K. BANERJEE, Compact AC modeling and performance analysis of through-silicon vias in 3-D ICs, *IEEE Trans. Electron Devices*, vol. 57, no. 12, pp. 3405–3417, 2010.

[16] H. A. WHEELER, Formulas for the skin effect, *Proc. I.R.E.*, vol. 20, pp. 412–424, 1942.

[17] S. RAMO, J. R. WHINNERY, and T. VANDUZER, *Fields and Waves in Communication Electronics*, John Wiley & Sons, New York, 1965.

[18] A. DEUTSCH, H. H. SMITH, C. W. SUROVIC, G. V. KOPCSAY, D. A. WEBBER, P. W. COTEUS, G. A. KATOPIS, W. D. BECKER, A. H. DANSKY, G. A. SAI-HALASZ, and P. J. RESTLE, Frequency-dependent crosstalk simulation for on-chip interconnections, *IEEE Trans. Adv. Packag.*, vol. 22, no. 3, pp. 292–308, 1999.

[19] J. R. CARSON, Wave propagation over parallel wires: The proximity effect, *Philos. Mag. Ser. 6*, vol. 41, no. 244, pp. 607–633, 1921.

[20] Ansoft HFSS, http://www.ansoft.com/products/hf/hfss

[21] D. G. FINK and D. CHRISTIANSEN, *Electronics Engineers' Handbook*, McGraw-Hill, New York, 1989.

[22] R. E. COLLION, *Field Theory of Guided Wave*, 2nd ed., IEEE Press, New York, 1991.

[23] C. PAUL, *Analysis of Multiconductor Transmission Lines*, 2nd ed., John Wiley & Sons, Hoboken, NJ, 2007.

[24] B. V. ZEGHBROECK, Metal-oxide-silicon capacitors, in chapter 6 of *Principles of Semiconductor Devices*, [online] http://ecee.colorado.edu/~bart/book

[25] S. M. SZE and K. K. NG, *Physics of Semiconductor Devices*, 3rd ed., Wiley-Interscience, Hoboken, NJ, 2006.

[26] G. KATTI, M. STUCCHI, K. DE MEYER, and W. DEHAENE, Electrical modeling and characterization of through silicon via for three-dimensional ICs, *IEEE Trans. Electron Devices*, vol. 57, no. 1, pp. 256–262, 2010.

[27] T. BANDYOPADHYAY, R. CHATTERJEE, C. DAEHYUN, M. SWAMINATHAN, and R. TUMMALA, Electrical modeling of through silicon and package vias, in *Proc. IEEE International Conference on 3D System Integration*, San Francisco, CA, September 2009, pp. 1–8.

[28] N. ARORA, *MOSFET Modeling for VLSI Simulation—Theory and Practice*, World Scientific Publishing, Singapore, 2007.

[29] E.-X. LIU, E.-P. LI, W.-B. EWE, and H. M. LEE, Multi-physics modeling of through-silicon vias with equivalent circuit approach, in *Proc. IEEE EPEPS*, Austin, Texas, October 2010, pp. 33–36.

[30] H. LI, E. LIAO, X. F. PANG, H. YU, X. X. YU, and J. Y. SUN, Fast electroplating TSV process development for the via-last approach, in *Proc. IEEE Electron. Comp. Technol. Conf.*, Las Vegas, June 2010, pp. 777–779.

[31] E.-X. LIU, E.-P. LI, W. B. EWE, H. M. LEE, T. G. LIM, and S. GAO, Compact wideband equivalent-circuit model for electrical modeling of through-silicon via, *IEEE Trans. Microwave Theory Tech.*, vol. 59, no. 6, pp. 1454–1460, June 2011.

Index

Electrical Modeling and Design for 3D System Integration: 3D Integrated Circuits and Packaging, Signal Integrity, Power Integrity and EMC, First Edition. Er-Ping Li.
© 2012 Institute of Electrical and Electronics Engineers. Published 2012 by John Wiley & Sons, Inc.